Wisconsin Land and Life

Edited by

Robert C. Ostergren and Thomas R. Vale

THE UNIVERSITY OF WISCONSIN PRESS

The University of Wisconsin Press
2537 Daniels Street
Madison, Wisconsin 53718-6772

3 Henrietta Street
London WC2E 8LU, England

All maps and figures have been prepared by Richard Worthington,
University of Wisconsin–Madison Cartography Laboratory

Library of Congress Cataloging-in-Publication Data
Wisconsin land and life / edited by Robert C. Ostergren and Thomas R. Vale.
    584 pp.   cm.
    Includes bibliographical references and index.
    ISBN 0-299-15350-9 (cloth: alk. paper).
    ISBN 0-299-15354-1 (pbk.: alk. paper).
    1. Wisconsin—Geography.   I. Ostergren, Robert Clifford.
II. Vale, Thomas R., 1943–   .
F581.8.W57   1997
917.75—DC20   96-36447

For Clarence Olmstead and Francis Hole,
Professors Emeriti of Geography,
in recognition of their long commitment
to the study of Wisconsin

Contents

Contents

Figures

Wisconsin Land and Life

Introduction
A Celebration of the Land and Life of Wisconsin

Robert C. Ostergren and Thomas R. Vale

Geographers are interested in places. We study the natural environment—climate, vegetation, soils, water, and landforms—in order to understand how processes that operate in nature produce the natural characteristics of places, and we assess how humans interact with those characteristics. We analyze human habitation of locales, often from a historical perspective, in order to appreciate how human occupance helps to create individual or group senses of place. We explore both people-nature interaction and human habitation as forces that, when expressed spatially, may create distinctive regions, economies, or landscapes. In each of these various pursuits, we geographers seek to understand the richness of the surface of the earth and the human enterprise that creates a home out of that surface. The expression of that richness and that home-building for a particular area may be called the land and life of a place.

This book is a geographic exploration of the land and life of Wisconsin. It attempts to capture the personality of our home state through chapters that analyze the natural environment, the historical processes of human habitation, and the regionalization of nature and people. It cannot be complete in the sense of covering all possible topics, for all peoples, for all times; rather, it offers a selection, a sampling, of major land and life themes of the place we call home. Its authors are

3

Wisconsin geographers who are linked by ties to the state's academic geography departments. Some are current faculty; others are retired professors; still others are individuals who completed, or are completing, their doctoral degrees.

It is not surprising that the Department of Geography on the University of Wisconsin–Madison campus is so important in the affiliations of the authors who have contributed to this volume. From the first offering of geography courses in the 1860s, to the emergence of a geography curriculum in the 1890s, to the establishment of a separate Department of Geography in 1928 (formerly, geography was included within a joint Department of Geology and Geography), the Madison campus has long been home to geographers who have studied the state. Lawrence Martin, for example, first published his *Physical Geography of Wisconsin* in 1915, and, through re-publication, it remains a standard reference of the state's landform characteristics and landform history almost a century later. Vernor Finch engaged in field research in the Montfort area (60 miles due west of Madison) in the 1920s, creating a detailed portrait of the land use of a 46-square-mile area, an effort that remains a model of intensive surveying and field sampling of an agricultural region. A number of geographers, including Richard Hartshorne, participated in the painstakingly detailed field surveys of land cover and land use for individual townships in the Wisconsin Land Economic Survey of the 1930s. In ensuing decades, a constant flow of Ph.D. dissertations focused on Wisconsin topics.

One name in particular, though, seems to epitomize the historical association of the Madison geography department and Wisconsin studies: Glenn Trewartha. A native of the state, Trewartha earned his Ph.D. from Madison in 1925 with a dissertation on the dairy industry of Wisconsin. After joining the Madison department, he went on to become one of its most notable academics. Trewartha's work on Wisconsin topics included a broad analysis of the Driftless Area, including major publications on landform development, historical settlement, and human alteration of environments. Also a climatologist, Trewartha wrote in 1981, as one of his last papers, a study of the annual distribution of rainfall over Wisconsin and the upper Midwest. His was a lifetime of contributions to the understanding of the state—its natural environment, history of human habitation, and people-nature interactions.

In a sense, the demands of specialization make it difficult for any individual today to encompass the breadth of professional interests illustrated by Trewartha. Instead, modern geographers become competent in some area of particular interest, nonetheless maintaining appreciation for, and some ability in, contexts and perspectives broader than those in other academic traditions. We geographers enhance our particular specialties by interactions with other geographers (and certainly with those in sibling disciplines) who have expertise in different areas of inquiry, an interaction that helps to create a sense of intellectual community. Overall, in spite of distinctive specializations, we geographers are united by a common appreciation for the land and life of places.

This book represents such a blending of particular expertise within unifying themes. The first group of chapters on natural environments and wild landscapes explores processes and patterns in the natural world of the state. As is often the case with geographic treatments of nature, each chapter combines purely natural phenomena with some theme that links people with nature; these linkages, which represent different human interpretations of the natural world, include nature as a constraint, nature as resource, nature as altered environment, and nature as object of affection. The second group of chapters focuses on specific settlement processes and cultural patterns that have figured prominently in shaping the highly distinctive and varied human geography of the state. All these chapters are strongly historical, placing emphasis on the ways in which people of diverse backgrounds and interests have occupied and organized Wisconsin places and spaces over time to meet their particular economic, social, and cultural needs and, in so doing, have often created a particular sense of place or identity. The third theme, regional economies and landscapes, ties together a group of chapters that feature people-nature interactions, whether they be of an active or highly economic kind, such as dairying, lumbering, and tourism, or of a more perceptual kind, such as local boosterism, nature appreciation, and simple recognition of the individuality of places. Most of these chapters focus on highly identifiable places or regions around the state and, in particular, bring out the distinctive landscape expression of whatever form of human habitation or people-nature interaction has been an important force in that place or region. Nonetheless, the final chapter in this section reminds us that however distinctively we may wish to view the various landscapes of our state, we must be cognizant of the fact that there is a constant erosion of all differences by the homogenizing forces of our modern everyday life.

We hope that readers will find, within the pages of this volume, information and ideas that will excite their minds and encourage them to learn more about our home state. It is our wish that these chapters will celebrate the land and life of this place we know as Wisconsin.

PART ONE

NATURAL ENVIRONMENTS AND WILD LANDSCAPES

Wisconsin's natural endowment is bountiful. Some of this richness results from its spatial character, the geographic situation of our state, a location where the drier, midcontinental environment sweeps up from the southwest, where the moister, eastern continental environment edges in from the southeast, and where the cold, high-latitude environment comes down from the north. As a result, within the state boundaries, we find our lives enriched by a varied blend of natural characteristics, both living and inanimate—howling winter snow and (occasionally) desiccating summer drought, gently rolling plains and steep-sided valleys, forests of pine and prairies of grasses, snowy owls from the Arctic and opossums from the Deep South.

Wisconsin's nature, however, is also enriched by its temporal character, a history that has seen glaciers grind over most of the state and then retreat. The impacts of these now-absent environmental agents extend to nearly all aspects of the natural world—landforms, soils, hydrology, vegetation, and wildlife. We can better appreciate many of these glacier-modified attributes because the great tongues of ice did not invade the southwestern quarter of the state, leaving for us a nonglaciated Driftless Area.

The rich natural bounty of Wisconsin is celebrated in the chapters in this first section of the book. The chapters vary in approach: some provide generalizing overviews, whereas others explore in detail a case study of a particular character or theme. Several stress the spatial patterning of some natural phenomenon, and others emphasize a temporal view. In spite of focusing on a particular part of the natural world, all the chapters see the connectedness that binds those parts together.

Although primarily dedicated to providing information and insight on nature, the chapters also illustrate the richness of the links between the human and the natural worlds, the endless ways that people find meaning in nature. These meanings include nature as resource, to be used for utilitarian purposes; nature as hazard, to be respected for its potential force; nature as scientific phenomena, to be understood; nature as artifact, to be seen as altered by human activity; nature as object of affection, to be cared for, even treasured.

The chapters provide illustrations of what each of us might seek in our intellectual explorations of Wisconsin's nature. They can only hint at both the possible topics and the potential interactions that are open to us, as geographers interested in experiencing the bounty that is the wild Wisconsin landscape.

1

From End Moraines and Alfisols to White Pines and Frigid Winters
An Introduction to the Environmental Systems of Wisconsin

Thomas R. Vale

On a morning in late September, I look out of my second-floor office window on the Madison campus. The sky overhead, uniformly dark with cloud and intermittently loud with thunder, pours down a shower of rain. The sheets of water partly obscure the columnar conifers and more globular broadleaf trees (a few showing the yellowing of the oncoming fall season) that grace the rounded hillslope, itself the depositional product of a past glacier. Much of the moisture disappears on the ground surface, apparently soaking into the soil, but near the base of my building, water spills off the edge of a sloping lawn into a concrete gutter and then gushes onto the asphalt parking lot below. The watery symphony does not drown the familiar cawing of crows, calling from somewhere beyond my sight.

At first glance, the natural landscapes of Wisconsin appear to lack diversity. No mountain ranges rise from level lowlands, as they do in Wyoming, nor have rivers carved canyons, as they have in Utah. Vegetation does not vary from short-grass prairie to alpine tundra over a score of miles, as it does in Colorado. We Wisconsinites cannot move from chilling cold to desiccating heat in a few minutes, as San Franciscans can on a California summer afternoon.

Yet, first impressions can be deceiving. By looking closely and by considering all the environment, we would find that the natural characteristics of our home state are indeed varied. The totality of nature includes both the subtle and the spectacular in a rich array of features—landforms, vegetation, weather and climate, soil, animal life, and hydrology. Moreover, because any one of these phenomena varies spatially in ways different from the others, the combination at any one location is unique: No two places are ever the same.

We can celebrate this diversity in environmental characteristics, but the richness presents problems for generalization: How can the natural landscapes of Wisconsin be discussed without both simplifying and omitting much that is relevant and enjoyable to know? The solutions adopted in this chapter are two. First, three particular places are discussed, not only to suggest the state's natural diversity, but also to illustrate particularly widespread components of natural Wisconsin. Second, the components of landscape in these three locations are presented as parts of interacting environmental systems. In these pages, we generate a perspective that links parts of the natural world together, a viewpoint that necessarily precludes completeness of coverage on any single topic but that encourages a way of looking at the world that can be applied elsewhere in the state or, indeed, in the world. Through this integrative outlook, which is at the heart of physical geography, we come to understand the natural world, and from that understanding comes appreciation, even affection.

Near Dodgeville, Iowa County

From a ridgetop a few miles east of the town of Dodgeville, I look southward, down a farm road which falls steeply into a ravine and then rises up and over a gently rounded upland. The fence of barbed wire bordering the pavement does not impede the brisk south wind, but one of its steel posts does provide a perch for a rigidly clinging meadowlark. Curving stands of alfalfa and corn, carefully plowed on the contour, intermixed with grassy pastures and fallow fields, extend beyond asphalt and fence. Scattered oaks mark the field borders. Toward the northwest, beyond a weedy stand of grass and thistle, a bold face of a road cut rises as a vertical cliff of tawny rock. Still farther around, to the northeast, the distant mass of Blue Mounds breaks the horizon. . . .

Today's view from the ridge both resembles and differs from that which was seen a century and a half ago. Still sloping gently to the south is the Galena dolomite, a layer of calcium and magnesium car-

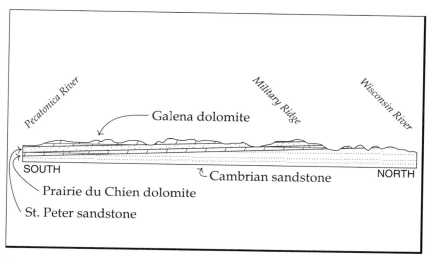

Figure 1.1. The sedimentary rock layers exposed in the Driftless Area dip to the south (Modified from Martin, 1965)

bonate rock, similar to limestone, that was laid down in a shallow sea over 400 million years ago just before Wisconsin emerged above the marine waters (it is this rock that we see in the road cut).[1] Streams have incised into the dolomite and, more deeply, into the slightly older, underlying St. Peter sandstone; this downcutting has produced the valleys of southwestern Wisconsin. The narrow valley over which we look is a tributary to the Pecatonica River, a stream that drains southward, down the regional dip of the dolomite and sandstone, from Military Ridge, a broad, rounded, continuous east-west upland capped by the Galena dolomite (Figure 1.1). No glaciers flowed here, in the Driftless Area, and so the stream-cut valleys and intervening ridges still remain unmodified by the action of flowing ice. Nonetheless, conditions during the Ice Age, the Pleistocene, were different from those of today, leaving their imprints on the contemporary scene. Particularly notable among these were the clouds of silt, lifted into the air during late Pleistocene times from the barren, exposed floodplain of the Mississippi River and blown eastward on westerly winds, to rain down over southern Wisconsin and, in fact, over most of the state. Several feet of such silt, called loess, probably accumulated on the ground surface at our vantage point east of Dodgeville.

These landform characteristics remain from presettlement times— the lay of the rock layers, the generalized form of the valleys, and the

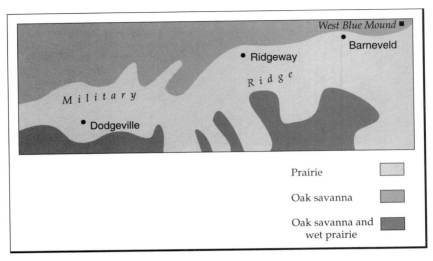

Figure 1.2. Prairie vegetation occupied the broad upland of Military Ridge, with oak savanna and wet prairie on the adjacent slopes. (Modified from Stroessner and Habeck, 1966)

presence of the loess blanket (although reduced by erosion associated with agricultural clearing and plowing)—but the vegetation covering the ground surface has changed drastically. Today, we see fields of corn and alfalfa, pastures, and forest. Before the advent of contemporary land uses, however, a prairie of grasses and flowering, nonwoody plants grew along all of Military Ridge and its adjoining spurs, including the ridge east of Dodgeville (Figure 1.2).[2] The prairie was made possible by the rich loess layer and a humid climate, both of which permitted plants to grow closely together in a dense, summer-green mat. But, in addition, the broad, undissected upland of the ridge encouraged wild fires, whether ignited by lightning or Native Americans, to spread unhindered by rough topography or bodies of water. The high frequency of burning, perhaps almost every year, discouraged the survival of upright, woody trees and shrubs and encouraged the prairie plants, for which the pyric consumption by fall or early spring fires was merely a substitution for the biological decomposition that would otherwise have occurred during the warm, moist summer months. The prairie vegetation itself encouraged this burning, with its annual production of a thick mass of leaves that, before or after the summer growth, was dead and dry and thus easily ignited.[3]

Prairies covered large areas of southern Wisconsin (Figure 1.3). In places, this carpet, called true prairie, lacked trees or shrubs, as was true on Military Ridge. But elsewhere the tall prairie plants concealed the sprouts of small trees, often bur oaks (*Quercus macrocarpa*), that grew up after each burn. The oaks in these brush prairies did not develop into upright tree form until the fires were eliminated with European settlement. Still more common than either true or brush prairies were the oak openings or oak savannas, prairie vegetation with an

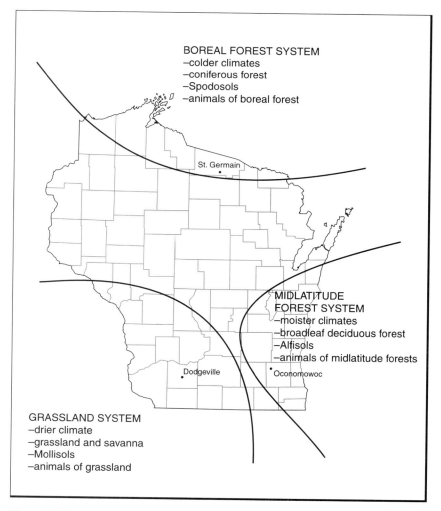

BOREAL FOREST SYSTEM
–colder climates
–coniferous forest
–Spodosols
–animals of boreal forest

St. Germain

MIDLATITUDE
FOREST SYSTEM
–moister climates
–broadleaf deciduous forest
–Alfisols
–animals of midlatitude forests

Dodgeville

Oconomowoc

GRASSLAND SYSTEM
–drier climate
–grassland and savanna
–Mollisols
–animals of grassland

Figure 1.3. Three major environmental systems meet in Wisconsin: grasslands from the southwest, midlatitude forests from the southeast, and boreal forest from the north.

open stand of widely scattered bur oaks; these, like the prairies, were also maintained by fire.

The presence of prairie on our site east of Dodgeville strongly influenced two other aspects of the natural environment. Beneath the waving flowers of little bluestem (*Andropogon scoparius*) and yellow cone flower (*Ratibida pinnata*), a soil characteristic of midlatitude grasslands developed. The prairie vegetation's yearly production of nonwoody leaves and roots readily accumulated in the upper soil as partly decomposed organic matter, or humus; the underground material was particularly important in this regard. The humus was rich in mineral nutrients because the prairie plants vigorously assimilated such elements from the soil, thereby protecting the nutrients from being leached out of the vegetation-soil system by downward percolating water. The richness of the humus, in turn, created a soil chemistry that discouraged the creation of hard clay layers in the subsurface, thus maintaining a loose texture in the soil. Finding the nutrient-rich, nonhardened soil to their liking, small burrowing organisms, such as earthworms and ants, churned through the soil, mixing the organic humus with the inorganic loess and increasing the depth to which the two soil elements were combined. This blend of a nutrient-rich humus and tiny rock particles is the distinctive feature of the midlatitude grassland soil, or Mollisol, a word derived from the Latin *mollis*, meaning "soft" (Figure 1.4).[4]

The development of the Mollisol, so strongly linked to vegetation, also ties to other environmental characteristics, particularly climate. The midlatitude continental climate of southern Wisconsin, sufficiently moist and warm to allow the growth of a thick prairie vegetation, is also sufficiently cool, particularly over the cold winter season, to permit the accumulation of humus material. In warm and wet climatic environments, by contrast, the organic material is rapidly decomposed biologically; thus, grasslands in the tropics have little humus accumulation.

Agricultural clearing and plowing has altered the Mollisols. Increased surface erosion has stripped off much of the humus material, and the work of tractor and plow has reduced the vertical differentiation of the soil layers. Fortunately, the soil remains able to generate the growth of crop plants so important to the local economy and so central to our aesthetic sense of this part of the state.

A second feature of the environmental system at our site near Dodgeville, also coupled with the prairie vegetation, is the population of wild animals that inhabited the skies overhead, the plants on

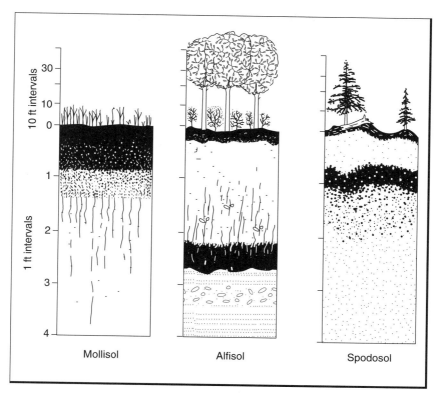

Figure 1.4. Each major environmental system is represented by a regional soil cross-section, or soil profile. The soil in any specific locale, depending upon environmental characteristics, may or may not resemble the regional type. (From Hole, 1976)

the ground surface, and the soil below.[5] Birds that depended upon the openness of the prairie grassland included the aerial displayers, such as the bobolink (*Dolichonyx oryzivorus*) and upland sandpiper (*Bartramia longicauda*); the ground-nesting grassland sparrows, such as the vesper (*Pooecetes gramineus*) and the grasshopper (*Ammodramus savannarum*); and those species that otherwise required the grassland for breeding, such as the greater prairie-chicken (*Tympanuchus cupido*) and the sandhill crane (*Grus canadensis*). Among larger mammals on the prairie, bison (*Bison bison*) grazed and wolves (*Canis lupus*) pursued them. Prairie voles (*Microtus ochrogaster*) darted among the grasses, and Franklin's ground squirrels (*Citellus franklinii*) dug their burrows in the Mollisol soil. Least weasels (*Mustela rixosa*) chased the voles, while badgers (*Taxidea taxus*) dug after the ground squirrels. More than any other single factor, the prairie vegetation determined what animals lived on the rounded ridge near Dodgeville.

This central importance of vegetation to wild animals is suggested by the fate of prairie species today: Of all the wild animals that roamed Wisconsin in pre-European times, none have so generally suffered a decline as the grassland birds and mammals. A few have succeeded in modern equivalent plant covers, in pastures and hayfields, or in areas that structurally resemble the prairie, such as road right-of-ways, but many more find little suitable habitat in the contemporary landscape.[6]

Still another environmental characteristic that reflected the vegetation in the past but that also integrated other features of the system, and continues to do so today, is streamflow.[7] Our upland ridge that falls away steeply to valley-bottom creeks contributes rainfall and snowmelt to the main channel of the Pecatonica River. Several characteristics of the drainage of the Pecantonica, including our ridgetop, enhance water runoff from the landscape generally and thereby enable quick delivery of much water to the stream channel below: steep slopes, hard bedrock relatively close to the ground surface (rather than bedrock covered by thick accumulations of sand and gravel deposited by glaciers or glacier-fed streams), and lack of forest cover (whose leaves would intercept the raindrops, causing them to drip more slowly down to the ground surface). Proportionately, then, water tends to run off the slopes in the Driftless Area, producing high flood flows in the downstream rivers, such as the Pecatonica.

Floods, a regular part of the regional environmental system, tend to occur at two times of the year. On an annual basis, the most likely flood usually develops in the early spring, when a combination of factors encourages surface runoff—snowmelt, cool temperatures (which reduce evaporation of surface water back into the atmosphere), and dormant vegetation (which, if biologically active, would transpire water to the air). The highest floods over longer time scales, however, are often summertime events, associated with particularly intense thunderstorm rains.

All these characteristics of our ridgetop area—landform, vegetation, soil, animal life, and hydrology—are linked either directly or indirectly to weather and climate.[8] The climate near Dodgeville is typical for much of southern Wisconsin, with an annual precipitation of about 33 inches (with a marked warm-season maximum), less than 40 inches of annual snowfall, a normal growing season (the length of time bracketed by the last night of 32° Fahrenheit in the spring and the first 32° night in the fall) of about 140 days, or nearly 5 months, and a high incidence of days with maximum temperatures above 90°. These average characteristics, which, of course, vary enormously from year to year and decade to decade, result from southern Wisconsin's presence in the middle of the North American continent and beneath the motion system of the Westerlies (a band of winds aloft—20,000–60,000 feet off the surface—that bathes the area with air from the west). The mid-latitude, continental location dictates a strongly variable temperature regime, hot in summer and cold in winter, and, the Westerlies, with their storms passing overhead all year, produce a climate without a dry season. The greater moisture content of the warmer air of summer, however, compared with the dry, frigid air of winter, means that more precipitation occurs in the mild half of the year.

Taken together, these characteristics of the ridgetop southeast of Blue Mounds form an environmental system that, in both fundamental nature and individual features, replicates nearby locales (Figure 1.5). Other parts of the state, however, have particular traits and a collective composition that are different.

Near Oconomowoc, Waukesha County

Seven or eight miles northeast of Oconomowoc, with a brisk north wind ushering in the fall season, I stand beside another farm road in a hummocky upland above the lowland of the Little Oconomowoc River. A dense stand of sugar maple (*Acer saccharum*) and basswood (*Tilia americana*), elm (*Ulmus americana*) and box elder (*Acer negundo*), lines the pavement and shields my view of the marsh beyond the trees. The still-green leaves of the trees catch the wind, transforming the motion into song and masking the delicate sounds of birds in the swaying thicket. On the other side of the road, farm pastures and new suburban homes sit atop the rolling landscape of well-drained soil. . . .

The general characteristics of climate near Oconomowoc resemble those near Dodgeville: annual precipitation of about 30 inches, seasonal peak in precipitation in the warm season, annual snowfall of about 40 inches, growing season of about 5 months. Yet, differences in details, reflecting more general statewide patterns, distinguish the two locales. First, the seasonal patterns of precipitation vary. Areas in southeastern Wisconsin tend to receive slightly less summer rainfall than places in the southwestern part of the state, a consequence of the more gentle topography in the southeast (the ridges in the southwest help to generate updrafts that may develop into thunderstorms) and the cooling effect of Lake Michigan on the lower atmosphere (which

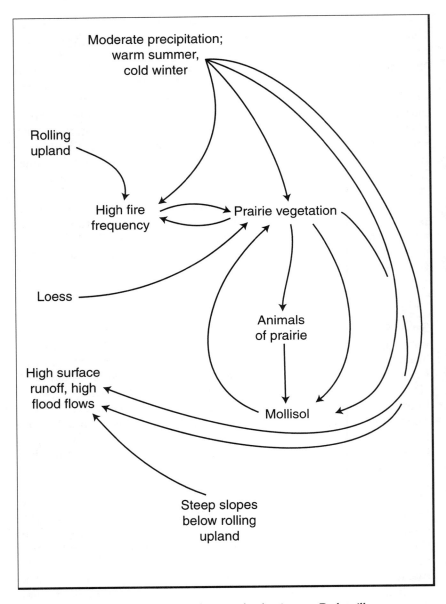

Figure 1.5. Environmental system for the site near Dodgeville

tends to create a more stable vertical temperature profile upward from the earth's surface). Winter precipitation, by contrast, tends to be slightly greater near the lake, a result of greater moisture availability. Second, winter temperatures are slightly warmer in the southeast, compared with the southwest, because of the moderating influences of Lake Michigan. Third, maximum temperatures on summer afternoons and, maybe more important, the frequency of particularly hot days, with temperatures above 90°, are lower in the southeast than in the southwest, again because of the presence of the cool lake waters.[9]

The cooler summer, by reducing evaporation and thus moisture stress, may have contributed to the presence of a moisture-demanding forest, rather than a drought-enduring prairie or oak savanna, on our upland site near Oconomowoc (Figure 1.6). The forest here, as in most forested areas of southern Wisconsin, is one of broadleaf, deciduous trees, a consequence of climatic conditions. The loss of leaves in the fall season permits the trees to become dormant and thereby survive the cold winter months, when soil water cannot be absorbed by roots. With the warming temperatures of spring, the trees break their dormant state, produce new leaves, photosynthesize, successfully flower, and set seed—all possible in a long warm, wet growing season. On our hummocky upland, moreover, this forest of broadleaf trees includes sugar maple and basswood, species that require moist soils, rather than the bur oaks of the drier savannas.

The slightly cooler summer temperatures may have also encouraged the presence of forest cover by reducing the likelihood of fires. The sugar maple–basswood forest itself would similarly discourage burning by creating plant material that is not highly flammable and by casting heavy shade that precludes the establishment of low-growing ground vegetation. In addition, the lakes and wetlands in the Oconomowoc area, and perhaps the broken topography (both a consequence of the Pleistocene ice), may have also acted as barriers to the fires borne on westerly winds.[10]

The animals that inhabit the forest are different from those of the prairie (Figure 1.3). Instead of bobolinks and upland sandpipers singing above the sea of bluestem, forest birds sing and nest among the branches of the tall trees—red-eyed vireos (*Vireo olivaceus*) and rose-breasted grosbeaks (*Pheucticus ludovicianus*), scarlet tanagers (*Piranga olivacea*) and northern orioles (*Icterus galbula*). Rather than herds of bison grazing the grasses, clusters of white-tailed deer (*Odocoileus virginianus*) forage along the forest edges, beside the marsh below the hummocky upland, and in the oak savanna farther south, in earlier times bounding back into the heavy cover at any sign of approach by

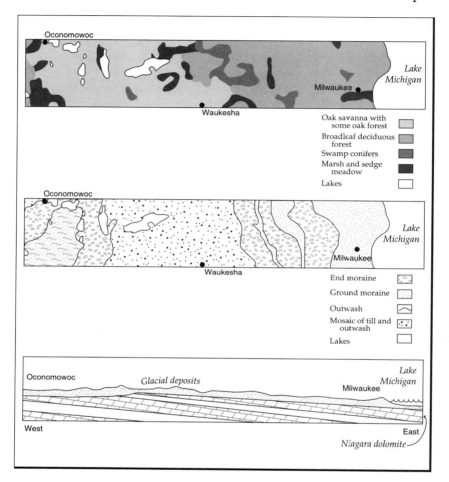

Figure 1.6. Vegetation, glacial landforms, and bedrock structure from Oconomowoc to Milwaukee. The broadleaf deciduous forest is best developed on the end moraines close to Lake Michigan. The end moraines near Oconomowoc are part of the Kettle Moraine, which includes much sand and gravel deposited by glacial meltwater streams. Beneath the glacial deposits, the sedimentary rock layers dip to the east, down into the Lake Michigan lowland. (Modified from Finley, 1976; Alden, 1918; Martin, 1965)

wolves or a mountain lion (*Felis concolor*). The southern flying squirrel (*Glaucomys volans*) nests in holes in the tree trunks, and eastern chipmunks (*Tamias striatus*) scamper about on the forest floor, sometimes pursued by long-tailed weasels (*Mustela frenata*).[11]

The forest vegetation has created a soil unlike the Mollisol of the

prairie. Trees do not produce as much nonwoody, below-ground organic material as the prairie plants do, instead concentrating their
biological production as wood above the earth. As a consequence, beneath the surface accumulation of logs, limbs, and twigs, only a shallow humus layer develops. Nevertheless, sugar maple and basswood
actively absorb many mineral nutrients into their roots, and thus the
leaves that drop to the ground surface in the fall return these materials to the soil. This cycling of nutrients, while maintaining soil fertility,
is not as complete as it is in the prairie, and thus the soil chemistry
is more conducive to the movement of certain soil elements by downward percolating water. These include some nutrients, which may be
washed entirely out of the soil to become part of the dissolved material
in groundwater and surface streams, and tiny clay particles, which
typically move downward to create a layer of concentrated clay in the
subsurface of the soil. The resulting soil, called an Alfisol (a reference
to *al*uminum and iron, the Latin for which is *ferris,* two elements that
are conspicuous in the upper part of the soil), is associated with the
broadleaf forests of much of the eastern United States, including those
of sugar maple and basswood in eastern Wisconsin generally and our
low upland near Oconomowoc specifically (Figures 1.3 and 1.4).[12]

The Alfisol is a relatively young soil, which, given millions of years,
develops into a less fertile soil type. Appropriately, then, our particular Alfisol is developed on 1–2 feet of late–Ice Age loess that overlies
deposits of unconsolidated rock material laid down by the last of the
Pleistocene ice flows, those present as recently as 13,000 years ago, during what is called the Wisconsin advance. In this part of the state, one
great tongue of ice (the Lake Michigan lobe) flowed southward into the
lowland that is now occupied by Lake Michigan; another (the Green
Bay lobe) extended southward just to the west of the first, covering
much of eastern Wisconsin, from Green Bay and Wausau to Oconomowoc and Madison. These side-by-side glaciers transported and deposited masses of rock material to their snouts and flanks, including
the linear area between them, thereby building what are called end
moraines. Meltwater streams also laid down accumulations of sand
and gravel, called outwash, in this zone between glacial lobes. Blocks
of ice, broken from the main glacier as it retreated, sat in the loose
rock material, eventually melting to create depressions, called kettles,
which subsequently filled with water. The resulting area of north-
south-trending ridges, rich with lakes, west of Milwaukee and Sheboygan, has been named the Kettle Moraine. The city of Oconomowoc sits
within this complex, and our hummocky upland northeast of the city
is a piece of an end moraine of the Green Bay lobe (Figure 1.6).

The wetland below us, however, sits not on moraine but on out-wash. In fact, the string of wetlands and lakes south of our site, those between Waukesha and Oconomowoc, occupy kettle depressions in outwash rather than moraines. Consisting of relatively large rock-particle fragments, sand, and gravel, the outwash material drains easily (except, of course, where the groundwater intersects the earth's surface within kettle holes, an intersection that results in marshes and lakes), and thus the outwash supports a vegetation, oak savanna, that tolerates droughty soils. The smaller particles of clay and silt in the end moraine retain moisture, thereby permitting the development of the sugar maple–basswood forest. These relationships are further supported by the differing frequency of fires—less frequent in the moist forest, more frequent in the dry savanna. Within a half-mile of our upland site, then, the changing nature of the glacial deposits interacts with the groundwater to create a rich diversity of environments—from lakes to marshes to moist forest to dry savanna.

The glacial deposits of loess, moraine, and outwash bury the bedrock (Figure 1.6). The most prominent rock layer so covered is the Niagara dolomite, the same stratum that caps West Blue Mound to the west, that outcrops as the cliff along the east shore of Lake Winnebago to the north, and that forms the lip over which Niagara Falls drops far to the east. Near Oconomowoc, the Niagara dolomite tilts, or dips, to the east, creating a west-facing escarpment just east of town but west of Waukesha (buried beneath glacial deposits, of course). The combination of hard dolomite and accumulated glacial material explains why this area reaches elevations of slightly more than 1,000 feet above sea level, among the highest altitudes between Madison and downtown Milwaukee.[13]

The depth of glacial deposits is thinnest atop the outcropping Niagara dolomite but reaches more than 200 feet in the nearby lowlands, including those over which the Oconomowoc River flows. These deep, unconsolidated materials absorb and store large volumes of rainwater and snowmelt, which means that water readily infiltrates the earth's surface, even when the ground is already wet. As a consequence, surface runoff of water is minimal, and water movement from the landscape to stream channels is mostly through the groundwater reservoir; this pathway of water delivery is slow, meaning that streams do not respond quickly or sharply to precipitation or snowmelt. In addition, the numerous lakes store water on the earth's surface, also slowing stream response. Floods in the area, in fact, rise relatively little, compared with average streamflows.[14]

This hydrological behavior is enhanced by other characteristics of

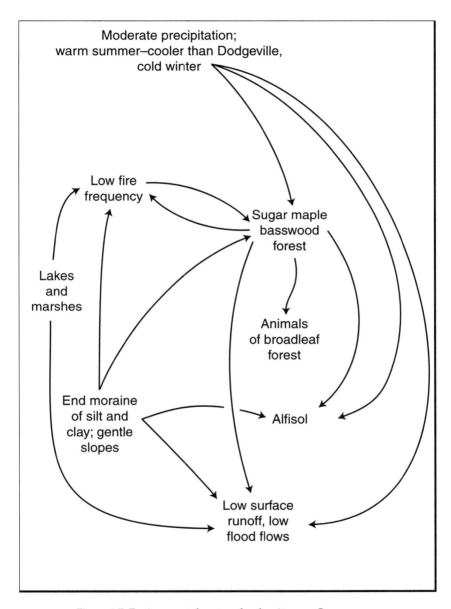

Figure 1.7. Environmental system for the site near Oconomowoc

the environment. The gentle slopes of the glacial deposits encourage water to soak into the ground rather than running off the surface (in contrast with the steep slopes of the Driftless Area, which promote surface runoff and thus "flashier" stream response to precipitation and snowmelt). The presence of forest (even if not everywhere), with its dense vertical structure of leaves and twigs, also slows the delivery of raindrops to the earth, and thus of water from slopes to stream channels.

The combination of environmental characteristics on the upland near Oconomowoc describes a system in sharp contrast with that on the ridgetop near Dodgeville (Figure 1.7). Still other areas of the state are different, in varying degrees, from either of these two locales.

Near St. Germain, Vilas County

Four miles north of St. Germain, I stand on the lower flank of an irregular upland of deep hollows and short, steep-sided ridges, all covered by a forest of tall, swaying white birch (*Betula papyrifera*) and aspen (*Populus tremuloides*), leaves rustling in the breeze, with an understory of aromatic balsam fir (*Abies balsamea*). A grove of red pine (*Pinus resinosa*) sits on a steep slope, and a single, multiple-trunked white pine (*Pinus strobus*) towers over the birch and aspen. Through the boles of the trees I see the quiet, blue waters of a lake, shimmering beneath rising wisps of the still morning air. Somewhere beyond my sight, a loon (*Gavia immer*) calls. . . .

The frequency and magnitude of flooding flows on streams near St. Germain, like those near Oconomowoc, are low, and the contributing reasons are similar in both locales: Gentle slopes, deep unconsolidated glacial deposits, and thick forests all encourage infiltration of rainwater and snowmelt into the earth, thereby reducing surface runoff and delaying the delivery of water from the landscape generally to stream channels. In addition, the abundant lakes in Vilas County store large volumes of water, a storage that also slows the rate of water movement downstream, thus retarding flood peaks.[15]

The glacial material that covers the bedrock (here, close to the southern edge of the Precambrian crystalline rock that forms the continental core, the shield of North America) consists mostly of outwash sands and gravels. Beyond the lake, the outwash material forms a level plain, and in this respect, it resembles the typical depositional outwash surfaces in southern Wisconsin. But the stream deposits of the undulating, irregular upland upon which we stand has a different history.

Here, the sands and gravels were laid down by streams on the surface of the glacial ice, probably especially during the waning stages of the Pleistocene (maybe between 10,000 and 15,000 years ago), when meltwater was abundant and the ice itself was stagnant or nearly so. As the underlying ice liquefied, the accumulation of stream deposits slumped downward, collapsing in uneven heaps, and even flowing as masses of rock debris and water. The resulting hummocky terrain characterizes not only our particular upland locale but also much of central Vilas County (Figure 1.8).[16]

The irregular topography created by collapsing stream sediment contributes to the abundance of lakes for which the county is famous. And the fame is deserved; the 346 lakes and ponds in Vilas County cover 140 square miles, or 15 percent of the county area. Perhaps only parts of Minnesota, Ontario, and Finland have as many lakes per square miles as Vilas County.[17]

Rooted in the sands, forests of red and white pine—often associated with broadleaf species such as red maple (*Acer rubrum*) and sugar maple—prospered in pre-European times (Figure 1.8). The abundance of pines reflects a general association of such trees with sandy, or otherwise dry and infertile, soils. Pines, with their waxy needles, which reduce water loss, tolerate the dryness of the sands; moreover, their retention of leaves (needles) all year helps in the conservation of nutrients that might otherwise be lost from the vegetation-soil system by rapidly percolating water if leaves were to fall to the ground surface all at one time. Today, this presettlement forest has been largely replaced—because of the heavy logging of the last century—by stands of white birch and aspen, or by mixtures of these broadleaf trees and pines or balsam fir.[18]

Environmental factors other than soil also favor conifers, trees like the pines, fir, and eastern hemlock (*Tsuga canadensis*), which bear their seeds in woody cones. The cool climate, with its long winter and short growing season (compared with southern Wisconsin), promotes evergreen trees. Spared the need to produce an entirely new crop of leaves each spring, the trees may begin photosynthesis as early as possible in the growing season. Moreover, the year-round presence of leaves on the trees allows the process to continue well into the waning daylight hours of autumn, or even during warm periods of winter. The pines, fir, and hemlock of northern Wisconsin, in fact, represent the southern edge of the great boreal forest, a forest of conifers, that extends northward into Canada and circumscribes the North Pole in the cold regions south of the Arctic tundra (Figure 1.3). In addition to a cool climate, the pines (but not the fir or hemlock) were also favored by

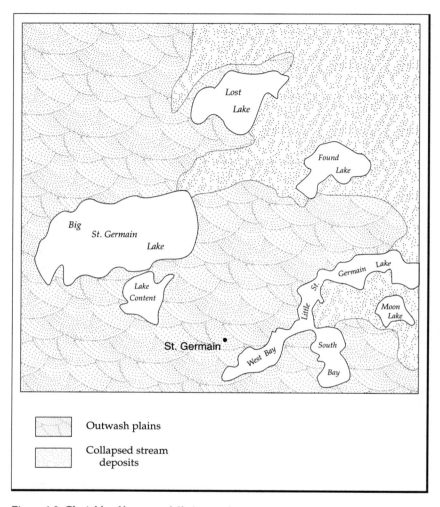

Figure 1.8. Glacial landforms, and *(facing page)*, vegetation near St. Germain. Sands and gravels deposited by glacial meltwaters cover the entire area. Pines dominated the pre-European vegetation. (Modified from Attig, 1985; Finley, 1976)

occasional fires, which created the sunnier ground surface needed for good survival of seedlings; burning, in turn, was promoted by the dry soil and the flammability of the trees themselves.[19]

The climate of our site in Vilas County is indeed cooler than the area of Wisconsin farther south. The length of the growing season, about 100 days, or 3 months, is only two-thirds as long as what is

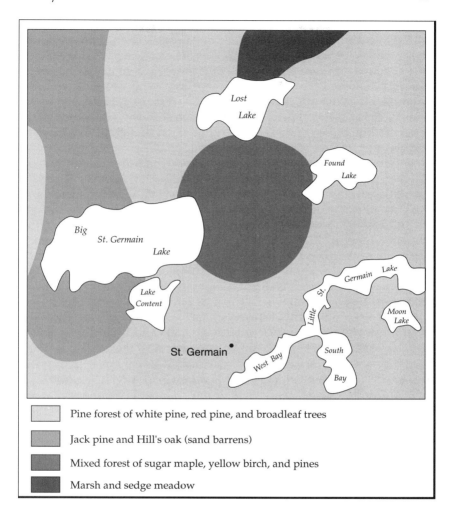

Pine forest of white pine, red pine, and broadleaf trees

Jack pine and Hill's oak (sand barrens)

Mixed forest of sugar maple, yellow birch, and pines

Marsh and sedge meadow

enjoyed near Oconomowoc, and the number of summer afternoons with temperatures at 90° or higher, fewer than two, is lower than in most of the rest of the state. The higher latitude explains the generally cooler conditions in the north, although the chilling effects of Lake Superior also depress summer temperatures. The annual precipitation, about 32 inches, represents the average for the state as a whole, but the seasonal distribution is distinctive: Winters tend to be drier than the other places that we have sampled, a reflection of the colder winter air carrying less moisture, and summers tend to be a little wetter, because slightly higher elevations force uplift and thus precipitation from rising air. More distinguishing of the Vilas County locale is the high annual snowfall, about 70 inches, a consequence of the higher latitude and its colder winters, as well as the possible contributions from lake-effect snows.[20]

The sandy substrate, coniferous trees, and moist climate have combined to favor development of a distinctive soil. The sands of the collapsed sediments encourage rapid infiltration and little retention of water. Under these conditions, the potential exists for the loss of nutrients from the vegetation-soil system as the soil water drains quickly into the groundwater and to lakes and streams. The conifers accentuate this nutrient-poor soil: these trees, compared with prairie plants and most broadleaf trees, do not absorb soil nutrients strongly, and thus the organic matter that they create is not as rich in nutrients, resulting in relatively little return of these elements to the soil. The cool climate also contributes to the weak cycling of nutrients in the vegetation-soil system by slowing the rate of decomposition of organic matter. The result of these interactions is the development of a soil that lacks certain basic nutrients, a condition described as acidic.

The acid condition, in turn, renders the soil vulnerable to the downward movement of tiny rock particles (clays) and organic molecules (from the plant humus). In a matter of centuries, these materials are transported out of the upper and redeposited in the lower part of the soil. This type of development is in direct contrast with the nutrient-rich soil of the prairie, recall, which creates a soil chemistry that resists such downward movement of clay. The soil in the outwash sand–coniferous forest–cool climate system is called a Spodosol (from the Greek, *spodos*, meaning "wood ash," a reference to the ashen or gray color of the portion of the soil out of which the clays and organic molecules are removed) (Figure 1.4).[21]

The birds and mammals that inhabit the cool, coniferous forests are, in part, different from those of deciduous forests farther south. Certainly, many species grace the wooded lands in both areas of Wisconsin, but the northern parts of the state also support many species of the great boreal forest (Figure 1.3). Among the mammals, the moose (*Alces alces*) and woodland caribou (*Rangifer caribou*) once extended their ranges southward from Canada into northern Wisconsin, although today both are extirpated from the state. The peppery red squirrel (*Tamiasciurus hudsonicus*), with its familiar scolding chatter, takes the place of the gray squirrel. The red-backed mouse (*Clethrionomys gapperi*) and woodland deer mouse (*Peromyscus maniculatus*) both scamper on the forest floor, pursued by the very rare lynx (*Lynx canadensis*) and

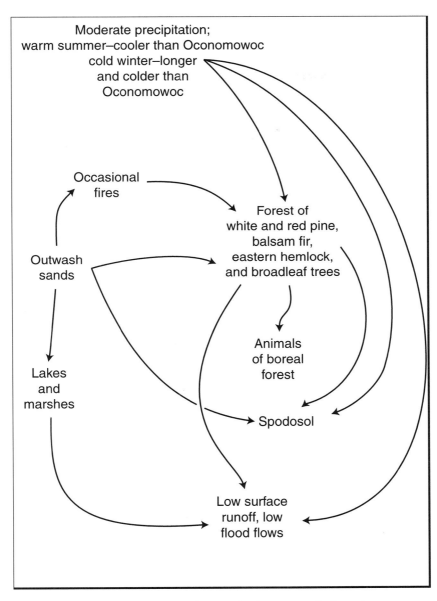

Figure 1.9. Environmental system for the site near St. Germain

pine marten (*Martes americana*). The birds that glean insects from the branches and leaves of the trees include the diverse group of wood warblers — yellow-rumped (*Dendroica coronata*) and black-throated blue (*Dendroica caerulescens*), northern Parula (*Parula americana*) and black-burnian (*Dendroica fusca*), Cape May (*Dendroica tigrina*) and magnolia (*Dendroica magnolia*). Ruby-crowned kinglets (*Regulus calendula*) forage with the warblers; gray jays (*Perisoreus canadensis*) patrol the forest edge; common loons sing their haunting calls from nearby quiet lakes.[22]

Taken together, the spot near St. Germain is an environmental system with similarities and differences from places elsewhere in Wisconsin (Figure 1.9). Its strong affinities to the northern boreal region of cold climate, spruce forest, and Spodosol soil allow us to claim a bit of the great northwoods as our own.

Wisconsin's environmental systems, as represented in this chapter by three locales in disparate parts of the state, vary locally as well as regionally. For example, by moving only a short distance from any one of the three specific sites, down from the upland and into an adjacent lowland, we would quickly encounter a freshwater marsh, with its own set of system characteristics: Groundwater drains poorly in the bottom of the valley or kettle hole; cattails (*Typha*) with hollow stems permit gas exchange even though their roots lie in a saturated soil; organic material from the marsh plants accumulates as undecomposed muck to form a Histosol (from the Greek *histos* for "tissue," a reference to plant tissue); sedge wrens (*Cistothorus platensis*) chatter from hidden perches, and otters (*Lutra canadensis*) swim after meals of fish; evaporation of water from the exposed marsh surface reduces the size of outflowing streams during periods of drought; cold-air drainage renders nighttime temperatures lower than those of the nearby upland. Other local excursions away from our initial starting points would take us into still more diverse combinations of environmental characteristics.

Whether across the state or across a valley, then, the environmental diversity of Wisconsin is rich. To learn about our natural endowment and to understand how its features are interconnected are first steps in the appreciation of and emotional attachment to our natural world, the landscape we call home.

Notes

1. Martin, *The Physical Geography of Wisconsin*, 69–77. The classic, and unmatched, reference to the landforms of and landform development in Wisconsin is L. Martin, *The Physical Geography of Wisconsin* (Madison, 1965). His perspective — extremely long time scales and evolution of the general topog-

raphy—is, however, different from the contemporary focus in geomorphology on shorter time scales, finer resolutions, and physical processes. See J. Knox et al., *Quaternary History of the Driftless Area,* Field Trip Guide Book Number 5 (Madison, 1982). For treatments of landform processes, see such works as R. Chorley, S. Schumm, and D. Sugden, *Geomorphology* (London, 1984), and numerous publications of the Wisconsin Geological and Natural History Survey.

2. W. J. Stroessner and J. Habeck, "The Presettlement Vegetation of Iowa County, Wisconsin," *Transactions of the Wisconsin Academy of Sciences, Arts and Letters* 55 (1966): 167–180. The best treatment of Wisconsin vegetation remains J. T. Curtis, *The Vegetation of Wisconsin: An Ordination of Plant Communities* (Madison, 1959).

3. Curtis, *Vegetation of Wisconsin,* 295–305. Other works that link vegetation structure to environmental setting include H. Walter, *Vegetation of the Earth* (New York, 1973); J. Vankat, *The Natural Vegetation of North America* (New York, 1979); M. Barbour and W. Billings, eds., *North American Terrestrial Vegetation* (New York, 1988). A particularly strong and detailed linkage between vegetation, soil, and landform is provided by F. G. Goff, "Upland Vegetation," in *Soil Resources and Forest Ecology of Menominee County, Wisconsin,* Bulletin 85, Soil Series 60 (Madison, 1967), 60–89.

4. S. R. Eyre, *Vegetation and Soils: A World Picture* (Chicago, 1963), 109–113. The soils of Wisconsin are treated in F. Hole, *Soils of Wisconsin* (Madison, 1976), and F. Hole, *Soil Guide for Wisconsin Land Lookers,* Bulletin 88, Soil Series 63 (Madison, 1980). For more detailed discussions of soils and soil development, see S. W. Buol, F. D. Hole, and R. J. McCracken, *Soil Genesis and Classification* (Ames, Iowa, 1980); P. Birkeland, *Soils and Geomorphology* (New York, 1984).

5. The standard references to birds and mammals of Wisconsin are S. D. Robbins, Jr., *Wisconsin Birdlife: Population and Distribution Past and Present* (Madison, 1991); and Hartley H. T. Jackson, *Mammals of Wisconsin* (Madison, 1961). Additional sources include S. A. Temple and J. R. Cary, *Wisconsin Birds: A Seasonal and Geographical Guide* (Madison, 1987); P. Ehrlich, D. Dobkin, and D. Wheye, *The Birder's Handbook* (New York, 1988).

6. Robbins, *Wisconsin Birdlife,* includes accounts of the modern status of bird species.

7. No single reference makes readily accessible a treatment of the hydrological characteristics of Wisconsin streams. However, a series of publications of the United States Department of the Interior and the Wisconsin Geological and Natural History Survey are easily readable. For extreme southwestern Wisconsin, see S. M. Hindall and E. L. Skinner, "Water Resources of Wisconsin: Pecatonica-Sugar River Basin," *Hydrologic Investigations Atlas HA-453* (Washington, D.C., 1973). See also R. P. Novitzki, *Hydrology of Wisconsin Wetlands,* United States Geological Survey Information Circular 40 (Madison, 1982). For more complete treatments of water, see L. Leopold, *A View of the River* (Cambridge, Mass., 1994); T. Dunne and L. Leopold, *Water in Environmental Planning* (San Francisco, 1977).

8. For a discussion of the weather and climate of Wisconsin, see R. Palm and A. deSouza, *Wisconsin Weather* (Minneapolis, 1983). For discussions of

atmospheric processes, see such books as Arthur Strahler and Alan Strahler, *Modern Physical Geography* (New York, 1992); R. Muller and T. Oberlander, *Physical Geography Today* (New York, 1984). A particularly good treatment of atmospheric processes in the middle latitudes is provided by J. Harman, *Synoptic Climatology of the Westerlies: Process and Patterns* (Washington, D.C., 1991).

9.   Palm and deSouza, *Wisconsin Weather*, 45–85.

10.   In addition to Curtis, *The Vegetation of Wisconsin*, see the map published by the U.S. Forest Service's North Central Forest Experiment Station, by R. Finley, *Original Land Cover of Wisconsin* (St. Paul, Minn., 1976).

11.   Species accounts are found in Robbins, *Wisconsin Birdlife*; and Jackson, *Mammals of Wisconsin*.

12.   Hole, *Soils of Wisconsin*, 61–70. See also the U.S. Department of Agriculture, Soil Conservation Service, *Soil Survey of Milwaukee and Waukesha Counties* (Washington, D.C., 1971).

13.   Martin, *The Physical Geography of Wisconsin*, 227–230, 278. See also W. C. Alden, *The Quaternary Geology of Southeastern Wisconsin*, U.S. Geological Survey Professional Paper 106 (Washington, D.C., 1918); D. Mickelson et al., *Late Glacial History and Environmental Geology of Southeastern Wisconsin*, Field Trip Guide Book, Number 7 (Madison, 1983).

14.   R. D. Cotter, et al., "Water Resources of Wisconsin: Rock-Fox River Basin," *Hydrologic Investigations Atlas HA-360* (Washington, D.C., 1969).

15.   E. L. Oakes and R. D. Cotter, "Water Resources of Wisconsin: Upper Wisconsin River Basin," *Hydrologic Investigations Atlas HA-536* (Reston, Va., 1975).

16.   J. Attig, *Pleistocene Geology of Vilas County, Wisconsin*, Wisconsin Geological and Natural History Survey Informational Circular 50 (Madison, 1985), 19–21.

17.   Martin, *The Physical Geography of Wisconsin*, 413–418.

18.   Curtis, *The Vegetation of Wisconsin*, 218–220.

19.   Vankat, *The Natural Vegetation of North America*, 104–106; Goff, "Upland Vegetation," 64–68.

20.   Palm and de Souza, *Wisconsin Weather*, 45–85.

21.   Eyre, *Vegetation and Soil*, 51–55. See also Hole, *Soils of Wisconsin*; and F. Hole, *Soil Survey of Vilas County, Wisconsin* (Washington, D.C., 1986).

22.   Species accounts are found in Robbins, *Wisconsin Birdlife*; and Jackson, *Mammals of Wisconsin*.

2

Wisconsin's Glacial Landscapes

David M. Mickelson

What would the Wisconsin landscape look like if it had not been glaciated? It would be very different, to be sure, but different in what details? Would it be flatter, or more deeply incised by stream valleys? Would bedrock ridges rise steeply above flat-floored bottomlands? Would soils look the same—our familiar dark soils revealed by plow and bulldozer—or would they be red clay residual soils like ones that can be seen in Missouri, Kentucky, and Tennessee? Would most of the state resemble the unglaciated southwestern part of Wisconsin? Clearly, our Wisconsin landscape has been much modified by the action of glaciers and the glacial climates of the recent past. These modifications are the theme of this chapter.

History of Glaciation

Throughout the Mesozoic (225 million to 70 million years ago) and much of the Cenozoic (the last 70 million years) the earth was without extensive glaciers. Temperature began to decline about 50 million years ago (Figure 2.1), presumably in response to continental plates changing position, particularly the movement of the Antarctic continent into a polar position. It is also hypothesized that mountain chains built along colliding plate margins changed atmospheric circulation

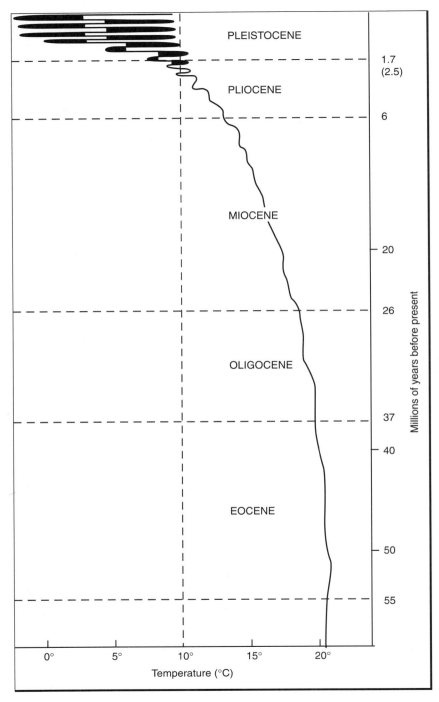

Figure 2.1. World temperatures began to cool about 50 million years ago, but the sharp cooling that initiated Pleistocene glaciers occurred about 2 million years before present.

patterns, which ultimately led to the cooling of the earth as a whole. Mountain glaciers became common, and probably by 20 million years ago glaciers of the western mountains of North America were as extensive as they are today. The great continental ice sheets, however, required further cooling in order to initiate and grow, eventually to spread southward to Wisconsin.

There is no record of when the earliest advance of ice took place in Wisconsin, but it was probably more than 700,000 years ago. How much earlier is not known, but there is evidence of glaciation in the western plains, as far south as Nebraska and Kansas, by 1.8 million years ago. The conclusion that early glaciations entered Wisconsin more than 700,000 years ago is based on lake and glacial sediment west of Eau Claire that has reversed magnetic polarity.[1] The polarity of the earth now (north is north, south is south) is considered "normal" polarity, but for considerable periods of earth history this pattern has been reversed (the north arrow of a compass would have pointed south). Because the earth has had a normal polarity for nearly all of the last 700,000 years, it is likely that these deposits in northwest Wisconsin are older than that.

It was recognized before the turn of the century by many glacial geologists (including Wisconsinites T. C. Chamberlin and R. T. Salisbury) that there had been multiple glaciations in the midcontinent region. They suggested (probably mostly on the basis of what was at the time becoming accepted in the European Alps) that there were four glaciations, each separated by an interglacial period. Without modern dating techniques it was impossible for them even to guess the length of the Ice Age or the age of the last glacial advance. The four glaciations were named the Nebraskan, Kansan, Illinoian, and Wisconsin, after states where deposits of each glaciation were well exposed. Unfortunately, the continental record of glaciation that they worked with is far less complete than that preserved in marine cores. Since the 1960s abundant evidence from fossil and isotope studies has demonstrated that huge continental ice sheets may have formed between 12 and 15 times during the last 2 million years. How many of these actually entered Wisconsin is not clear, the record is too fragmentary and we cannot date most deposits. The Wisconsin landscape is analogous to a chalk board that has been written on and mostly erased many times. Only the most recent scribings on the landscape are preserved for us to read.

Early Wisconsin Glaciation

The Badger State's namesake, the Wisconsin glaciation, was the final—and in some places in southern Wisconsin the most extensive—ice cover of all. We know the Sangamon interglacial ended about 110,000 years ago on the basis of the isotope record of deep sea cores and, more recently, ice cores from Greenland. Although there is evidence of early Wisconsin ice advances near Toronto, Ontario, there is little evidence of the extent of ice in the western Great Lakes area. That lack of evidence probably does not mean that glaciers failed to enter the region; rather, almost everywhere earlier deposits probably were mostly eroded and the remnants covered by younger materials laid down by the late Wisconsin glaciers. Ice almost certainly entered the Mississippi River basin because windblown silt, or loess, associated with glaciation blew off the Mississippi Valley and into southwestern Wisconsin during this time.[2] It was evidently a very cold period, with permafrost and barren, windswept tundra plains in northern and central Wisconsin.

Late Wisconsin Glaciation

Another ice sheet began to grow as early as 27,000 or 28,000 years B.P. (before present) and probably entered the state by 23,000 years B.P.[3] Because ice flow was controlled by the shape of the underlying topography, the southern margin of the ice sheet was lobate, or divided into lobes. The Lake Michigan lobe extended far down the Lake Michigan basin into central Illinois, but westward only a few tens of kilometers into Wisconsin, it joined the Green Bay lobe (Figure 2.2). The Green Bay lobe was funneled down the Green Bay and Lake Winnebago lowland to just south and west of Madison, and ice of this lobe actually flowed toward the northwest in Portage and Marathon counties, and in Langlade county, where it met the southwest-flowing Langlade lobe. The interlobate zone between the Lake Michigan and Green Bay lobes (over what is now the Kettle Moraine) was lower than the ice in the axis of the lobes on either side and formed a trough that sloped southwestward toward Walworth County.

Because of the east-west trend of the Lake Superior trough and the upland south of Lake Superior, ice flowing from the north was diverted toward the southwest to beyond Minneapolis and toward the southeast into the Langlade lobe. The extent of ice from this Superior lobe in Wisconsin is limited because of that topographical control. In fact, this diversion is probably why the Driftless Area of southwest Wisconsin was never glaciated. The east-west-trending Lake Superior basin has always diverted glaciers to the east and west, and allowed only lim-

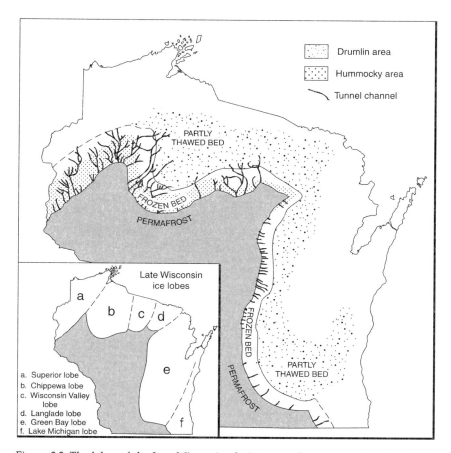

Figure 2.2. The lobes of the Late Wisconsin glaciers moved over ground surfaces that were either frozen or thawed. (Adapted from J. W. Attig, D. M. Mickelson, and L. Clayton, "Late Wisconsin Landform Distribution and Glacier-bed Conditions in Wisconsin," *Sedimentary Geology* 62.)

ited ice to cross the highland of the Upper Peninsula and northwest Wisconsin and to flow into the Wisconsin Valley and Chippewa lobes.

It was the late Wisconsin glaciation that had the most profound impact on the landscape. Numerous landforms such as moraines, drumlins, and eskers dominate the landscape of the glaciated part of the state. Thousands of lakes, all products of glaciation, dot the glaciated area. Our agricultural land is rich because the grinding action of glaciers produced silt-sized particles that weather easily, releasing nutrients for plant growth. Our roads and buildings are structurally strong in part because of the high quality aggregate that is often derived from glacial outwash. Deposits associated with glaciation also hold huge amounts of groundwater that in most places is of very good quality. Other glaciated deposits are the repositories of waste generated by our ever-increasing human population.

Glaciers

Glaciers, whether large or small, exist because more snow falls in the winter than melts in the summer. From this zone of accumulation the glacier expands to lower elevations (as in the case of a valley glacier) or southward (as the ice sheet that came into Wisconsin did), until melting is sufficient to limit its growth. Subsequent advances and retreats of the ice margin are caused by changes in the balance between accumulation and melting. A period of cold or snowy weather causes the margin to creep forward, and periods of warmer or less snowy weather cause the margin to retreat. The ice itself flows under its own weight from areas of thick ice to areas of thin ice, thus, always toward the margin. The flowing ice carries rock and soil to the edge in "conveyor belt" fashion, where it is deposited directly by the ice as till,[4] off the ice surface as debris flows, or is picked up by meltwater streams.

Effects on the Landscape

The overall effect of continental glaciers is to erode hilltops and fill valleys, thus reducing relief. That difference in relief on either side of the glacial border can be clearly seen in the state's topography, and it should also be obvious to anyone who has hiked or biked in much of the southern part of the state. Glaciation, of course, produces its own topographic forms, sometimes dramatic and sometimes subtle. Moraines are hummocky ridges that form parallel to the ice margin during times when the ice margin position is stable. Like a conveyor belt, the glacier brings sediment to the ice margin, where it accumulates. Where the ice edge remains in one position the "conveyor belt" delivers a greater thickness of sediment than between moraines, where the ice margin retreats fairly rapidly (Figure 2.3). Moraines are kettled, with a typical internal relief in southern Wisconsin of about 10 meters and in northern Wisconsin up to nearly 100 meters. Large erratics, rocks carried by the glacier from a distant source, often litter moraine surfaces.

The so-called Kettle Moraine, however, is not really a moraine at all, because it consists almost entirely of sand and gravel instead of

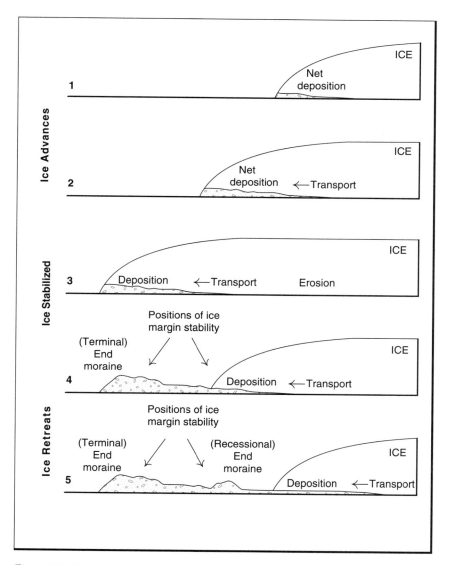

Figure 2.3. The various landforms created by glaciers depend, in part, on whether the ice front is advancing, stable, or retreating.

glacial deposits, or till. It is an accumulation of gravel carried by a river that ran under and on top of ice in the trough that formed when the Green Bay and Lake Michigan lobes pulled apart.[5] The many deep kettles, although not unique to the Kettle Moraine, are one of its dramatic landforms, as are large conical hills such as Lapham Peak and Holy Hill, west of Milwaukee. These conical hills probably formed as gravel in supraglacial streams tumbled down through openings in the ice and piled up at the glacier bed as "moulin kames."

Wherever meltwater from the ice surface found its way to the base of the glacier, it collected into tunnels and moved toward the ice margin. In most places these subglacial streams carried sediment to the ice margin and, when ice had melted, left no trace. Some, however, deposited sand and gravel, filling the tunnel. These were left as narrow winding ridges of sand and gravel called eskers. These are common in the Kettle Moraine and in other places where a substantial amount of coarse sediment was available to subglacial streams.

An airplane passenger flying between Madison and Milwaukee or Fond du Lac cannot help notice the streamlined landscape below. Over 5,000 drumlins, elongate hills paralleling the ice-flow direction, cover the landscape of the southern part of the Green Bay lobe. Smaller drumlin fields are present along the west edge of the Lake Michigan lobe deposits in Waukesha and surrounding counties, along the east and west sides of the former Green Bay lobe, and in the Langlade and Chippewa lobes in northern Wisconsin. Drumlins formed at the base of the glacier by the combined processes of erosion and deposition. From the ground they appear as individual elongated hills, but from the air one has the impression of a serrated knife having been scraped across the landscape. Drumlins have a peculiar distribution in that they are not present in the late Wisconsin age deposits of Ohio, Indiana, or Illinois, yet thousands occur in Wisconsin, Minnesota, and New York State. This may be due to differences in the temperature regime at the bottom of the glacier. More specifically, it seems likely that the outer 10 kilometers or so of the glacier had a frozen bed with no water present while the ice was advancing (Figure 2.2). We postulate that the glacier produced drumlins in Wisconsin, and only flat till plain 200 kilometers south in Illinois because part of the bed behind this frozen zone of the glacier in Wisconsin was frozen and part was wet, resulting in different processes from those of the unfrozen, wet bed of the glacier farther south.

The wealth of glacial features in Wisconsin has led to the development of Wisconsin's Ice Age Scientific Reserve, a series of nine parks jointly administered by the Wisconsin Department of Natural

Resources and the National Park Service. The purpose of these parks is to preserve and illustrate features produced by glaciation, and they are the first of their kind in the United States. The Ice Age Trail, which follows the outermost moraine of the late Wisconsin glacier throughout most of its length, crosses many of the features just described. Hiking the trail can provide a wonderful addition to one's knowledge of our glacial heritage.

Wisconsin Lakes

A prime attraction for Wisconsin residents and those of other states is the large number of lakes in the glaciated part of Wisconsin. There are no natural lakes in the Driftless Area, because millions of years of stream incision have removed the necessary dams and integrated the drainage. By far the most abundant lakes in Wisconsin are kettleholes. These forms are depressions created when masses of glacial ice are covered by sediment deposited directly from the glaciers or outwash deposited by meltwater streams during glacial retreat (Figure 2.4). Subsequent melting of the buried ice causes collapse of the overlying material and the development of a depression. In high parts of the landscape, kettles are often dry, but wherever they intersect the water table or where there has been accumulation of fine sediment in the bottom (which reduces infiltration), they are water filled. Thousands of kettleholes are present in Wisconsin.

Why are there so many kettles in northern Wisconsin and so few in areas farther south, particularly in Illinois? Most kettles form where ice along the retreating glacier margin becomes stagnant. Through melting, masses of ice become separated from the main part of the glacier. If burial does not take place, the ice block simply melts without leaving a trace. In many cases, however, sediment deposited by streams flowing over debris-rich ice buries the ice masses. As the ice continues to melt, sediment over and around the ice block collapses into depressions created by melting. Thus, an abundance of sediment on the glacier surface in its marginal areas is essential to the formation of kettles. Three conditions created this abundant supraglacial sediment in Wisconsin. One of these was the direction of ice flow relative to the regional slope; the second, the presence of a buttress of ice or frozen ground at the ice margin; and the third, the same presence at some lobe junctions. All of these created an upward (compressive) flow of ice near the margin, dragging material from the glacier bed to the glacier surface. This happened wherever ice flowed up the regional slope, such as along the west side of the Green Bay lobe and also in the interlobate zone

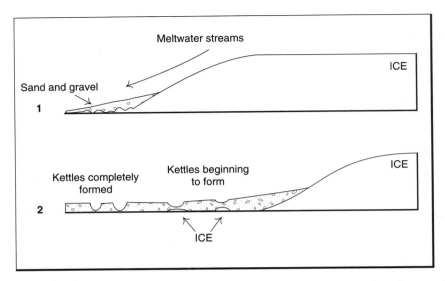

Figure 2.4. Kettleholes are formed when buried ice melts and the overlying sediment collapses.

between the Lake Michigan and Green Bay lobes, where intense compressive flow carried sediment to the glacier surface. Here deposits as much as 50 meters thick formed an insulating blanket over the uneven ice surface below. Differences in thickness of debris and localized concentrations of crevasses led to differential melting and the isolation of individual masses of ice that later became kettles.

The presence of frozen ground contributed to the creation of many lakes throughout the state. In the area previously occupied by the Chippewa and Wisconsin Valley lobes (Figure 2.3) in northern Wisconsin, ice flowed down the regional slope from the Northern Highland. Here, permafrost in front of the advancing glacier slowed the advance, causing compressive flow and intense upward movement of sediment from the base of the ice to the surface. The existence of permafrost conditions as the glacier retreated was also important in northern Wisconsin and in the Green Bay lobe because it allowed the accumulation of thick outwash over parts of the ice margin before much melting took place. Probably many of the kettlehole lakes in Wisconsin formed, not at the glacier margin, but long after the active ice margin had retreated, when temperatures rose and permafrost disappeared about 13,000 years ago.[6]

Lakes are also produced by glacial scour. Glaciers erode their bed, but, unlike streams, they have the ability to scour deep, broad basins into the underlying rock, which, when no longer occupied by ice, are water filled. Classic examples of this overdeepening by glacier ice are the Great Lakes basins, Green Bay, and Lake Winnebago. These basins are broad and long and carved into the underlying rock.

Devils Lake, on the other hand, is a moraine-dammed lake. It was a through-flowing river valley before the Green Bay lobe advanced to its maximum position, blocking the north and south ends of the valley and building a moraine.[7] When the ice retreated, Devils Lake was isolated from the rest of its former valley by the moraine and outwash deposits. In other places, lakes formed because preglacial valleys were only partly filled with glacial sediment. The isthmus area of Madison lies over an 80-meter-deep valley that was part of an integrated stream network before the advance of the Green Bay lobe.[8] Deposits from the glacier partly fill this main valley and its tributaries, but depressions like Lakes Mendota, Monona, Wingra, and Waubesa, not filled completely, remain as basins.

Economic Effects

Economically, glaciation has enriched Wisconsin in several ways: by producing the accumulation of aggregate sources, by creating favorable conditions for waste disposal, and by improving conditions for agriculture. Each of these will be treated individually.

Aggregate Resources

All concrete that is used for roads, house foundations, buildings, and other structures and all asphalt that is used for roads and parking lots and driveways require vast amounts of aggregate—the coarse particles that form their framework. Aggregate typically comes from two sources: sand and gravel deposited by glacial meltwater and dolomite rock that lies beneath. In 1993, of the aggregate used in Wisconsin, about half came from each source and had a combined value of over $66 million. This is by far the largest mining activity in the state, and the tax revenues on the raw aggregate as well as on the finished concrete and asphalt are substantial. Unlike metals, crushed stone and sand and gravel have a relatively low value per unit weight; thus, location is an important criterion for their usefulness. Generally, beyond 10 miles of transport, every doubling of the distance of hauling doubles the price of the aggregate, considerably increasing the cost to the private and public sectors of any construction project. Fortunately, good-quality sand and gravel were deposited in several settings. Outwash is particularly important. It accumulated very quickly (in modern streams, up to 0.5 meter per year) and during ice retreat was deposited in many low areas of the landscape. The best-quality aggregate for concrete contains very little silt and clay. Thus, braided outwash streams, whose channels continually cut back and forth across the floodplain winnowing silt and clay from the gravel, are perfect environments for the accumulation for this sediment. Other more localized deposits of sand and gravel were left at the point of contact with the glacier ice. Eskers were an early source of sand and gravel because they were easily recognized and easily mined. Thus, many of the eskers that were in southern Wisconsin have been mined away and used for roads or buildings. In places where sand and gravel were deposited on the glacier margin or over masses of separated ice, hummocks composed of this sand and gravel occur, and these are also sources of aggregate in many parts of the state.

Waste Repositories

At the other end of the grain-size spectrum, silt and clay are abundant in till of some glacial advances. Particularly in the Great Lakes and Green Bay–Lake Winnebago basins, and in Lakes Michigan and Superior, silt and clay accumulated on the lake bottoms between each of the late-glacier advances. Each subsequent ice advance incorporated these fine sediments and deposited clayey till. Although not ideal, these deposits are the best location for sanitary landfills in the state. The clayey sediment has relatively low hydraulic conductivity (transmits water slowly), and where it is thick, it retards the movement of contaminants into aquifers below. Fortunately, many of Wisconsin's population centers and concentrations of industry are in eastern Wisconsin, within the area covered by these clayey tills and fine-grained lake sediment.

Impact on Agriculture

Glaciation is directly reponsible for the gently rolling terrains of central and eastern Wisconsin, which are so productive agriculturally. In addition to having more gently rolling land, glaciated landscapes have soils that are naturally productive. During glacial transport, sediment is worn and broken into smaller sizes. In particular, large amounts of silt are produced. When exposed to water and air in the soil, these readily break down chemically, releasing nutrients that can be used in plant growth. In addition the clay produced by this weathering process is effective at holding moisture and nutrients for use by plants.

Even the Driftless Area of southwest Wisconsin reaped this benefit of glaciation, because windblown silt, or loess, was picked up from the outwash plain of the Mississippi River and blown eastward. It accumulated up to 10 meters thick in places close to the Mississippi, to about 0.5 meter thick near Madison, and to less still along Lake Michigan. If it were not for glaciation, much of the Driftless Area would be covered with thin residual soils depleted of nutrients, and they would be much less productive than they are today. Loess is thin or absent in much of northern Wisconsin, although areas in front of the ice margin, for instance, the Antigo "flats," received more than a meter.

Summary

Much of Wisconsin's landscape has been molded by glaciation. Coarse-grained outwash and ice-contact-stratified deposits provide millions of dollars worth of aggregate each year, and the fine-grained sediments that accumulated in lakes along the glacial margin, or as till at the ice margin, provide waste-disposal sites that would have a much higher risk factor if these sediments were not present. Freshly ground clay and silt also produce natural fertility in our soils, and the smoothing effect of continental glaciation on the land surface has made much more land available for cultivation than would have existed without glaciation. The thousands of lakes left by the retreating ice provide recreation for hundreds of thousands of summer tourists from Wisconsin and other states. Finally, the aesthetically pleasing glaciated landscape contains landforms such as moraines, kettles, drumlins, and eskers that provide recreational and educational opportunities in many of Wisconsin's county and state parks. Wisconsin, both economically and aesthetically, is a land made richer by its history of glaciation.

Notes

1. R. W. Baker et al., "Pre-Wisconsin Glacial Stratigraphy, Chronology, and Paleomagnetics of West-Central Wisconsin," *Geological Society of America Bulletin* 94 (1994): 1442–1449.

2. D. S. Leigh and J. C. Knox, "Loess of the Upper Mississippi Valley Driftless Area," *Quaternary Research* 42 (1994): 30–40.

3. J. W. Attig, L. Clayton, and D. M. Mickelson, "Correlation of Late Wisconsin Glacial Phases in the Western Great Lakes Area," *Geological Society of America Bulletin* 96 (1985): 1585–1593.

4. *Till* is a genetic term used for sediment deposited directly by the gla-

cier. The term *diamicton* is a descriptive term to describe poorly sorted sediments with a large range in grain size deposited in a variety of environments. It includes till, mudflow and landslide deposits, and so on.

5. For a discussion of the Kettle Moraine, see J. W. Attig, "Glacial Geology of the Kettle Moraine," *Wisconsin Natural Resources* 10 (1986): 17–20.

6. J. W. Attig, D. M. Mickelson, and L. Clayton, "Late Wisconsin Landform Distribution and Glacier-bed Conditions in Wisconsin," *Sedimentary Geology* 62 (1989): 399–405. For a more regional view of chronology and conditions see D. M. Mickelson et al., "Late Glacial Record of the Laurentide Ice Sheet in the United States," in H. E. Wright, Jr., ed., *Late Quaternary Environments of the United States*, Vol. 1: *The Late Pleistocene*, ed. S. C. Porter (Minneapolis, 1983), 3–37.

7. J. W. Attig, et al., *Ice Age Geology of Devil's Lake State Park*, Wisconsin Geological and Natural History Survey Educational Series 35 (Madison, 1990). See also K. I. Lange, *Ancient Rocks and Vanished Glaciers: A Natural History of Devils Lake State Park* (Madison, 1989).

8. D. M. Mickelson, *A Guide to Glacial Landscapes of Dane County*, Wisconsin Geological and Natural History Survey, Field Trip Guide Book 6 (Madison, 1983).

3

Challenges of Wisconsin's Weather and Climate

Waltraud A. R. Brinkmann

No natural characteristic of Wisconsin is more important to its land and life than its weather and climate. The blankets of snow and numbing arctic cold of winter, the increasing warmth and daylight in awakening spring, the hazy skies and thunderstorm downpours of summer, the crisp and colorful afternoons in fall—all are familiar parts of the Wisconsin landscape, welcome ingredients to a Wisconsinite's sense of place. To describe and understand these atmospheric components of the Wisconsin scene are the purposes of this chapter.

The Climate Controls

No two places on earth have precisely the same climate, but similar climates can be grouped into types. Wisconsin's is a "temperate continental" type. Only a few regions in Eurasia have a climate similar to Wisconsin and surrounding states. There is nothing like it in the Southern Hemisphere. The reason is the relatively unique combination of climate controls.

Wisconsin's location in the interior of the second largest continent, North America, is one important controlling factor. Land surfaces heat and cool more rapidly than water. Oceans, on the other hand, are great stabilizers of temperature. By virtue of its interior location, Wisconsin is subject to greater climatic variations than might be found in more coastal locations, such as western Oregon along the Pacific coast. That area lies in the same latitude belt and receives the same amount of solar radiation as Wisconsin, but has a maritime climate with milder winters and cooler summers. Lakes Superior and Michigan do moderate Wisconsin's climate, but their effect is not as pronounced as the Pacific Ocean's effect on Oregon.

The state's location relative to the general circulation pattern of the atmosphere is another important climate control. In middle latitudes, a broad westerly current circles the globe at upper levels and westerly winds prevail at the surface. The upper-level current, which has a wavelike structure, guides the movement of surface weather systems such as extratropical cyclones (or storms). The convergent nature of extratropical cyclones causes air masses from different directions to be drawn into their centers.

Mild Pacific air masses are the most frequent in Wisconsin. These air masses are relatively dry because moisture is wrung out of them as they cross the Rocky Mountains to our west. Because the terrain east of the Rockies is quite flat, Wisconsin is open to invasions by extremely cold arctic air masses from the north and very warm and humid tropical air masses from the Gulf of Mexico and the tropical Atlantic. This is very different from what happens in similarly centrally located places in the other great Northern Hemisphere land mass: Eurasia. In Eurasia, the main mountain systems run east-west. As a result, the moderating influence of the Atlantic Ocean is carried farther inland, and humid tropical air masses are lacking. Warsaw, Poland, for example, is located at about the same latitude as Winnipeg (nearly 300 miles north of Wisconsin's northern border), yet its climate is similar to Wisconsin's, although its winters are slightly milder and its summers are cooler as well as drier.

Climate varies from year to year. In a warm year, Wisconsin's climate is more like that of Illinois to the south; in a cold year, it is more like that of western Ontario to the north. Large interannual climate variability is a characteristic of most midlatitude locations that are influenced alternately by arctic and tropical air masses. This variability has a profound effect on human society; it makes such things as designing structures and planning (of a food supply, for example) difficult. This problem was formally recognized early in the century. The 30-year period 1901–1930 was adopted internationally as the climatic "normal" period. The hope was that averages for this period would provide a minimum error estimate that could be used in predicting future conditions. However, long-term climatic change was not taken into account. Today, by international agreement, the normal period is always the most recent 30 years and is changed every 10 years. For example, the normal period for 1995 is 1961–1990.

The climate of a region is the sum total of all the weather experienced there. It consists of the averages, or normals, and the extremes. These weather conditions are best portrayed by describing the march of the season.

Wisconsin Weather and Climate Round the Year

Temperature

Seasonal temperature changes result from seasonal changes in both incoming solar radiation (due to the inclination of the earth's axis in relation to the sun) *and* the frequency and character of air masses. In winter, the low amounts of energy from the sun mean that the air over Wisconsin will be cold. Pacific air prevails during about three-quarters of the winter days, keeping winters from being even more frigid than they actually are. Cold, dry arctic outbreaks occur on the remaining days. Arctic outbreaks are colder and more frequent over northern Wisconsin, which is closer to the source of arctic air, than in the south. Late January is generally the coldest time of year. The map of normal daily minimum temperatures for January (Figure 3.1) shows how Lake Michigan moderates the minimum temperatures over eastern Wisconsin. During spring, solar radiation intensity increases, and the general circulation patterns of the atmosphere migrate poleward. The intrusion of arctic air masses becomes less frequent. This air also becomes warmer because the source, arctic Canada, warms up. Tropical air masses begin to reach northward into Wisconsin.

In summer, mild Pacific air masses and warm, humid, tropical air masses are about equally frequent in Wisconsin. It is the tropical air masses of July and August that bring our warmest summer days. Normal daily maximum summer temperatures range from the upper 80s over the southern counties to the lower 80s over the northern counties. The cooling effect of the Great Lakes, particularly of Lake Superior, is quite pronounced (Figure 3.2). As the environmental temperature approaches body temperature, a growing number of people feel physical discomfort, particularly when the humidity is high. Over the southern counties, temperatures rise into the 90s on a number of days each summer, and occasionally exceed 100° F. At those times, the cooler northern counties become especially attractive. The lower temperatures and humidities, plus the many forests, lakes, and streams, draw thousands

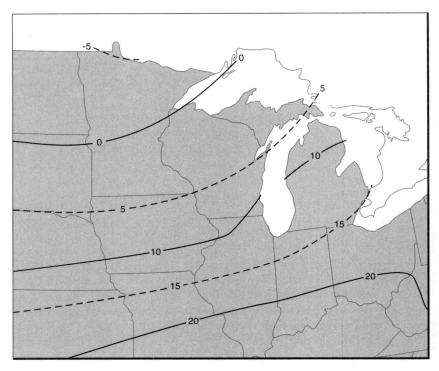

Figure 3.1. Normal daily minimum temperature (°F) for January

of tourists from southern Wisconsin, Illinois, and Minnesota to the northwoods. Tourism is, in fact, one of the state's three major sources of income.

In fall, the general circulation patterns migrate toward the equator. Humid, tropical air becomes less common, and declining solar radiation cools the air. Arctic air masses begin to invade Wisconsin more frequently, and they turn colder. When the minimum temperature drops below 32° F, frost threatens agricultural crops. This, however, is not a major climate hazard in Wisconsin, except for cranberry growers, because farmers select crop varieties that correspond to Wisconsin's growing season. Damaging early frosts can occur though. One of the most devastating freezes occurred on September 22, 1974. Widespread frost damaged 20–80 percent of the corn crop; 90 percent of the soybean crop in the southern counties was destroyed. Total crop loss was estimated at $200 million.

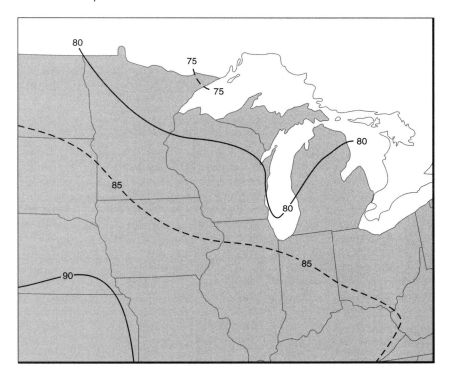

Figure 3.2. Normal daily maximum temperature (°F) for July

Precipitation

While air mass characteristics and frequencies help explain seasonal temperature variations, the nature and location of air mass boundaries explain seasonal precipitation and such weather extremes as hail and tornadoes. Along air mass boundaries, or fronts, the warmer of the two air masses is forced to rise. Since rising air cools and loses its ability to hold moisture, clouds often form within it. If the clouds become sufficiently tall, precipitation is produced.

Winter. In winter, when air temperatures are below freezing, precipitation is usually in the form of snow. Since colder winter air contains less moisture, Wisconsin's winter precipitation is relatively light, and, in fact, January and February are the driest months of the year. Most of the state receives the liquid-water equivalent of only 1–2 inches per month. However, winter precipitation totals appear impressive, because it takes about 10 inches of snow to equal 1 inch of rain. Most of the state receives about 40–50 inches of snow during an average winter season. Twice that much is received in the Lake Superior snow belt, a narrow belt across the extreme north that is affected by the lake. The lake-effect snowfalls occur when cold, dry arctic air passes over the lake. The air picks up water vapor and then drops it as prodigious amounts of snow along the immediate downwind shores.

Winter storms are sometimes accompanied by strong winds, which can form snowdrifts, occasionally in excess of 10 feet. We remember the "Blizzard of 1978" for its enormous 15-foot drifts. Stranded travelers filled up schools, churches, and motels. Numerous schools closed and the University of Wisconsin suspended classes at the Madison and Milwaukee campuses (an almost unheard of event).

On rare occasions, the air temperature aloft is above freezing and rain forms. This freezes instantly when it hits cold objects near the ground, such as power lines or trees. One of the most destructive ice storms ever to hit the state occurred in March 1976. Ice accumulations up to 5 inches thick and winds in excess of 50 miles per hour snapped trees and utility poles and downed power and telephone lines. Many roads were completely blocked by fallen trees, poles, and wires. Up to 100,000 people were without power at the height of the storm; some were without power for more than 10 days. Many counties declared a state emergency, and the total damages exceeded $50 million.

Few lives are lost as a direct result of winter storms (from exposure, for example). More frequently, snow and ice storms are an indirect cause of death. People die from heart attacks while shoveling snow or from skidding on icy roads. Property damage and social disruption can, however, be substantial, particularly in the more urbanized southern counties.

In rural areas, in parks, and in forests, snow is considered a valuable gift from nature, permitting various types of winter recreation. This is particularly true for the northern counties. There, the snow cover season is one of the longest in the country (Figure 3.3). Furthermore, millions of acres of forest have been set aside for public use in northern Wisconsin because they are unsuitable for agriculture. This makes the northwoods one of the best winter playgrounds.

Spring. In spring, the still relatively cold arctic air begins to clash with the tropical air masses that are reaching northward, and thunderstorms become more frequent in Wisconsin. Thunderstorms form when the air is humid and unstable. In spring and early summer, solar radiation becomes more intense, resulting in strong surface heating. This, in turn, heats the air near the surface, causing it to become unstable and rise. It is this rising moist, unstable air that tends to produce

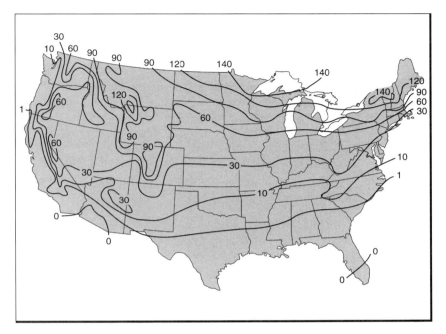

Figure 3.3. Number of days per year with snow cover

thunderstorms. The most intense thunderstorms are those in which the rising air receives additional uplift from the presence of a front. In fact, lines of severe thunderstorms often form ahead of cold fronts, the leading edge of masses of cool air. Thunderstorms and associated weather systems move across the state guided by the upper-level winds. These tend to travel southwest to northeast, reflecting the direction in airflow aloft.

Summer. Thunderstorms tend to be most numerous in summer. Their frequency of occurrence is usually reported as the number of days per year in which thunder is heard at least once at a location. In the tropics, where the sunlight is intense year-round, almost every day is a thunderstorm day; at the poles there are none. Wisconsin, because it is more poleward than most other parts of the country, has relatively few days with thunderstorms. Only the West Coast and the New England states have fewer, because, in those locations, the cool coastal waters are a stabilizing influence (Figure 3.4). In Wisconsin, the number ranges from about 40 in the south to about 30 in the north. Almost all these storms occur during the warm season, but damaging thunderstorms and lightning have been reported in all calendar months.

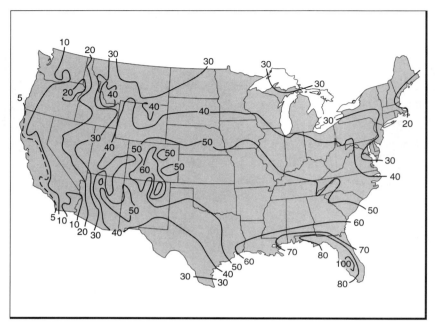

Figure 3.4. Number of days per year with thunderstorms

Lightning forms within tall clouds when positive electrical charges accumulate in the upper portion of a cloud and negative charges accumulate in the lower portion. Air is a poor conductor, but when the accumulation reaches a critical value, a spark or lightning strike occurs. Most lightning takes place within the cloud. About once in every four strikes, negative charges from the cloud bottom move downward, seeking the shortest distance to the ground (i.e., the tallest object) or a good conductor of electricity, such as metal.

Farm buildings and churches are usually the tallest objects in rural areas, and much property loss is due to lightning-caused fires in rural buildings. Another source of property loss, particularly in Wisconsin, is livestock killed while standing in open fields, near wire fences, or under trees. People, too, are in danger in such open spaces, as well as in parks and on golf courses. In Wisconsin, 38 percent of all injuries and deaths from lightning have occurred in such locations, 6 percent on boats and near water, and 3 percent near heavy equipment. There also have been several reported injuries when lightning struck telephone lines. An average of six or seven injuries and fatalities occur per year, about 2 percent of the national total. This is comparable to the state's population size, which is 2 percent of the U.S. population.

Florida, with 5 percent of the U.S. population, accounts for 11 percent of all lightning-caused deaths and injuries.

All Wisconsin thunderstorms produce rain. They are the source of most of the state's precipitation, particularly in summer. An individual storm usually lasts 6 hours and affects an area of less than 400 square miles. A 1.5-inch rainfall total from such a storm is fairly common in Wisconsin. A five-inch or greater downpour in 6 hours is another matter. Such high-intensity rainfalls are the cause of flash floods, but they are relatively rare, occurring only about once in a 100 years. (This expression is a probability at a single particular point—like the number of days with thunderstorms—and several reports of such an event somewhere within Wisconsin are to be expected within a 100-year period.) Because thunderstorms cover relatively little land, flash floods are very localized and total property losses tend to be small. The potential for loss of lives is high, however, since flash floods develop so rapidly that timely warnings are difficult.

Hail is another potentially damaging component of thunderstorms. Hailstones grow when strong updrafts within the cloud hold the stone up in the cold upper regions of the cloud or carry it repeatedly into that region of the cloud. The stone falls out of the cloud when it becomes too heavy for the strength of the updraft. Small stones, produced by weak updrafts, may melt on the way down and reach the ground either as rain or as harmlessly small stones. Only the most severe thunderstorms have strong enough updrafts to produce damaging hail.

Wisconsin lies along the northeastern edge of the North American region of maximum hailstorm occurrence. In eastern Wyoming and eastern Colorado, the center of hailstorm activity, the average number of days a year with hail is six or more (Figure 3.5). In Wisconsin, it ranges from about three in the south to two in the northeast. Hail damage to property is relatively small; it involves damage to buildings (roofs, windows, aluminum siding) and to automobiles. Hail damage to crops is considerable and is one of the most important agricultural hazards.

Sometimes a giant twisting funnel extends down from the base of a thunderstorm cloud. The tornado, which is a funnel cloud reaching the ground, is the most violent weather phenomenon. Its core is an area of intense low pressure. Air is drawn inward and upward at speeds estimated to reach 300 miles per hour. The destructive power comes from the strong winds and from the explosive effect of the rapid pressure drop when a tornado moves over a structure. The path of destruction is generally less than 50 yards wide and 3 miles long. Some tornadoes are extremely violent, causing severe damage, while others are weak.[1]

No part of the world experiences more tornadoes than the central

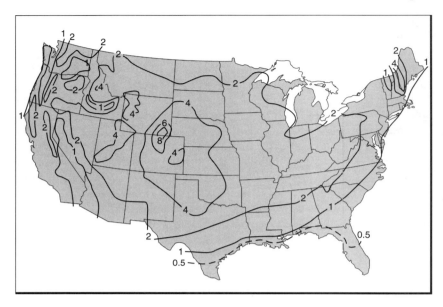

Figure 3.5. Number of days per year with hail

United States. Wisconsin lies along the northern edge of this region (Figure 3.6). Annually about 19 tornadoes are reported in Wisconsin, or 3 or 4 per 10,000 square miles, compared with 7 or 8 per 10,000 square miles in Oklahoma, in the heart of the U.S. tornado alley. One of the worst recent tornado disasters in Wisconsin struck Barneveld in 1984. It killed 9 people and injured 197. It leveled 90 percent of the village, causing $25 million in damages. There were two reasons for this incredible destruction: It was an intense tornado, and it traveled northeasterly along Main Street, systematically destroying structures along the linear business district. Another great tornado in 1878 passed just 10 miles south of Barneveld. It stayed on the ground for an incredible 150 miles, killed 19 and injured at least 45 people. Examination of tornado tracks, particularly the long-track tornadoes, suggests that there are two Wisconsin tornado alleys: a southern alley and a more pronounced west-central alley. These two alleys may be related to the preferred tracks of extratropical cyclones and associated fronts, but they may also reflect the gentle topography that tornadoes prefer.[2]

Most thunderstorms are beneficial because they are the main source of precipitation. Only about seven or eight of all the storms that occur in Wisconsin in a year cause more than a million dollars in damage;

these are most often *lines* of thunderstorms. Of these, one is usually a major hailstorm (i.e., most of the damage is done by hail), one or two are major tornadic storms, one is a windstorm, and another generates high-intensity rainfall and causes a major flash flood.

Every so often an extremely disastrous event occurs. One of the most devastating lines of storms in recent years moved through west-central Wisconsin in July 1980. Intense lightning, large hail, locally heavy rains, strong winds, and tornadoes devastated the area. At the Eau Claire Municipal Airport winds of 112 miles per hour were recorded before the anemometer blew away. Total damage exceeded $240 million.

Severe weather tends to be most common in early summer. June has the largest number of hailstorm days and the largest number of tornado days in most of Wisconsin. June is also the wettest month. Four to five inches are normally received over most of the state. In July, severe weather and precipitation begin to decrease as the northern landmasses warm up and the contrast between arctic and tropical air masses decreases. The intensity of extratropical cyclones decreases, and they drift farther poleward.

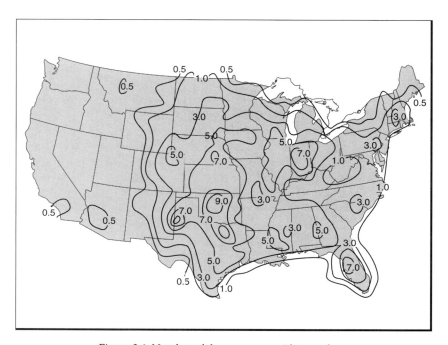

Figure 3.6. Number of days per year with tornadoes

Wisconsin's agriculture benefits from the precipitation maximum in early summer, the time when plant growth is at its peak. The warm, dry days of late summer are ideal for maturing and harvesting. There is, however, the danger of drought as temperatures continue to increase into July and August, the warmest months of the year. Very dry summer months can result in serious crop losses, since temperature regulates the water needs of plants.

Fall. Solar radiation intensity begins to decrease in fall. The general circulation pattern begins to shift southward, and temperature contrasts between air masses intensify once again. Frequently, there is a brief increase in thunderstorms in Wisconsin in early fall before the activity retreats southward toward the Gulf Coast.[3] Rain, too, often increases briefly to a secondary maximum.

Interannual Variability

One year is never like the next. Year-to-year variations occur because of variations in the location and strength of the upper-level winds and associated changes in storm tracks and air mass frequencies. Cold winter months are the result of a frequent occurrence of arctic air masses. Warm winter months are warm, not because of days with unusually high temperatures, but because of few occurrences of cold arctic air (Figure 3.7). In summer, the reverse is true: The frequency of the occurrence of warm air masses makes the difference between warm and cool summer months.[4]

The summer of 1988 was one of the warmest in recent years. June 1988 was not the warmest June on record; neither was July or August. But the mean temperature for all three was significantly above normal, and this three-month-long sequence of above-normal monthly means made the summer of 1988 the warmest on record for most of the state.

Warm summer months bring drought conditions, in part because the high temperatures increase the water need of plants, and in part because the rain-producing frontal boundary is farther north. Serious agricultural droughts are not generated, however, by just one or two months with below-normal precipitation but rather by a prolonged dry period.

Summer precipitation in 1988 was below normal, but the year's drought actually began to develop in spring. The April to August precipitation total that year for most of the state was one of the lowest on record. Communities were forced to restrict water use. Emergency bans on outdoor campfires and smoking were widespread. Numerous Fourth of July firework celebrations were either canceled or severely

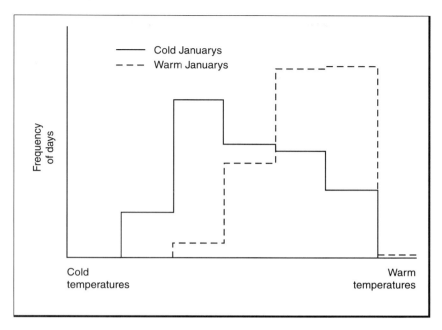

Figure 3.7. Frequency distribution of daily minimum temperatures for three cold and three warm Januarys

curtailed. Hydroelectric power generation was down. All 72 counties were declared disaster areas.

This drought's many indirect and delayed effects on farmers are a good example of why it is almost impossible to estimate the total economic impact of a drought. The 1988 drought resulted in the loss of about half of the state's hay and corn crops. The resulting higher feed prices were somewhat mitigated, however, by increased milk prices. Some farmers reduced the size of their dairy herds; others ceased operation. The drought had its greatest impact on farmers who were already under economic stress, rather than on farmers located in areas of the most severe conditions.[5]

An earlier drought of similar magnitude, the drought of 1976, is noteworthy because it stirred an interest in weather modification in the state. A common approach to rainfall enhancement involves silver iodide, which when introduced into a cloud can trigger formation of precipitation particles. Cloud seeding therefore requires the presence of suitable clouds. Evidence supporting the effectiveness of cloud seeding in certain weather situations has grown in recent years, but its effectiveness in a climate like Wisconsin's has yet to be established. The lack of clouds during drought periods makes it unlikely to work well.

The early 1960s were another noteworthy period of relative dryness in the Great Lakes region. This period is not generally referred to as a drought, because we tend to define a drought in terms of its impact on vegetation. The dry years of the early 1960s caused relatively little damage to vegetation because it was a "cold" drought. It was the result of a southward displacement of the upper-level flow and the predominance of cool, dry, arctic air over the region. The low summer temperatures reduced the water needs of plants and thus had a compensating effect. The water levels of the Great Lakes, however, which in summer respond strongly to precipitation, were significantly reduced and record low levels were established for Lake Michigan.[6]

Just as a major drought takes more than one dry month to develop, major regional flooding happens after several wet months. Such flooding usually results from a frequent passage of extratropical cyclones and associated fronts. Rainfall intensities do not have to be high. A drizzle every other day will eventually saturate the soil, and any further rainfall will enter streams as runoff and cause river levels to rise. Such floods take weeks and months to develop and are therefore relatively easy to predict. Few lives are lost from such floods, but damages are high because a large area is affected. The "Flood of 1993" is a good example of such regional flooding. Each summer month of 1993 was wet, but generally not the wettest on record. It was the combination of a preceding wet winter and spring *and* a very wet summer that produced the record or near-record flood stages on many rivers. Losses (including reduced crop yields due to wet weather) are estimated to have been near $1 billion.

These recent extreme events highlight one of the characteristics of Wisconsin's weather and climate: its variability. Going back in time, we find records of unusual events of equal or even greater magnitude. The worst multiple-year drought occurred in the 1930s. The cold wave of January 1922 produced the all-time record low for the state (−54° F at Danbury). The New Richmond tornado of 1899 killed 117 people. In the "Great Snowstorm of 1867" Milwaukee's "horse cars did not attempt to run" and travel ceased "on both the wagon and railroads leading out of the city."[7]

Climate Change

Estimates of global temperatures for the past 100 years or so, based on instrumental records, show that the current decades, the 1980s and

1990s, have been the warmest of the century. The 1930s were another warm decade. The 1800s, by contrast, represent the end of a several-centuries-long period of coldness known as the Little Ice Age. Temperature trends for Wisconsin have generally been in tune with these global trends. In the mid-1800s, average temperatures were perhaps 1° or 2° F lower and winters were snowier.[8] A warming trend into the 1900s culminated in the 1930s. Many record high temperatures were established during that decade. The all-time high for the state, 114° F, was recorded in July 1936 at the Wisconsin Dells. The 1960s and 1970s were cold,[9] but a renewed upward trend began in the 1980s. Several record high temperatures established in the 1930s were broken in the summer of 1988. The decreased snow cover over southern and central Wisconsin in the 1980s and 1990s has had a considerable effect on winter recreation.

What kinds of trends will the future bring? Climate change is difficult to predict because we do not completely understand the nature and relative importance of such causes of change as greenhouse gases, volcanic activity, or industrial and agricultural dust. Greenhouse gases prevent much of the upward ground radiation from escaping to space. The effect is to raise the temperature of the lower atmosphere and the earth's surface. Volcanic activity and dust decrease the amount of incoming solar radiation and thus have a cooling effect. To what degree these and other phenomena may cancel each other is currently a major topic of research. Furthermore, as the global temperature changes, so does the general circulation pattern, which has significant regional effects. How all of this translates into climate change for Wisconsin is difficult to say, since computer models of the atmosphere are still too coarse to represent such fine detail adequately. However, if the growing concentrations of atmospheric greenhouse gases turn out to be the most important influence in the near future, then we could see a year-round warming, which would harm some activities in the state and benefit others. For example, higher temperatures would result in more evaporation, higher water needs by plants, and thus more frequent agricultural drought. On the other hand, the danger from flooding might decrease. Lowered Great Lakes water levels due to increased evaporation would reduce shoreline erosion, but would hamper the movement of ships through shallow connecting channels. It would take less energy to heat buildings in winter but more to cool them in summer. Much of the winter precipitation might fall in the form of rain, and this would curtail winter sports activities, but road transportation would benefit from reduced ice and snow hazards. Wisconsin's weather and climate will continue to be challenging.

Acknowledgment

The author thanks the "Global Climates" class of 1995 for helpful comments.

Notes

1.   To distinguish strong from weak tornadoes, the Fujita Scale was developed to classify these storms in six intensity categories, from F0 (light damage) to F5 (incredible damage).

2.   R. G. Gallimore, Jr., and H. H. Lettau, "Topographic Influences on Tornado Tracks and Frequencies in Wisconsin and Arkansas," *Transactions of the Wisconsin Academy of Sciences, Arts and Letters* 58 (1970): 101–127.

3.   W. A. R. Brinkmann, "Severe Thunderstorm Hazard in Wisconsin," *Transactions of the Wisconsin Academy of Sciences, Arts and Letters* 73 (1985): 1–11.

4.   W. A. R. Brinkmann, "Variability of Temperature in Wisconsin," *Monthly Weather Review* 111 (1983): 172–180.

5.   J. A. Cross, "1988 Drought Impacts among Wisconsin Dairy Farmers," *Transactions of the Wisconsin Academy of Sciences, Arts and Letters* 80 (1992): 21–34.

6.   W. A. R. Brinkmann, "Comparison of Two Indicators of Climate Change: Tree Growth and Lake Superior Water Supplies," *Quaternary Research* 32 (1989): 51–59.

7.   D. M. Ludlum, *Early American Winters II, 1821–1870* (Boston, 1968), 174.

8.   E. W. Wahl and T. L. Lawson, "The Climate of the Midnineteenth Century United States Compared to the Current Normals," *Monthly Weather Review* 98 (1970): 259–265.

9.   W. A. R. Brinkmann, "Secular Variations of Surface Temperature and Precipitation over the Great Lakes Region," *Journal of Climatology* 3 (1983): 167–177.

4

Some Patterns in the Earth Beneath Our Feet

Francis D. Hole

On April 15, 1994, large silver maple trees (*Acer saccharinum* L.) showered bright red bud scales into the intersection of Dunning Street and Sommers Avenue in east Madison, Wisconsin. The resulting short-lived red carpeting on the gray pavement was beautiful. By comparing the colors of pavement and of bud scale–carpet with standard color chips of an international scientific Munsell color chart (with which soil scientists categorize the colors in soils), I learned that the pavement showed a precise range of neutral grays from light to dark gray (in the categorization scheme from the Munsell chart: 7.5YR 7/0 to 4/0) and that the bud scale crumbs were dark red to dusky red (10R 3/6 to 4/6), among the brightest colors in the soil color book. A new neighbor, who happened to see me crouched at the street curb with my chart, stopped his car and asked: "What are you measuring?" It turned out that he was a professional "conservator of fine arts," who uses numerical color charts in monitoring the condition of surfaces of walls in galleries and historic buildings. He was delighted to meet a fellow student of the identification and spatial arrangement of colors. I was persuaded on that welcome spring day that patterns which trees imprint in soils, and even on streets, are among the "cunningest patterns of excelling nature," to adopt a phrase from Shakespeare's Othello.

The rain of bud scales was a dramatic, singular event in the natural formation of patterns on and in Wisconsin's earth, on and in its soils. By the word *soil,* I mean the root domain of lively darkness and silence. This unorthodox definition appreciates the stillness and biological activity of the perpetual nightscape in which most plant roots and associated organisms function. Soil scientists like myself, being sighted (unlike most denizens of the soil), quickly open up soil bodies for brief examination, as if we were surgeons doing exploratory operations on the life-supporting earth. We describe colors and shapes of components and discern the genesis and function of soils and their parts. The uniqueness of our perspective comes not only from our mental dexterity but also from our ability to see.

Soil patterns reflect the spatial arrangements of a large number of phenomena, whether or not visible: materials, both inanimate (air, water, solid particles) and animate (living cells and multicellular organisms); forms (shapes, sizes, and orientations of entities); and processes and interactions, including phenological events (migrations, flowerings). Our focus is on *soil patterns* in Wisconsin. We will first describe soil characteristics "scientifically" and, second, react to that science experientially.

Some Vertical Patterns in Soils

How are soils classified? It is on the basis of vertical patterns, as is the case with the familiar plug taken out of a watermelon that exhibits a sequence of layers: green, white, and red. Each of the hundreds of thousands of soils that have been classified scientifically on lands of this planet has a unique vertical pattern of soil layers, referred to as *soil horizons.*

Vertical movement of materials in soils brings about this layering. For millennia, rainwater has percolated down through soils, carrying particles of clay and plant nutrients from surface soil into subsoil. Plants have taken up nutrients from soil and air, and then, through leaf-fall and root-fall, have recycled nutrients back into soil and air. In deciduous forest soils (Figure 4.1, left) these vertical movements have laid a layer of leafy litter on the ground. Biotic activity of roots and soil animals has mixed the decomposing litter into the surface soil (0–15 cm), forming a dark top layer. In the next layer below that (15–30 cm), soil particles have been washed clean of dark stains so that the soil there is paler than above or below. The subsoil (30–60 cm) has dark brown clay coatings. The entire column of soil has been leached of lime (carbonates) to a depth of 90 centimeters by percolating rainwater over thousands of years.

A vertical cross-section of soil under a hemlock stand (*Tsuga cana-*

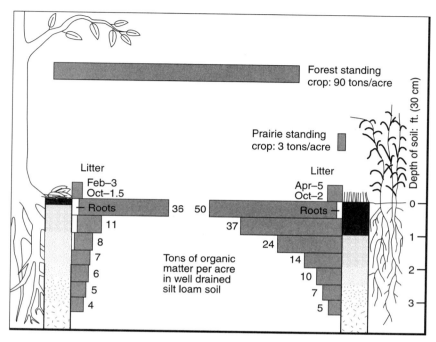

Figure 4.1. The distribution of organic matter (tons/acre) in forest (white oak, black oak) and prairie (big bluestem, Indian grass) ecosystems in south-central Wisconsin (Adapted from a figure prepared by G. A. Nielsen and F. D. Hole, 1963; adaptation by H. D. Foth, *Fundamentals of Soil Science*, 6th ed. [New York, 1978], 162, fig. 7.3. Copyright © 1978 John Wiley & Sons, Inc. Reprinted by permission of John Wiley & Sons, Inc.)

densis) in northern Wisconsin (Figure 4.2) is more dramatic than the deciduous forest soil (Figure 4.1, left). The hemlock soil shows a double pattern.[1] The upper unit (sequum) starts at the top with a hemlock litter layer labeled 01 for fresh organic matter and 02 for decomposed litter. This litter was too acid for large earthworms and other soil mixers. Rains wash some acid compounds out of the 0 horizons and through the upper 10 centimeters of the sandy soil. These acid compounds strip the coatings and stains of iron oxides (called sesquioxides, s) and organic matter, leaving the surface mineral soil (the horizon labeled E) white, which is the natural color of a clean grain of quartz. The abundant mites and springtails, fungi and bacteria, which easily escape the eye, decompose hemlock needles without mixing action. The hemlock forest floor seems very still and lifeless as compared with the floor of the oak forest (Figure 4.1, left). The stains removed from the white layer (horizon E, Figure 4.2) end up coating and even cementing together particles in the coffee-brown subsoil (labeled Bs).

In many areas this sequential pattern of 0-E-Bt soil horizons rests directly on the C horizon, which is the parent material. But in the double-sequence shown in Figure 4.2 another pale horizon (E'x) and a brown subsoil (B'tx) appear above the C layer. It seems likely that, before the hemlock stands developed, a hardwood forest prevailed with a less acid litter layer, below which formed a dark A horizon, now gone, and also a thicker, pale horizon, of which the E'x layer is the lower surviving part. The lower sequum (E'x-B'tx) is brittle; x is the symbol for brittleness; t is the symbol for clay accumulation. The soil in Figure 4.2, then, reveals development under two different vegetation covers.

Under prairie (Figure 4.1, right) the perpetual growth and death of fibrous roots and numerous soil organisms for several thousand years have darkened the soil, forming humus-rich surface soil (0–30 cm) and somewhat less darkened subsoil (at 30–90 cm), with carbonates (lime) leached to 90 centimeters. Fires set frequently by lightning and by native peoples have for centuries suppressed invading trees and have released plant nutrients abruptly into prairie soil. Ants and burrow-

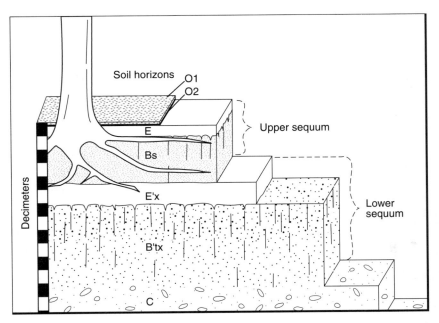

Figure 4.2. Diagrammatic cross-section of a Spodosol that is a bisequal soil (Reproduced with permission of the Iowa State University Press from *Soil Genesis and Classification* by Buol, Hole, and McCracken, 1989, p. 240)

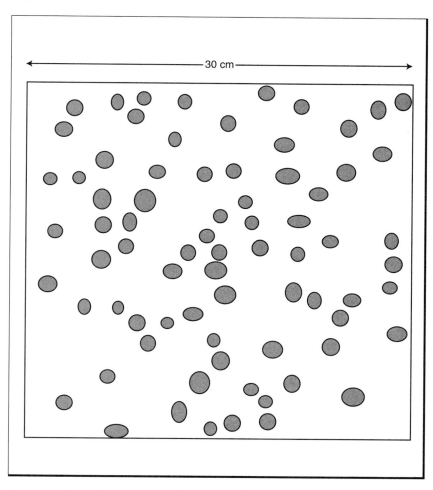

Figure 4.3. Exit holes of the periodic cicada at the surface of the soil. Under a single large tree as many as 40,000 holes were made by these emerging insects. (Traced from C. L. Marlatt, *The Periodical Cicada*, Bureau of Entomology Bulletin 71 [1907], figure 40: a photograph of ground with 1 square foot delineated, inside which 75 cicada holes are seen. F. D. Hole traced this quadrat, 1994.)

ing animals have dug in these soils (Figure 4.3), mixing materials up and down.

Some Horizontal Patterns in Soils[2]

Emphasis on the role of up and down movement of materials in producing vertical patterns in the earth beneath our feet need not lead

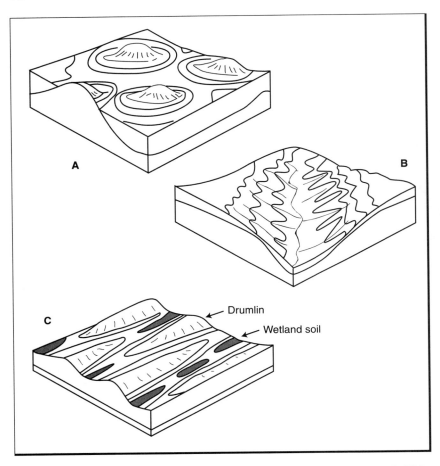

Figure 4.4. Three soilscape fabrics shown schematically: (A) circular or spotted, (B) irregularly striped, (C) simply and discontinuously striped. (Reproduced from F. D. Hole, *Soil Guide for Wisconsin Land Lookers*, Bulletin of Geology 2822 [1980], fig. 16)

us to disregard the horizontal component (Figure 4.4). For example, horizontal contraction and expansion of clay in the B'tx horizon in Figure 4.2 has created a small-scale polygonal pattern of tonguing of white soil into the cracks.

Patterned ground[3] polygons, measuring 10–50 meters across (Figure 4.5) formed 13,000 years ago in fluvial sediments in Adams County[4] (central Wisconsin), just west of the Johnstown Moraine, when the glacier stood on it. Today field crops grow greener and taller along the hidden polygonal lines during an unusually dry summer, when the roots reach down a meter to moist soil in the buried, collapsed troughs of the former permafrost-affected patterned ground. It was a time of glaciation (13,000 years ago) when low winter temperatures locally contracted the ground in the Driftless (unglaciated) Area to a depth of about 4 meters. The ice wedges, which slowly enlarged the shrinkage cracks, took centuries to grow to a width of 2 meters at the top of the permafrost. Geologists have recently observed the "fossil" ice-wedge casts by excavating the ground and by using ground-penetrating radar. These hidden horizontal patterns are well preserved, in contrast with the ephemeral red carpet of bud scales mentioned at the start of this chapter.

Other examples illustrate the importance of horizontal patterns in soils. The spacing of the mounds of the western mound-building ant

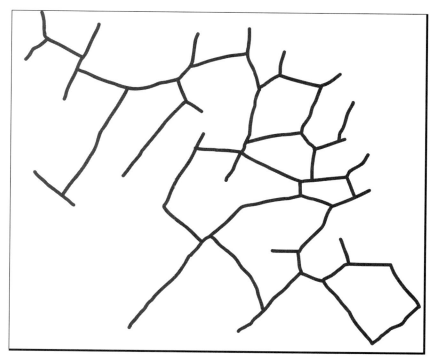

Figure 4.5. Ice-wedge polygons on stream sediment deposited beyond mouths of small valleys cut after Lake Wisconsin drained, in the SW ¼ SW ¼ SW ¼ Section 4, T. 14 N., R. 6 E., southwestern Adams County. Traced from U.S. Department of Agriculture aerial photograph AJA-IFF-284. Area shown is 170 meters wide. (Reproduced from Clayton, 1987)

may average several meters in undisturbed prairie. The ants of a given colony move out to collect secretions from aphids on plants within a radius of a meter or so of a mound, and bring nutrient-rich materials back to the mound. In this way, high levels of available phosphorus and potassium are maintained in the mound during the growing season.

Under a single specimen of American beech (*Fagus grandifolia*) an annular soil pattern develops in eastern Wisconsin. The smooth bark of the tree facilitates stem flow of water during rains, such that the soil adjacent to the tree receives up to five times as much water in the course of a year as the soil at a horizontal distance of only 2 meters from the bole of the tree. The two soils may be as different as if they were from different climatic zones.

Erosion of soil creates horizontal and vertical patterns of soil loss and of sediment accumulation in both open and closed hydraulic systems. An open system is exemplified by soil erosion that moves sediment through dendritically converging tributaries into valleys of major rivers. Closed systems are represented by kettles, depressions without surface drainage, that were formed by the melting of blocks of glacial ice buried under river-laid sand thousands of years ago. In both open and closed systems, soil patterns of exposed brown subsoil on cultivated uplands and sheets of resulting dark sediment in lowlands developed. Presettlement soil erosion had proceeded at the slow pace of about 1 ton of soil lost per acre (2.224 metric tons/hectare) of upland each year. This rate has increased 10-fold from disturbance on fields and by urbanization.

In windstorms, single strong trees are blown over each year. Roots and attached soil are pulled vertically out of the ground and displaced horizontally a distance of 1–2 meters. The tipped-up root-plate gradually sheds adhering soil and decomposing roots to form a mound as much as 60 centimeters high, adjacent to an equally large pit from which the soil was jerked during the storm. Wisconsin's forested landscapes are dotted with many such tree-tip mounds and hollows, creating horizontal as well as vertical patterns in the soil.

Getting Acquainted with Soil Patterns at Representative Sites

The people of the United States had a love affair with soil maps published in color as early as 1899. In that year, President McKinley himself called to the attention of the Senate and House of Representatives some new, "large scale" (1:63,360; 1 inch = 1 mile or 1.6 kilometers) soil maps just issued by the U.S. Department of Agriculture. The maps were large enough for farmers to distinguish soil patterns on their properties. The secretary of agriculture pronounced these maps "the most important work of this character ever undertaken in any country."[5] Today, published soil maps with aerial photographic background are available for 54 of Wisconsin's 72 counties.[6] At least half of each county's book of soil surveys consists of maps with patterns of bodies of soil, water, and bedrock clearly delineated. Soil scientists who did the fieldwork mapped at a rate of about a half square mile (1.29 km²) per day per worker. The total length of all those soil boundaries is about a million miles (1,609,000 km) from the state of Wisconsin. A single representative section (640 acres, 259.2 hectares), near Marshall, will now be examined.

The Soilscape Pattern in a Glaciated Landscape[7]

A contour map reveals several glacially streamlined hills called *drumlins,* and lowlands between, in the Marshall landscape (Figure 4.6). The hills and some of the lowlands were being built 16,000 years ago (on the Johnstown Moraine[8] Clayton et al, 1992) under a slowly moving sheet of ice that was about a half-mile (792 m) thick. Deposits of sand and gravel *outwash* were made by rivers of meltwaters 15,000–12,000 years ago. Meltwaters flowed from the glacier the year around, but at a much reduced rate in winter. Winds blew dust off the outwash plains, leaving a foot (30 cm) or more of new soil material on both uplands and flats in the landscape. With the establishment of vegetation, all the deposits became stabilized. The published soil map (Figure 4.6) shows 72 soil bodies, entire and partial, one body of water, and one body of disturbed land (C.F.L., or cut and fill land).

The boundaries between these units total about 24 miles (38.6 km) in length; they are not uniform in breadth, as the even width of soil boundaries on maps suggests. The actual width of a soil boundary on the ground varies along its length. For example, the boundary is a few meters wide between soil bodies marked Ho and KrE2 in the southeast quarter of the section. Much wider is the boundary between soil bodies SaA and VrB in the northeast quarter of the section. The boundary of the pond at the east edge of the section lengthens in wet seasons and shortens in dry seasons. Many boundaries of wetland soils "vibrate" with slow motion as wetlands shrink and expand. Some landowners report that beaver dams account for expansion of some ponds and wetlands.

The soil boundaries in the Marshall landscape are not of uniform importance, aside from variation in width. For example, the line be-

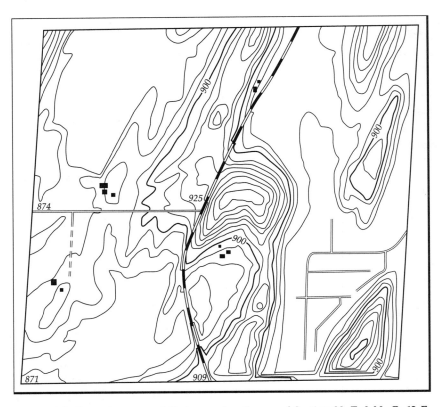

Figure 4.6. Contour map, and *(facing page)*, soil map of Section 33, T. 8 N., R. 12 E., Deerfield Quadrangle, Wisconsin (Contour map from U.S. Geological Survey, 1962; soil map from Glocker, Patzer, et al., 1978)

tween soil bodies KdD2 and KdC2 at the center of the section signifies a change in slope (C to D: 6–12 percent slopes to 12–20 percent slopes) only. On the other hand, the boundary in the southeast quarter between Ho and KrE2 signifies changes in slope (0 to E: no slope to 20–35 percent slopes), degree of erosion (0 to 2), and kind of soil (a wet muck, Ho, to a well-drained, stony, sandy soil, Kd).[9]

A Soilscape Pattern in the Driftless Area

In the Driftless Area of southwestern Wisconsin, the landscape has been more dissected by dendritic stream systems than the landscape of glacial deposits near Marshall. The terrain is underlain by Cambrian sandstone capped with Prairie du Chien dolomite (Ordovician).

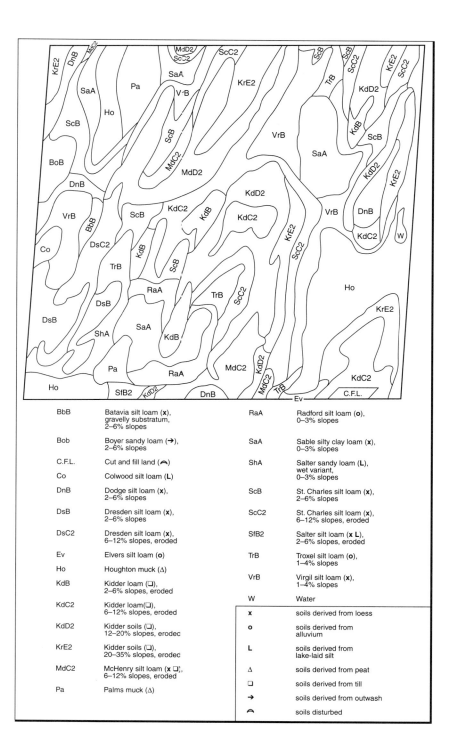

BbB	Batavia silt loam (**x**), gravelly substratum, 2–6% slopes	RaA	Radford silt loam (**o**), 0–3% slopes
Bob	Boyer sandy loam (→), 2–6% slopes	SaA	Sable silty clay loam (**x**), 0–3% slopes
C.F.L.	Cut and fill land (⌒)	ShA	Salter sandy loam (**L**), wet variant, 0–3% slopes
Co	Colwood silt loam (**L**)		
DnB	Dodge silt loam (**x**), 2–6% slopes	ScB	St. Charles silt loam (**x**), 2–6% slopes
DsB	Dresden silt loam (**x**), 2–6% slopes	ScC2	St. Charles silt loam (**x**), 6–12% slopes, eroded
DsC2	Dresden silt loam (**x**), 6–12% slopes, eroded	SfB2	Salter silt loam (**x L**), 2–6% slopes, eroded
Ev	Elvers silt loam (**o**)	TrB	Troxel silt loam (**o**), 1–4% slopes
Ho	Houghton muck (Δ)		
KdB	Kidder loam (□), 2–6% slopes, eroded	VrB	Virgil silt loam (**x**), 1–4% slopes
KdC2	Kidder loam (□), 6–12% slopes, eroded	W	Water
KdD2	Kidder soils (□), 12–20% slopes, erodec		
KrE2	Kidder soils (□), 20–35% slopes, erodec		
MdC2	McHenry silt loam (**x □**), 6–12% slopes, eroded		
Pa	Palms muck (Δ)		

x	soils derived from loess
o	soils derived from alluvium
L	soils derived from lake-laid silt
Δ	soils derived from peat
□	soils derived from till
→	soils derived from outwash
⌒	soils disturbed

A loess cover, laid down from about 15,000 to 12,000 years ago,[10] was deposited at varying thicknesses over the bedrock and the cherty reddish brown clay that typically caps the dolomite. Erosion has been dominant in southwestern Wisconsin since about 350 million years ago, when the deposition of Silurian dolomite and some later strata was completed in the last sea that covered Wisconsin. The streams in this section are working in the Ordovician dolomite and the underlying sandstone.

The published soil map shows 50 soil bodies, whole and partial, and about 32 miles (51.5 km) of soil boundaries. The number of bodies is fewer and the length of boundaries is greater than in the glaciated section, where lowlands are more extensive than uplands and where the variety of parent materials for soils is greater. The increase in total length of soil boundaries from 24 to 32 miles (38.6 to 51.5 km) results from the greater extents of steep land. Erosion of silty material has been from upland soil bodies down to floodplains occupied by silt loams.

Building a Lasting Friendship with the Soil

I recommend that each of us citizens take the trouble to photocopy the portions of the modern soil maps that show the places in which we live, work, and enjoy recreation regularly. I also suggest that, on the basement walls of our favorite buildings, we place murals at true scale and in color of cross-sections of the native soil of the locality, with enlarged images of native soil organisms, including root structures, worms, insects, and vertebrates, and of microorganisms that fix nitrogen from the air. We can be ready to converse with our friends about the soils that support us. These maps and murals will help us to become acquainted with the faces of what soil scientist Hans Jenny (1899–1992) called "my friend, the soil."[11] This "friend" feeds us out of its own substance. My own face, which looks with appreciation at soil patterns, soil maps, and soil murals, is just one of the faces of the soil.

Increased awareness of soils helps us not only as individuals but also as communities. To facilitate this communication we may form outdoor soil clubs that correspond to the chapters of the National Audubon Society. I suggest that collective soil club be named the Jenny Society. Dr. Hans Jenny grew up in Switzerland and spent most of his career on the faculty of the Department of Soil Science at the University of Berkeley, California. His soil research was global in scope. He and his wife, Jean Jenny, were leaders in a movement to set aside soil sanctuaries. In these areas, ecologically important soils are preserved for society to practice observation and contemplation. Professor Jenny was aware of the importance of aesthetics as well as science in soil appreciation and conservation.

The emphasis of the Jenny Society would be on combining objective information about soil patterns with subjective experience of patterns of "flora, fauna, people and of soil."[12] More specifically, the society would have three purposes: One is to raise our level of awareness of the origin and nature of soil communities that support life. Another is to help us reclaim our natural capacity for enjoyment of soilscapes as we walk and otherwise move about over them. The third purpose is to enroll ourselves as informed and committed stewards of the soil.

Local chapters of the Jenny Society might be named in honor of individuals who display unusual kinship with the soil. For example, Frank Leverett (1839–1943)[13] impresses me as particularly appropriate. It was my good fortune to visit briefly with this master mapper of glacial patterns (which are soil patterns, too), in his home at Ann Arbor, when in 1941 I brought for his review a new glacial map of Wisconsin by geology professor F. T. Thwaites of the University of Wisconsin. I saw Frank Leverett again at subsequent meetings of the Geological Society of America, in Chicago.

As a young man, Leverett had walked from his home in Iowa to the offices of the Wisconsin Geological and Natural History Survey at the University in Madison to apply for a job with the U.S. Geological Survey, of which T. C. Chamberlain was an administrator in Madison. Over the next six decades Frank worked in 20 states of our country and in several provinces of Canada, mapping glacial features. His field seasons involved 200 days on foot each year. He walked 20–30 miles (32–48.3 km) a day for a total estimated distance of 100,000 miles (160,900 km) during his career. He wrote field observations on 45,000 pages of 300 notebooks. He observed and drew all map boundaries while in the field, a procedure also followed by Chamberlain and hundreds of soil surveyors who later recorded detailed soil patterns. Since exposures of glacial materials were rare, Leverett talked to local people to learn about the stratigraphy. He used a color code on his maps: red for end moraines, brown for fluvial materials (outwash), and blue for till plains. He recorded locations of rock outcrops, orientations of striae, presence of buried channels, and sequences of buried drift sheets and soils. This modest, indefatigable man probably contributed more than any other person to our understanding of the work of continental glaciers in the Great Lakes region. He taught at the University of Michigan at Ann Arbor for 20 years. His home was a popular, informal classroom to which young geologists and soil scientists came for instruction and inspiration. Perhaps in Wisconsin and Michigan, then, we might have the Frank Leverett Chapter of the Jenny Society to remind us all of a most desirable marriage—understanding of and affection for the soil.

Conclusion

The root domain, called soil, is both an integrative part of environmental systems and the basis of land life on this planet. Soil patterns change in response to the geological factors of parent material and landforms, climatic factors, biological factors, and the age of soil.[14] The intricacies of soil bodies on landscapes are clearly shown on published soil maps of Wisconsin. The range in size of soil bodies is exemplified in Rock County, Wisconsin,[15] in which the smallest soil body reported is 2 acres (0.8 ha) in size. Near Janesville a single body of a prairie soil (Plano silt loam) measures 12.9 kilometers (8 mi.) long and 3.6 kilometers (2.5 mi.) wide, for a total area of 17.5 square miles (11,200 acres or 4,536 ha). Patterns of soil bodies vary widely. Soils are subject to burrowing by animals, displacement by plant roots, heaving by freeze-thaw cycles and the shrink-swell action of clay, shifting by air and water and earthquakes, and to both disturbance and protection by human beings.

We people have a natural capacity for enjoyment of soil, which is, in fact, Mother Earth. Soils are as interesting to study out-of-doors as are the plants and birds that soils support. Enjoyment of soil involves observation of the earth beneath our feet and practice of its stewardship. We can become, in appreciation of Frank Leverett and Hans Jenny, amateur soil walkers, exhuming soil profiles for temporary admiration, watching for changes through the seasons, contemplating with affection "the root domain of lively darkness and silence."

Notes

1.   See description of the Iron River silt loam, in C. J. Milfred, G. W. Olson, F. D. Hole, F. P. Baxter, F. G. Goff, W. A. Creed, and F. Stearns, *Soil Resources and Forest Ecology of Menominee County, Wisconsin,* Wisconsin Geological and Natural History Survey, Bulletin 85, Soil Series No. 60 (Madison, 1967), 199. Also note S. W. Buol, F. D. Hole, and R. J. McCracken, chapter 13, "Spodosols: Soils with Subsoil Accumulations of Humus and Sesquioxides," in Buol, Hole, and McCracken, eds., *Soil Genesis and Classification,* 3d ed. (Ames, Iowa, 1989).

2.   For further information on topics in this section, see Buol, Hole, and McCracken, *Soil Genesis and Classification;* and P. Farb, *Living Earth,* (New York, 1959). F. D. Hole published "Effects of Animals on Soils," *Geoderma* 25 (1981): 75–112.

3. Information about patterned ground can be found in A. L. Bloom's *Geomorphology, a Systematic Analysis of Late Cenozoic Landforms* (Englewood Cliffs, N.J., 1978). Information about ice-wedge casts in Wisconsin is found in the following: L. Clayton, *Pleistocene Geology of Adams County, Wisconsin*, Wisconsin Geological and Natural History Survey, Information Circular 59, University of Wisconsin Extension, (Madison, 1987), 9, fig. 5; J. W. Attig, L. Clayton, K. R. Bradbury, and N. C. Blanchard, "Confirmation of Tundra Polygons and Shore-Ice Collapse Trenches in Central Wisconsin and Other Applications of Ground-Penetrating Radar," *Abstracts with Programs*, vol. 19 (1987), 21st Annual Meeting of the North-Central Section, The Geological Society of America, St. Paul; J. W. Attig and L. Clayton, "Conditions beyond the Ice Margin in Wisconsin during the Last Glacial Maximum," *Abstracts with Programs*, vol. 24 (1992), 26th Annual Meeting of the North-Central Section, The Geological Society of America, Ames, Iowa. R. F. Black, "Periglacial Phenomena of Wisconsin, North-Central United States," *Report of the VIth International Congress of Quaternary* vol. 4: *Periglacial Section* (Lódź, 1964). Black also authored "Periglacial Features Indicative of Permafrost: Ice and Soil Wedges," *Quaternary Research* 6 (1976): 3–26.

4. Patterns of soil bodies on landscapes of Adams County are reported by D. E. Jakel in *Soil Survey of Adams County, Wisconsin*, U.S. Department of Agriculture Soil Conservation Service (Washington, D.C., 1984). C. J. Milfred and R. W. Kiefer found that a sequence of aerial color photographs taken in late summer from a small aircraft flying low and with the use of a hand-held camera yielded images of patterns of green and yellow in a field of maize (corn) near Middleton, Wisconsin, that showed an intricate pattern of soil boundaries delineating 24 soil bodies. The published soil map of Dane County showed four soil bodies in this field. No patterned ground was detected in this area. See report by Milfred and Kiefer, "Analysis of Soil Variability with Repetitive Aerial Photographs," *Soil Science Society of America Journal* 40 (1976): 553–557.

5. See H. Jenny's *E. W. Hilgard and the Birth of Modern Soil Science* (Pisa, Italy, 1961), 94.

6. The reader will find helpful information in the book by F. D. Hole (with contributions by G. B. Lee, and M. T. Beatty), *Soils of Wisconsin* (Madison, 1976).

7. The term *soilscape* refers to the soil portion of a landscape. V. M. Fridland (1919–1983), a Russian soil scientist, made important contributions to the study of soilscape patterns. His works include the following: V. M. Fridland, "Structure of the Soil Mantle," *Geoderma* 12 (1974): 35–41; V. M. Fridland, *Pattern of the Soil Cover* (Moscow, 1972; English translation: Jerusalem, 1976). Two soil geographers in the United States have treated the subject of soilscape patterns: F. D. Hole and J. B. Campbell, *Soil Landscape Analysis* (Totowa, N.J., 1985). The reader is referred to C. L. Glocker, R. A. Patzer, et al., *Soil Survey of Dane County, Wisconsin*, (Washington, D.C., 1977; distributed by U.S. Department of Agriculture Soil Conservation Service in cooperation with the Research Division of the College of Agricultural and Life Sciences, University of Wisconsin, 1978).

8. Clayton, Lee, John W. Attig, David M. Mickelson and Mark D. Johnson,

1992. Glaciation of Wisconsin. Educational series 36, Wisconsin Geological and Natural History Survey. University of Wisconsin—Extension, Madison, Wisconsin.

9.   Hole and Campbell, in their book *Soil Landscape Analysis,* proposed nine ranks of soil boundaries to express degrees of contrast between adjacent soil bodies.

10.   Personal communication, Lee Clayton, 1994.

11.   This quotation is from H. Jenny, "My Friend the Soil." *Journal of Soil and Water Conservation* 39 (1984): 158–161. Professor Jenny's grasp of soil science and the "contract" between people and the soil is well summarized in H. Jenny, *The Soil Resource: Origin and Behavior* (New York, 1980); and H. Jenny, "The Making and Unmaking of a Fertile Soil," in W. Jackson, W. Berry, and B. Colman, eds., *Meeting the Expectations of the Land* (San Francisco, 1984), 42–55.

12.   Words from a soil song by F. D. Hole (1985).

13.   Information about Frank Leverett may be found in these articles: S. G. Berquist, "Memorial to Frank Leverett," *Science* 99 (1944): 312–313; R. L. Rieck and H. A. Winters, "Frank Leverett, Pleistocene Scholar and Field Worker," *Journal of Geologic Education* 29 (1981): 222–227; H. A. Winters and R. L. Rieck, "Frank Leverett: Michigan's Master Geologist," *Journal of Geologic Education* 64 (1980): 11–13.

14.   Professor Hans Jenny's elucidations of the concept of five factors of soil formation are presented in the following books: H. Jenny, *Factors of Soil Formation* (New York, 1941); and H. Jenny, *The Soil Resource, Origin and Behavior* (New York, 1980). Hans Jenny wrote a chapter, "The Making and Unmaking of a Fertile Soil," in Jackson, Berry, and Colman, eds., *Meeting the Expectations of the Land.* Buol, Hole, and McCracken published *Soil Genesis and Classification.* M. I. Harpstead, F. D. Hole, W. E. Bennett, and M. Bratz-Stevens published *Soil Science Simplified,* 2d ed. (Ames, Iowa, 1988).

15.   R. J. Engel, H. F. Gundlach, K. O. Schmude, C. L. Glocker, E. L. Weber, and F. L. Anderson, *Soil Survey of Rock County, Wisconsin,* U.S. Department of Agriculture Soil Conservation Service (Washington, D.C., 1974).

5

Eastern White Pine in Southwestern Wisconsin
Stability and Change at Different Scales

Susy Svatek Ziegler

As much as shimmering lakes or laughing loons, eastern white pine (*Pinus strobus* L.) is an icon of the great northwoods. The tree is widely distributed throughout the northeastern United States and adjacent parts of Canada (Figure 5.1). Here, where winters are long and cold, the climate favors the growth of coniferous trees that are able to tolerate high snowfalls, low temperatures, and short growing seasons, such as eastern white pine. In northern and central Wisconsin, in spite of vegetation changes associated with the thorough logging of pines that began in the 1840s and ended with the collapse of the lumber industry in the 1930s, white pine continues to thrive on well-developed loams or sandy loams, where its dark green boughs tower above the foliage of the shorter-growing deciduous trees. The species also grows, however, in southwestern Wisconsin. Here, close to the southern limit of its geographic range (Figure 5.2), white pine grows in small, isolated stands on steep sandstone bluffs, where it is surrounded by a mosaic of deciduous forest, prairie remnants, livestock pastures, and agricultural crops. On such sites, most other trees have a difficult time becoming established; thus, white pine has a competitive advantage over these species with less tolerance of drought-prone slopes and minimal soil development. Growing at its range limit, on dry sites, with potential competitors nearby, white pine in southwestern Wisconsin has a story

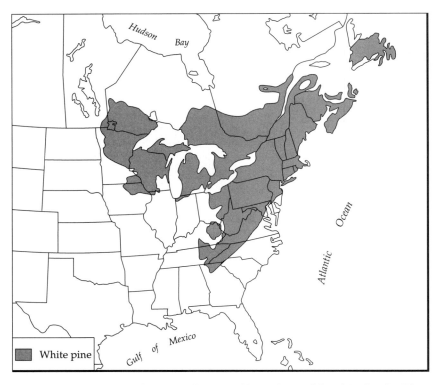

Figure 5.1. Native range of eastern white pine (From G. Wendel and H. Smith, *"Pinus strobus* L. Eastern White Pine," in R. Burns and B. Honkala, eds., *Silvics of North America,* vol. 1, U.S.D.A. Agricultural Handbook No. 654 [Washington, D.C., 1990], 476–488).

of persistence in the face of seemingly threatening change. Its narrative embraces many scales, the first of which is the long perspective provided by geological time.

The Distant Geological Past

The story begins some 590 million years ago—about 200 million years before the evolution of land plants—in the Cambrian period of the Paleozoic era, when advancing and retreating shallow inland seas spread well-sorted sediment over the ancient core of the North American continent. The area that is now southern Wisconsin was covered at least twice by inland seas. During these episodes, sediments eroded from the land surface were deposited in the sea to form sandstone

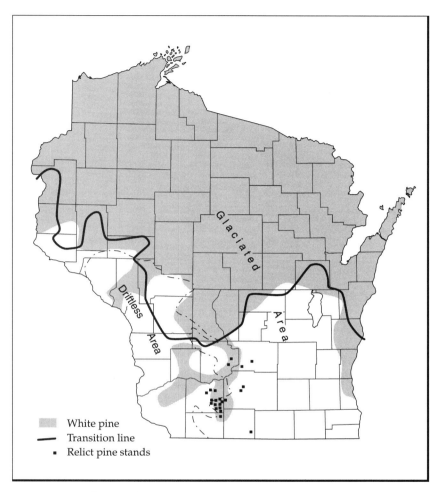

White pine
Transition line
▪ Relict pine stands

Figure 5.2. Distribution of eastern white pine *(shaded area)* in Wisconsin, as described by L. S. Cheney in 1897–1898 (modified from N. Fassett, "Preliminary Reports on the Flora of Wisconsin. V. Coniferales," *Transactions of the Wisconsin Academy of Sciences, Arts and Letters* 25 [1930]: 177–182). The line trending northwest to southeast marks the transition between northern conifer-hardwood forest, where pine was a common species at the time of European settlement, and the prairie–deciduous forest vegetation of the southern part of the state, as shown on the map by R. Finley, *Original Vegetation Cover of Wisconsin* (compiled from U.S. General Land Office Notes, North Central Forest Experiment Station, U.S. Department of Agriculture [St. Paul, 1976]). The squares indicate the location of 22 relict pine stands studied by McIntosh, 1948, referred to later in this chapter.

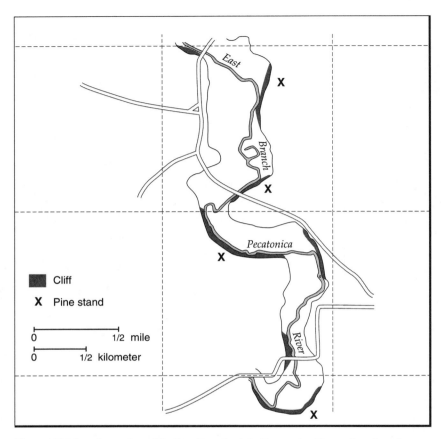

Figure 5.3. Map of a section of the East Branch of the Pecatonica River where it undercuts St. Peter sandstone to produce steep bluffs, upon which white pine stands dominate

and shale. During the Ordovician period, beginning about 500 million years ago, Wisconsin was inundated three or more times by shallow seas. In the middle Ordovician period, the St. Peter sandstone, consisting almost entirely of quartz, was deposited in what is now the central United States. Other sedimentary rocks such as dolomite, limestone, and shale were deposited over the St. Peter sandstone throughout the Silurian and Devonian periods. The large-scale deposition of sediment apparently ended about 360 million years ago, which incidentally was at least 150 million years before the evolution of conifers.

From this long-ago formation of rock layers, the ground was laid, literally, for white pine in southern Wisconsin. The unglaciated Driftless Area of southwestern Wisconsin is a hilly region that has been highly dissected by stream channels for several millions of years. As a result of the varied topography, the region is characterized by a diversity of plant habitats. In particular, steep slopes of exposed Cambrian sandstone of the Tunnel City Group and Ordovician sandstone of the St. Peter Formation are created where the streams, eroding laterally, undercut the sandstone (Figure 5.3). The resulting cliffs provide niches for plant species that tolerate poorly developed, well-drained soils, such as eastern white pine. On these nutrient-poor soils, white pine is able to persist in a region where deciduous trees typically dominate the forest vegetation. How long white pine has been here, however, is not known, but it is conceivable that the species has graced the region that we call southwestern Wisconsin for millions of years.

The narrative now jumps ahead in time to the last million years of the Pleistocene, during which northern and eastern Wisconsin were glaciated several times. The most recent of these cold episodes, the Wisconsin period of glaciation, which began 85,000 years before present, resulted in most of the state being covered by lobes of glacial ice that were as much as 2 kilometers thick. The unglaciated Driftless Area in southwestern Wisconsin, in contrast, may have served as a refugium for arctic and boreal plant species that migrated southward in front of the advancing ice. Henry Hansen and John Curtis believe the area could have supported northern tree species, presumably including eastern white pine, during this time.[1] However, evidence from drill cores taken throughout the area, and the presence of relict frost cracks and ice-wedge polygons that are 30 meters across, suggest that the climatic conditions in southwestern Wisconsin during the last glacial maximum may have been too harsh for tree growth.[2] Even arctic-alpine species may not have survived this severe periglacial climate. Certainly, though, the Driftless Area, during the retreat of the glaciers beginning about 12,000 years ago, must have provided suitable habitats for trees as their ranges shifted in response to climatic warming. At this time, eastern white pine migrated from the Appalachian highlands westward toward the Great Lakes region.[3] Then the range of the species shifted northward into Michigan, Wisconsin, and Minnesota (Figure 5.4). The question remains: Was white pine present in the unglaciated area of Wisconsin during the last glacial advance, or did it migrate into the region as the glaciers retreated?

Unfortunately, as yet, no clear answer is possible. The reconstruction of vegetation change in the Driftless Area during the Pleistocene is problematic. Throughout much of eastern North America, analysis of fossil pollen grains (palynology) and plant remains preserved in

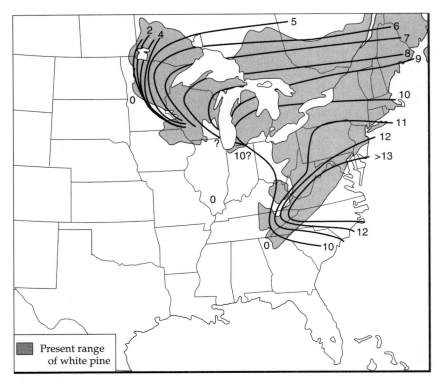

Figure 5.4. Postglacial migration of eastern white pine. Isolines indicate the time in thousands of years before present that white pine reached a given region. Stippled area indicates the present range of white pine in eastern North America. (Redrawn with permission from *Late-Quaternary Environments of the United States*, vol. 2, edited by H. E. Wright, Jr., University of Minnesota Press. Copyright © 1983 the Regents of the University of Minnesota.)

wetland areas may provide a detailed record of past vegetation and climate. However, in southwestern Wisconsin there are few wetlands or bogs because of the well-developed drainage of the unglaciated landscape. Pollen analysis is possible in the few bogs that formed in impounded meander cutoffs or where tributaries built up their valleys by increased deposition of sediment when river levels were higher during glacial times. Unfortunately, however, the record of late Pleistocene flora of southwestern Wisconsin is often incomplete because of poor pollen preservation in the existing bogs.

The earliest palynological evidence in the Driftless Area dates from 10,480 ± 100 radiocarbon years ago in a soil core extracted from the west branch of Blue Mounds Creek.[4] Because the record extends back only to the time of pollen preservation, the flora of the Driftless Area during the last glacial maximum is not yet known, if vegetation existed there at all. The only definitive thing we know is that white pine was abundant in southwestern Wisconsin when pollen began to accumulate in the few bogs that formed at the end of the last glacial period.

The Holocene

Pine has been present in the pollen record of Blue Mounds Creek throughout the Holocene, or the last 10,000 years. Because of its tolerance of poorly developed soils, white pine has grown on sandstone bluffs in the area for thousands of years, while deciduous genera, such as oak and hickory, and herbaceous species have occupied less extreme sites.[5] Early in the Holocene (9,500 years ago) rapid warming and increased dryness resulted in an abrupt change in dominance from spruce to pine.[6] This change presumably marks the postglacial transition from boreal to temperate vegetation. During the following warmer and drier Climatic Optimum (8,000–6,000 years ago), white pine migrated northward, and the mixed conifer–northern hardwood forest moved into southern Canada.[7] About 7,000 years ago, the prairie reached its easternmost position, extending into southeastern Wisconsin and northeastern Illinois.[8] During this period, white pine stands presumably persisted on the sandstone bluffs. Therefore, white pine presently found in southwestern Wisconsin is a relict of the late Quaternary period and represents a population that at one time was more extensive and is now disjunct because of shifts in range location throughout the past several thousand years.

Historical Times

The palynological evidence indicates that these pine relicts have persisted in the Driftless Area of southwestern Wisconsin from the late Pleistocene through the Holocene, implying successful adaptation of the species to the local environment. Nonetheless, the degree to which the species is maintaining itself at the southern margin of its present range is variable over both space and time. John Curtis reported that, "although these relic stands are considerably removed from the area of their climatic optimum, there is no indication that they are in a retreating or moribund condition."[9] A study of the forest vegetation of the Driftless Area in northeast Iowa, on the other hand, showed that white pine grew in relict communities primarily on drier sites, but that white pine seedlings were not a significant component of the herb layer, suggesting that the species was not reproducing well.[10]

In 1948, Robert McIntosh completed a taxonomic survey of species composition of 22 sites in six counties in southwestern Wisconsin where white pine, red pine, and/or jack pine, all species with northern affinities, were abundant.[11] He found representatives of diverse plant communities, including numerous species characteristic of northern pine stands. McIntosh indicated that, in a few of these stands, white pine was reproducing sufficiently well to persist in this area. His study also indicated that in several other stands white pine was not reproducing well, as marked by the absence of seedlings and saplings, with a trend toward local extinction.

Forty-five years after McIntosh completed his study, I relocated and assessed the dynamics of eight of the stands in which white pine is the dominant species.[12] These stands were selected to sample a range of topographic settings and to make comparisons with the patterns of regeneration that McIntosh noted 45 years before. The results of my study showed that contemporary patterns of reproduction of white pine in these stands are variable. At six of the eight sites white pine recruitment of seedlings and saplings is vigorous. These patterns of regeneration appear to be correlated with the aforementioned climatic, edaphic (related to soil), topographic, and biotic factors. The impacts of an additional factor, various agents of disturbance, however, are more difficult to tease apart.

Throughout the Holocene, the dynamics of white pine have been influenced by both natural and human disturbance factors. Individual tree fall caused by wind, especially of senescent individuals, has created canopy gaps throughout the history of the stands. Additional agents of disturbance include fire, logging, grazing, and browsing. I will discuss in turn the impact of these four factors on white pine regeneration.

The charcoal record provides evidence that fire disturbance has influenced vegetation patterns in the Driftless Area for most of the Holocene, with fingers of woodland on rarely burned steep slopes extending into fire-dependent prairie in low-lying and gently sloping areas.[13] John Curtis suggested that white pine stands on sandstone outcrops may have been protected from fire.[14] A long-standing silvicultural theory is that catastrophic fires or blowdowns are essential for white pine establishment and regeneration.[15] Yet, the fires that burned this area prior to European settlement apparently were of high frequency and low intensity. Furthermore, fire is no longer a common disturbance agent, because fire suppression practices were initiated by European settlers at the end of the nineteenth century. The trees in the present stands have no fire scars that could be used to reconstruct fire history, and it is unlikely that fire has disturbed the stands in southwestern Wisconsin for at least many decades. Nevertheless, white pine continues to persist here and it appears that most successful regeneration is associated with small disturbances such as individual tree fall, logging, and grazing.[16]

As European settlers made southwestern Wisconsin their new home in the mid- to late 1800s, they harvested what trees were available to use for building materials. The settlers also cleared land for agriculture. Widespread logging of white pine, predominantly for local use, is reflected today in the size and age structure of the stands; the oldest trees are no more than 100 years old. In the pollen record, too, the recent decline in percentage of pine in the Driftless Area coincides with logging associated with European settlement.[17] Apparently, however, a viable seed source remained for white pine to regenerate after logging. The present stands, then, are young compared with the 500-year maximum age that individual white pines can obtain. Although widespread logging continued throughout the first several decades of this century, today the demand for lumber is satisfied by cutting elsewhere. Some property owners still clear brush and cut logs for personal use, creating small clearings in the stands. Evidently, logging has allowed more sunlight to reach the forest floor, promoting the germination and rapid early growth of white pine seedlings, which are intermediate in shade tolerance.[18] The logging histories of the stands would be difficult, if not impossible, to reconstruct, but they would help explain local patterns of white pine regeneration.

In spite of the steepness of the terrain, many of the pine stands have served at one time or another as pasture for cattle grazing. In part, this grazing promotes the establishment of white pine seedlings, because cattle differentially remove grasses, weeds, and hardwood seedlings and saplings that would retard white pine development.[19] However, grazing also impairs the regeneration of white pine when cows uproot seedlings and trample saplings. The future of white pine in southwestern Wisconsin, however, is apparently not limited by grazing cattle.

Not all forms of disturbance are beneficial to white pine. While grazing actually may stimulate white pine growth, deer browsing is detrimental to the tree. Browsing deer remove the terminal buds of white pine seedlings and saplings, which can damage or kill them. Especially during harsh winters, deer browse results in extensive damage to seedlings and saplings. To make matters worse for white pine regeneration, deer populations in southwestern Wisconsin have been rising over recent decades. As with logging and livestock grazing, however, deer browsing does not seem to preclude white pine regeneration at all sites.

The Thompson Bluff Stand

The environmental relationships and disturbance factors that collectively influence the dynamics of white pine in the stands can also illuminate the regeneration characteristics and history at a particular site. Thompson Bluff is a white pine stand located in Iowa County in the valley of the east branch of the Pecatonica River. This property has been in the Thompson family for five generations, since it was homesteaded in 1852. The bluff is an outcrop of St. Peter sandstone that supports a wide range of size and age classes of white pine. Less dominant tree species at this site include red oak (*Quercus rubra* L.), white oak (*Quercus alba* L.), black cherry (*Prunus serotina* Ehrh.), and paper birch (*Betula papyrifera* Marsh.). The Thompsons no longer graze their dairy cattle in the pine stand, although they did until 1975. In spite of evidence of deer browse throughout the stand, white pine regeneration appears vigorous.

According to Steve Thompson, the current owner of the farm, the pine stand has not been logged for many years, although for personal use Steve removes trees that fall on their own. His great-grandfather, however, described the bluff as virtually clear of trees, presumably a result of logging by early settlers. Since then, white pine has become reestablished. A comparison of a pair of photographs taken of the stand circa 1900 and in 1994 indicates that the stand is denser today than at any time since the turn of the century (Figure 5.5). Robert McIntosh described the stand in 1947: "The conifers seem to be at least holding their own if not making local extensions, as indicated by large numbers of seedlings and the presence of younger trees. . . ."[20] Indeed, a comparison of tree diameter size classes shows a wider range today than 45 years ago, with no decrease in the percentage of stems in the smallest size class (Figure 5.6). This size structure suggests that white pine regeneration continues at this site, despite past population fluctuations.

Apparently, white pine is well established on this site. Without disturbance by fire, logging, and grazing, however, and with disturbance by browsing deer, can the species maintain itself? Do small canopy gaps created by senescent trees and windthrow provide sufficient sunlight and space for new white pine establishment? Is it not ironic that human activities may effectively substitute for the natural disturbance

Figure 5.5. Photographs taken of Thompson Bluff circa 1900 *(top)* and in 1994 *(bottom)* indicate that the pine stand is denser today than at the turn of the century. The sandstone outcrop in the foreground of the 1900 photograph is now tree-covered. The road was built circa 1950. (1900 photograph courtesy of Steve Thompson; 1994 photograph by the author)

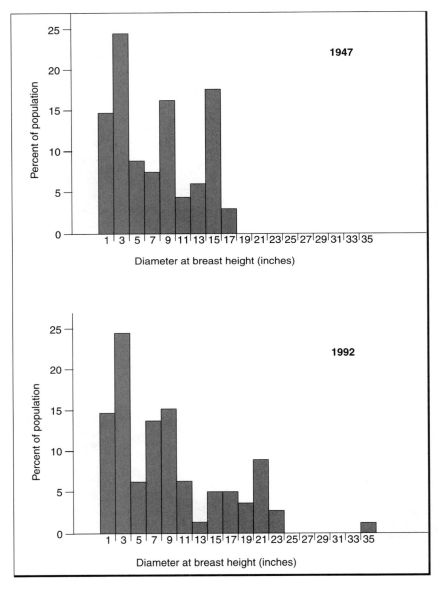

Figure 5.6. White pine size structures of the Thompson Bluff stand in 1947 and 1992 show an increase in the range of sizes.

regime that initially allowed the white pine stands to become established? On a human time scale, it appears that human disturbance over the last century has promoted white pine regeneration. However, the average natural life span of white pine may be four or five times that of the average person, so it may be too soon to draw conclusions about the effects of human impact. To stretch further the temporal scale over which we must view vegetation stability and change, white pine has been present in southwestern Wisconsin for over 10,000 years. Although population dynamics have fluctuated throughout the late Pleistocene, white pine regeneration continues. Under contemporary environmental conditions white pine likely will persist at Thompson Bluff, and in general at the southern limit of its range in Wisconsin, as it has throughout the Holocene.

Notes

1. See H. Hansen, "Postglacial Vegetation of the Driftless Area of Wisconsin," *American Midland Naturalist* 21 (1939): 760–761; J. Curtis, "Origin and Development of the Vegetation of Southwestern Wisconsin," contribution to the Appendix of R. Finley, *Geography of Wisconsin: A Content Outline* (Madison, 1976), 558; and J. T. Curtis, *The Vegetation of Wisconsin: An Ordination of Plant Communities* (Madison, 1959), 13.

2. James Knox, personal communication; see also D. Mickelson, J. Knox, and L. Clayton, "Glaciation of the Driftless Area: An Evaluation of the Evidence," in *Quaternary History of the Driftless Area*, Field Trip Guide Book No. 5 (Madison, 1982), 155–169.

3. M. Davis, "Biogeography of Temperate Deciduous Forests," *Geoscience and Man* 8 (1976): 21; M. Davis, "Holocene Vegetational History of the Eastern United States," in H. E. Wright, Jr., ed., *Late-Quaternary Environments of the United States*, vol. 2 (Minneapolis, 1983), 173; T. Webb III, E. Cushing, and H. Wright, Jr., "Holocene Changes in the Vegetation of the Midwest," in H. Wright, Jr., ed., *Late-Quaternary Environments of the United States*, vol. 2, 162.

4. This is one of three pollen cores that were used to document vegetation changes along the prairie–deciduous forest ecotone. See A. Davis, "The Prairie–Deciduous Forest Ecotone in the Upper Middle West," *Annals of the Association of American Geographers* 67 (1977): 204–213. Henry Hansen also analyzed pollen cores in the Driftless Area, but he did not radiocarbon date the sediment, so his findings cannot be compared with those of Davis.

5. Davis, "The Prairie–Deciduous Forest Ecotone in the Upper Middle West," 211; J. Knox, P. McDowell, and W. Johnson, "Holocene Fluvial Stratigraphy and Climatic Change in the Driftless Area, Wisconsin," in W. C. Mahaney, ed., *Quaternary Paleoclimate* (Norwich, England, 1981), 110.

6. A peak in the charcoal record at this time suggests that the decline of spruce, one species of a relict boreal community, may have been due to cata-

strophic fires during this dry period. See Davis, "The Prairie–Deciduous Forest Ecotone in the Upper Middle West," 208.

7.   P. Delcourt and H. Delcourt, "Vegetation Maps for Eastern North America: 40,000 YR B.P. to the Present," in R. Romans, ed., *Geobotany*, vol. 2 (New York, 1981), 150.

8.   A map illustrating the movements of the prairie-forest ecotone is shown in J. C. Bernabo and T. Webb III, "Changing Patterns in the Holocene Pollen Record of Northeastern North America: A Mapped Summary," *Quaternary Research* 8 (1977): 89.

9.   Curtis, *The Vegetation of Wisconsin*, 216.

10.   R. Cahayla-Wynne and D. Glenn-Lewin, "The Forest Vegetation of the Driftless Area, Northeast Iowa," *American Midland Naturalist* 100 (1978): 307–319.

11.   The results of this study are summarized in the following two sources: R. McIntosh, "Pine Stands in Southwestern Wisconsin," M.S. thesis, University of Wisconsin–Madison, 1948; R. McIntosh, "Pine Stands in Southwestern Wisconsin," *Transactions of the Wisconsin Academy of Science, Arts and Letters* 40 (1950): 243–257.

12.   S. Svatek, "Forest Dynamics of Relict Eastern White Pine Stands in Southwestern Wisconsin," M. S. thesis, University of Wisconsin–Madison, 1993; and S. Svatek Ziegler, "Relict Eastern White Pine Stands in Southwestern Wisconsin," *American Midland Naturalist* 133 (1995): 88–100.

13.   Davis, "The Prairie–Deciduous Forest Ecotone in the Upper Middle West," 206, 208.

14.   Curtis, *The Vegetation of Wisconsin*, 216.

15.   See, for example, D. Maissurow, "Fire as a Necessary Factor in the Perpetuation of White Pine," *Journal of Forestry* 33 (1935); and Curtis, *The Vegetation of Wisconsin*, 212.

16.   Peter Quinby also correlated successful white pine regeneration with small disturbance patches rather than catastrophic fire. See P. Quinby, "Self-Replacement in Old-Growth White Pine Forests of Temagami, Ontario," *Forest Ecology and Management* 41 (1991): 104.

17.   Bernabo and Webb, "Changing Patterns in the Holocene Pollen Record," 82; and Davis, "The Prairie–Deciduous Forest Ecotone in the Upper Middle West," 212.

18.   Peter Quinby ("Self-Replacement in Old-Growth White Pine Forests," 104) cites F. S. Baker, "A Revised Tolerance Table," *Journal of Forestry* 47 (1949): 179–181; and K. T. Logan, *Growth of Tree Seedlings as Affected by Light Intensity, II: Red Pine, White Pine, Jack Pine, and Eastern Larch*, Canadian Department of Forestry Publication 1160 (1966).

19.   K. Lancaster and W. Leak, *A Silvicultural Guide for White Pine in the Northeast*, U.S. Department of Agriculture Forest Service Technical Report NE-41 (Broomall, Pa., 1978).

20.   McIntosh, "Pine Stands in Southwestern Wisconsin," 41.

6

Wisconsin's Vegetation History and the Balancing of Nature

Duane Griffin

Vegetation, more than any other element in the landscape, reflects the environmental conditions of a place. In our classrooms and textbooks, we teach and learn that the tapestry of life-forms and species that we see in the landscape is the result of the complex interplay of climate, soils, topography, disturbance regimes, and the web of interactions among the organisms that live there.[1] On human time scales these factors may seem static. A forest might be felled for timber or burned by wildfire; a prairie might be plowed or paved. But in the absence of dramatic events such as these, the vegetation of the places in which we live may seem unchanging, so much so that the "forest primeval" and the "mighty oak" may come to represent the very essence of stability and timelessness.

The idea that there exists a balance of nature, a constant and stable point from which any deviation is an aberration and to which any alteration should return, is an old one. It is implicit in most cosmologies and a great deal of ecological theory. Within this view, the state of balance for the biophysical environment is analogous to the state of health for an organism. Actions or events that alter the biotic landscape and move it away from homeostasis are seen as unhealthy, while those that manage, conserve, preserve, or otherwise restore or maintain the balance are healthy.[2]

This is a comforting view. From a scientific perspective it suggests that we need only identify the state of balance for a particular community in order to predict how it will respond to a given change. On a practical level, it implies that finding an appropriate relationship between society and the natural environment might be a matter of defining the proper symmetry between our own consumption and the balance of nature. Psychologically, it promises that we might find in nature the timelessness, purity, even grace, that may be lacking in our daily, all too human, lives. One problem that this view poses, however, is that by invoking timelessness we deny the role and importance of history; and in nature, no less than in human affairs, history matters.

On the north side of Bascom Hill, the heart of the University of Wisconsin campus in Madison, is a small woods dominated by large red oak trees.[3] For decades, Bascom Woods has served as a place to take a quiet break from the daily academic grind and a hideaway for sweethearts. For many who pass and pause along its paths, the woods may represent a small bit of the "forest primeval," an informal monument to wild nature that has somehow escaped becoming another dormitory, library, or office building. Just to the south of the woods, at the steps to Bascom Hall, is a monument to history, a bronze statue of Abraham Lincoln gazing toward the state capitol dome. Few passersby realize that the man and the oak woods are linked by more than proximity.

Oak woods are common in southern Wisconsin, but forests dominated by red oak are somewhat uncommon. In his study of Wisconsin's vegetation, John T. Curtis notes that some such forests appear to be have been initiated by past disturbance events such as land clearing.[4] Tree ring counts from one of the old oaks in Bascom Woods reveal that it was about 130 years old when it was felled in a windstorm in 1992. Presumably, others of the same size are of similar age. What could have happened in the 1860s to initiate the woods that exist today?

The answer may lie elsewhere on Bascom Hill. A plaque on the wall of nearby North Hall, situated between the statue and the woods, recounts how the forced-air heating system in the building was shut down as an economy measure during the Civil War. Wood stoves were installed in the students' dormitory rooms for cooking and heating. The students were responsible for collecting their own firewood, "often," notes the plaque, "a tree from nearby Bascom Woods."

We know from the records of the public land survey (conducted in the 1830s) that steep, north-facing slopes in Dane County tended to be occupied by widely spaced white oaks, often with a shrubby understory that was periodically destroyed by fire.[5] By the 1860s this understory had probably grown into a dense thicket harboring the ancestors of the modern red oaks. By removing the larger trees and ignoring the small saplings, the students may have unknowingly shifted the balance of ecological favor toward the red oaks that now dominate the woods. In this sense, Bascom Woods is a living relic of the Civil War. A forest it is; primeval it is not.

Bascom Hill provides a tidy (if somewhat conjectural) illustration of the dynamic nature of vegetation in the landscape and the linkages that can exist between nature and history. Indeed, all of Wisconsin's modern vegetation bears the imprint of the past two centuries of changing human land use. Our biotic landscapes are different from what they were prior to European settlement and from what they would have been had that settlement not occurred. The changes that created Wisconsin's modern vegetation, in turn, took place in the context of the presettlement vegetation, which had been conditioned by human activity even deeper in the past.[6]

On broader spatial and temporal horizons, we can see, if somewhat dimly, that more dramatic changes have taken place in Wisconsin's vegetation over the past 12,000 years. Our primary source of information on vegetation in the deep past is fossil pollen preserved in lake, marsh, and bog sediments. Analysts extract cores of this sediment and process them to isolate the microscopic grains of pollen, which they then painstakingly identify and count. Radiocarbon dating of organic debris in the sediment provide dates that the analysts use to reconstruct the chronology of changes in the relative abundances of pollen taxa. The result is a view of the vegetation that existed in the area at the time the sediments were deposited. The view is somewhat skewed and obscured by the nature of pollen production, transport, deposition, and preservation. It is further clouded by the low temporal and taxonomic resolution of pollen data: a single sample from the core may represent many decades of pollen accumulation, and most taxa can be resolved only to the generic or family levels.[7] Despite these shortcomings, modern pollen spectra do reflect the vegetation assemblages that produce them, and by looking at the fossil spectra in light of these reflections, we can infer the general character of vegetation in the past.

What does the pollen record tell us about the vegetation history of Wisconsin? Published summaries provide regional descriptions of vegetation changes that have occurred: open spruce woodlands established as the glacier retreated; the closing of this woodland as black ash and birch became more common; the replacement of these forests as pine arrived from the east around 11,000 years B.P. (before present); the development of oak-hardwood forests in the south around 9,000 years B.P.; the sudden drying that occurred around 5,500 years B.P.;

Table 6.1. Common and scientific names of trees mentioned in the text

Common names	Scientific names
Ash, black	*Fraxinus nigra*
Apsen, large-toothed	*Populus grandidentata*
trembling	*P. tremuloides*
Basswood	*Tilia americana*
Beech	*Fagus grandiflora*
Birch, yellow	*Betula lutea*
river	*B. nigra*
white	*B. papyrifera*
Box elder	*Acer negundo*
Cedar, white	*Thuja occidentalis*
Cherry, black	*Prunus serotina*
Cottonwood	*Populus deltoides*
Elm, American	*Ulmus americana*
slippery	*U. rubra*
Hemlock	*Tsuga canadensis*
Hickory, shagbark	*Carya ovata*
Ironwood	*Ostrya virginiana*
Maple, red	*Acer rubrum*
silver	*A. saccharinum*
sugar	*A. saccharum*
Oak, bur	*Quercus macrocarpa*
black	*Q. velutina*
Hill's	*Q. ellipsoides*
red	*Q. rubra (Q. borealis)*
swamp white	*Q. bicolor*
white	*Q. alba*
Pine, jack	*Pinus banksiana*
red	*P. resinosa*
white	*P. strobus*
Spruce, black	*Picea mariana*
Tamarack	*Larix laricina*
Walnut, black	*Juglans nigra*
Willow, black	*Salix nigra*

and the return to moister conditions and the arrival of hemlock in the northern forests around 3,500 years B.P.[8]

Insightful as these summaries are, we can gain a still clearer understanding of Wisconsin's Holocene vegetation dynamics by focusing on two particular factors that drive them: climate and disturbance regimes. By viewing the pollen record through these two lenses and interpreting the results in terms of what we know about modern and presettlement vegetation distributions, new patterns emerge, new insights into the vegetation as it changed in the past.

Wisconsin is divided into two distinct floristic and vegetational

zones (Figure 6.1), separated by a southeast to northwest trending ecotone known as the Wisconsin tension zone.[9] The boundary between the
two zones is marked by the southern limit of pine as a common forest element and by the range limits of many plant species. The tension
zone corresponds to a number of climatic factors that are responsible
for the large-scale pattern of vegetation in the state. Within this larger
pattern, topography, soils, disturbance regimes, and biotic interactions
structure local vegetation patterns. In general, oak and pine species are

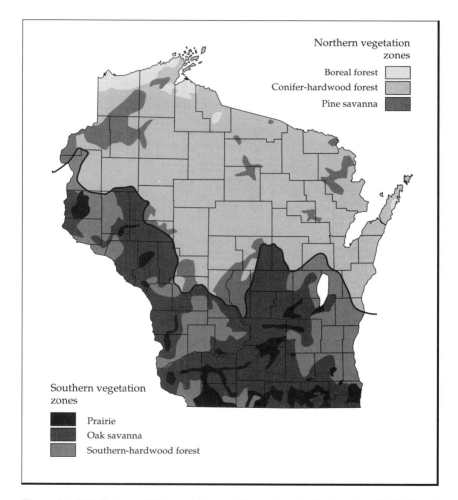

Figure 6.1. Distribution of Wisconsin's presettlement vegetation based on Public Land
Survey field notes (Adapted from the frontispiece to Curtis, 1959)

fire and drought tolerant but cannot reproduce well in the shade cast by a closed forest canopy. Hardwood species, in contrast, are generally shade tolerant but cannot withstand burning or prolonged drought. Thus, the particular combination of soil moisture, canopy cover, and disturbance regimes limits the type of vegetation that occurs at a site.[10]

North of the tension zone, climate is characterized by a high frequency of cool, dry, continental arctic air masses from Canada. Winters are longer, snowier, and colder than in the southern part of the state. Total precipitation is slightly less than in the south, but summer temperatures are cooler; thus, there is less evaporation, so more of the precipitation is available to plants.[11]

Four main vegetation "types" can be identified in the presettlement landscape of this cool, moist, northern region. On moister soils, the upland vegetation consisted of mixed conifer-hardwood forests. In these forests, the dense canopy produced high-moisture, low-light conditions that discouraged undergrowth and ground fires. Windstorms were the most common presettlement disturbance agents, and they sometimes affected thousands of acres. On sites with poorer soils, pine forests were the dominant vegetation. The more open canopy in these forests created drier surface conditions and allowed an herb or shrub understory to develop. During dry periods, the shrubs and the mat of dead pine needles provided highly combustible fuel for ground fires that served to kill off less fire-tolerant species that might grow and eventually shade out the pines. Pine forests also developed on sites where catastrophic disturbances destroyed hardwood forests. Pine barrens (savannas with jack pine as the dominant tree species) occupied sites with the poorest soils, especially sandy glacial outwash areas. Hardwood or conifer swamps occurred on waterlogged lowland sites.[12]

South of the tension zone, climate is controlled by the interplay of Pacific air, warmed and dried in its passage over the Rocky Mountains, and warm, moist, tropical air from the Gulf of Mexico. Precipitation and snow cover are more variable, and summers are generally warmer and longer than in the north. Prior to European settlement, extensive and frequent fires were common and served to maintain the oak savannas and tallgrass prairies that covered most of the region. Hardwood forests occurred only on moist soils where open water or topographic barriers acted as firebreaks. Oak woods developed on protected sites where soils were too droughty to support hardwoods, and lowland hardwood forests and sedge meadows occupied waterlogged sites. Young pine trees are susceptible to summer heat and drought and need an insulating blanket of snow to protect their roots from stress induced by alternate freezing and thawing in late winter and early spring. Because snow cover is so variable and drought is so common in southern Wisconsin, pines were restricted to rock outcrops in the Driftless Area of southwestern Wisconsin.[13]

The differentiation between the northern and southern vegetation zones is clear in the pollen record. Presettlement pollen spectra from northern sites are dominated by pine pollen. In contrast, the most abundant pollen from southern Wisconsin sites is oak. Consequently, an "ecotone index" (EI) can be derived from the ratio of pine to oak pollen for a given level of a core. By calibrating critical values for the index with the presettlement vegetation map and calculating indices for each level of a core, we can tell whether or not the vegetation was primarily northern or southern in character.[14] By mapping these values for 12 sites in and around Wisconsin, it is possible to determine the general location of the ecotone (Figure 6.2*a*).

Similarly, the ratio of hardwood pollen[15] (associated with little disturbance) to that of pine and oak (encouraged by disturbance) serves as an index (DI) of how prevalent disturbance was in the landscape (Figure 6.2*b*). The interpretation of past vegetation change based on these simplified pollen groupings is conservative, since changes in taxa that are not included may be important but are not recorded. Yet, even at this level of resolution, the changing index values at 1,000-year time steps for the 12 pollen sites reveal a complex pattern of change throughout the Holocene.[16]

Following the rapid demise of the spruce-fir forests between 11,000 and 10,000 years B.P., the ecotone between conifer and deciduous forests ran east-west across the southernmost part of the state. Low DI values for most of the state suggest that fires, windstorms, and other disturbances were relatively rare in the southeastern three-fourths of the state. This disturbance pattern fluctuates with regard to individual sites, but remains more or less constant until the onset of the mid-Holocene dry period that began, in Wisconsin, around 5,500 years B.P.

In contrast with the relatively stable pattern of disturbance regimes in the early Holocene, the ecotone index suggests that the conifer-deciduous forest boundary was quite dynamic. Between 10,000 and 9,000 years B.P., it moved northward in western Wisconsin and Minnesota and then in central and eastern Wisconsin between 8,000 and 6,000 years B.P. (Figure 6.3). Between 5,000 and 3,000 years B.P. the ecotone moved only slightly farther north, but all the sites shifted to high disturbance levels during this period (Figure 6.3). This increase in disturbance levels, rather than the position of the ecotone per se, reflects the effects of the mid-Holocene dry period in Wisconsin, when precipi-

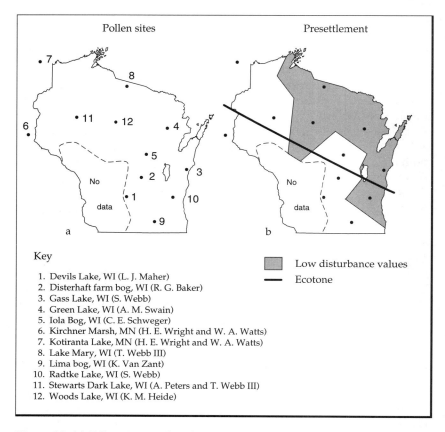

Figure 6.2. (*a*) Pollen sites used in this analysis; (*b*) Location of the ecotone between Wisconsin's northern and southern vegetation types and distribution of high and low disturbance regimes, based on EI and DI values from the presettlement pollen spectra. Compare the patterns shown in this map with those in Figure 6.1.

tation in the south was 10–20 percent lower than modern levels, and summer temperatures were about 0.5° C warmer than the present.[17]

There has been some confusion over the timing of the mid-Holocene dry period in the Midwest.[18] Pollen cores from sites in Minnesota clearly indicate that the dry period began 7,500 years B.P. and persisted there until 5,000 years B.P. The timing of this dry period was extended across the Midwest and was assumed to have held true for southern Wisconsin. Data from Devils Lake and other sites in south-central Wisconsin and eastern Iowa contradict this assumption and suggest that the dry period occurred much later, beginning around 5,500 years B.P. and ending by 3,500 years B.P.[19] The mapped index values shown in Figure 6.3 suggest that some warming occurred, at least in eastern Wisconsin, around 7,500 years B.P. The ecotone shifts northward, but the low-disturbance hardwood forests seem to have persisted, suggesting that there was something different about this earlier warming phase from the one that occurred at 5,500 years B.P.

The position of the modern ecotone is controlled by the balance of cold continental arctic and warmer Pacific and Gulf air masses. Where the arctic air dominates throughout the winter and, especially, early spring, constant snow cover insulates the roots of young pine trees. South of the tension zone, however, intrusions of Pacific and Gulf air in late winter and early spring melt the snow cover and create the freeze-thaw conditions that limit pine survival and reproduction. The northward shift in ecotone position between 8,000 and 7,000 years B.P. suggests that spring and summer intrusions of warm air increased in southern Wisconsin at that time. However, the persistence of hardwood forests at most sites until 5,500 years B.P. suggests that conditions were cool and moist enough to allow the survival and reproduction of their component species and discourage widespread fires. The change in disturbance regimes between 6,000 and 4,000 years B.P. suggests that intrusions of Pacific air during the spring and summer became strong enough to block the Gulf moisture, creating warmer and drier conditions throughout the area. These, in turn, would have promoted more frequent (and extensive) fires and favored the more drought- and fire-tolerant oak and pine species. As a result, the southern vegetation shifted from mesic deciduous forests to oak savannas and prairies. In the north, the mixed conifer-hardwood forests contracted their area, and the pine forests and savannas expanded theirs.

Whatever the ultimate causes of the mid-Holocene dry period, they weakened by 3,500 years B.P. and the ecotone gradually began to shift back toward the south.[20] By 2,000 years B.P., it had steepened along its east-west gradient and occupied approximately its modern position (Figure 6.3). After 3,000 years B.P. disturbance rates declined somewhat and hardwood forests became more common. Though it is not reflected in the maps, important compositional changes occurred in northern and eastern Wisconsin beginning around 3,500 years B.P. Hemlock, one of the dominant species in the modern and presettlement northern hardwood forests, migrated into northern Wisconsin from Michigan's Upper Peninsula, where it had been since 6,000 years B.P. Beech followed this migration into the Upper Peninsula and northeastern Wisconsin (along Lake Michigan) after 3,500 years B.P. but did

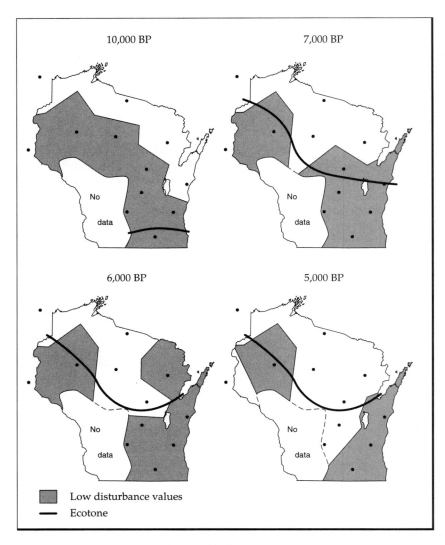

Figure 6.3. Ecotone position and areas of high- or low-disturbance regimes have varied throughout the Holocene. Because the pollen record for the Driftless Area is not suitable for this type of analysis, the area is not mapped.

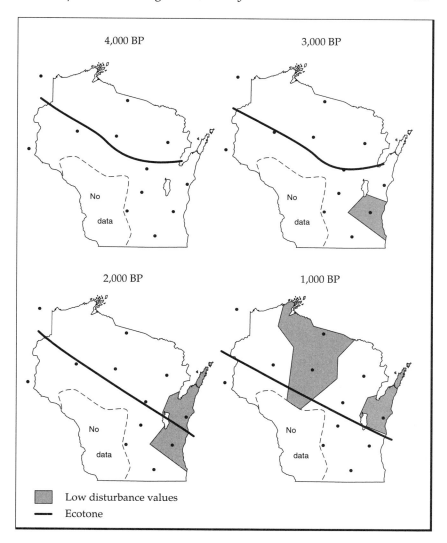

4,000 BP

3,000 BP

No data

No data

2,000 BP

1,000 BP

No data

No data

Low disturbance values
Ecotone

not extend its range southward until the Little Ice Age, a period from ca. A.D. 1450 to 1850, when conditions in much of the Northern Hemisphere were slightly cooler and moister.[21] The effects of the Little Ice Age are apparent in the presettlement maps (Figure 6.2) as the further decrease in disturbance levels in the northeastern half of the state.

Even at the relatively crude level of resolution available from the groups of taxa used in this analysis, it is clear that Wisconsin's vegeta-

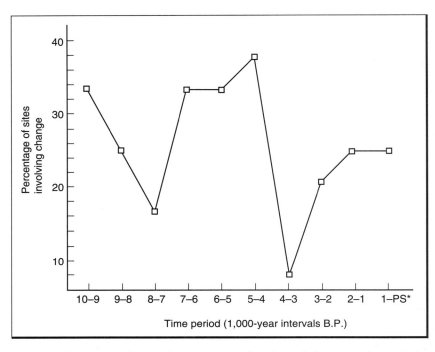

Figure 6.4. Percentage of sites whose ecotone or disturbance index status changed relative to the total number that could have involved change (i.e., 12 of 12 = 100 percent) in the time period indicated (*PS = presettlement, ca. 1830s)

tion has not been static over the past 10,000 years. In fact, the overall impression is one of constant change. We can see this most clearly by looking at the percentage of sites whose status changed in each 1,000-year time period relative to the maximum number that could have changed (i.e., a change in 12 out of 12 sites = 100 percent) (Figure 6.4). If we are generous and say that a change involving fewer than 20 percent of the sites is insignificant and represents stability, then only two periods, from 8,000 to 7,000 and from 4,000 to 3,000 years B.P., qualify as stable.

At the scale of individual sites there appear to have been longer periods of relative stability, though this is partly an artifact of the analysis.[22] If we start with the presettlement map and work backward, the longest period with no change that we can trace at a particular site involves those locales in southern Wisconsin that seem stable since the onset of the mid-Holocene dry period around 5,500 years B.P., when the oak savannas first appeared. Realistically, we can push that date up to the end of the dry period, around 3,500 years B.P., or even later. This is a more honest estimate of the age of Wisconsin's vegetation assemblages at the time of settlement. If we were able to look at even finer spatial and temporal scales, we would, no doubt, be able to see the effects of specific site histories, such as the one sketched earlier for Bascom Woods, embedded in the waves of broad-scale change.

What are we to make of the old notion of a balance of nature in light of Wisconsin's Holocene vegetation history? Does the long-term view suggest that it is a delusion and should be abandoned? There are no simple answers to these questions, because any answer depends less on nature itself than on our own wants and needs, our definitions of stability and change, and on the spatial and temporal scales at which we choose to apply our definitions.[23]

Had it not been for the hardy students of the 1860s, Bascom Woods would be a different forest from what it is today, but the regional climate, steep hillslope, and fire suppression would have all conspired to create a deciduous woods of some sort. To the sweethearts and passersby, it is structure, not species composition, that matters, and Bascom Woods has been a forest for at least a century. Its composition is changing as the old oaks die and are replaced by basswood, ash, and box elder trees, but these changes are not readily evident to the casual observer. In this context, Bascom Woods would appear to even the oldest homecoming alumni as the most stable element in the landscape, a fixed point of reference from which to judge the far more dramatic changes that have occurred in the built environment.

Even when species composition is important to people, change may be so subtle as to go almost unnoticed. The vegetation changes brought about by the onset of the mid-Holocene dry period seem dramatic and almost instantaneous in the pollen record. Yet, the archaeological record does not record a parallel shift in the material culture of the contemporaneous peoples whose daily livelihood and survival were immediately bound to their local environments. This suggests that the changes, which probably took a century or two to occur, were of a kind and magnitude that allowed people to continue living more or less as they had been.[24]

In a general sense, then, vegetation in the landscape can usually be viewed as more or less stable, and there is probably little harm in the idea of the balance of nature among sweethearts and passersby. Problems arise, however, when we uncritically extend this perspective to our scientific understanding of nature and then to management strategies based on this understanding.

The balance of nature implies that there must be an optimal ecological community, a single "best fit" solution, for a given set of environmental conditions. This idea of optimal assemblages of biota in delicate equilibria with their environments is an important component of the climax theory of vegetation advanced by Frederick Clements.[25] Ecological communities represent highly complex assemblages of interacting species, and Clements likened these assemblages to organisms. According to the theory, the community develops (grows) through a series of predictable stages to a predictable state of maturity, the climax, that is stable and self-replicating unless disturbed. Evolutionary theory implies that each climax community must have a long history as an intact unit in order for the complex coevolution of its component species to have occurred.

Clements' theories dominated ecological thinking for most of the twentieth century and provided the basis for many environmental management strategies that have been implemented in the United States. They also served as a basis for interpreting human impacts on ecological systems. From the Clementsian perspective, human activities alter the fine tuning of ecological communities and impinge not only on the communities but also on their entire evolutionary histories. Indeed, the fear that human impacts may upset these ancient and delicate harmonies and lead to the catastrophic collapse of environmental systems is implicit in much of the environmental mythology that has emerged in recent decades and has motivated a great deal of environmental activism.[26]

Clements' idea of similar climax communities repeated spatially across environmentally homogeneous regions was challenged early on by Henry Gleason, who noticed that the species composition of plant communities varied more or less continuously in space and time.[27] He proposed an individualistic theory of vegetation in which community composition is determined by the adaptations of individual species that allow them to survive and reproduce under continually varying environmental conditions. From this perspective, plant communities are better viewed as recent congregations of species with different biogeographic and evolutionary histories that have been brought together (temporarily) by chance. This conception contrasts strongly with the deterministic models of tightly knit assemblages of coevolved species with predictable successional pathways and end states that Clements and his followers envisioned.

Gleason's ideas formed a minority viewpoint for many years but have gradually eclipsed Clements'. In paleoecology, the view of Holocene vegetation change as the migration of discrete, intact formations has given way to more complex views based on the individualistic responses of particular taxa.[28] Most ecologists have abandoned the notion of a single, stable "balance of nature" in favor of more dynamic perspectives that acknowledge the role of history, chance, and change and are more sensitive to issues of spatial and temporal scale. Gradually, management strategies are beginning to follow suit.[29]

The vegetation patterns of Wisconsin's modern landscapes, even those that are relatively free of direct human interventions, do not exist because the species composing them have coevolved to exist in some optimal balance under modern environmental conditions.[30] They arose because there is an inherent flexibility in nature that allows viable, even complex, biotic communities to organize fairly quickly in the face of even rapid change. If there is a lesson we should learn from ecological history it is, perhaps, that there is not some elusive balance but, rather, a continual *balancing* of nature.

This lesson does not imply that we should become Polyannas about our impacts on the biotic landscape, but neither should we be Cassandras. Future biotic landscapes in Wisconsin will certainly bear the imprint of what we do today, but our descendants may or may not find them desirable, useful, or even livable. If we are to avoid this situation we will better serve ourselves and our landscapes by seeking to understand the mechanisms of change in nature than by chasing after chimeras of our own invention.

Notes

1. In the context of vegetation science, the term *disturbance* refers specifically to "an environmental change that makes plant resources available (such as water or light) that were formerly fully utilized by the pre-disturbance vegetation" and thus initiates vegetation change (T. R. Vale, *Plants and People: Vegetation Change in North America* [Washington, D.C., 1982], 3–4). This strict usage should not be confused with more popular understandings of the term that imply unwanted disruption. The presence of regular disturbances, such as periodic fires, is perfectly "natural" and may serve to maintain community composition.

2. F. N. Egerton, "Changing Concepts of the Balance of Nature," *Quarterly Review of Biology* 48 (1973): 322–350, provides a thorough history of the idea of the balance of nature.

3. Latin names of trees are listed in Table 6.1 and follow H. A. Gleason and A. Cronquist, *Manual of Vascular Plants of Northeastern United States and Canada*, 2d ed. (New York, 1991).

4. J. T. Curtis, *The Vegetation of Wisconsin: An Ordination of Plant Communities* (Madison, 1958), 110–111.

5. S. Will-Wolf and T. Montegue, "Landscape and Environmental Constraints on the Distribution of Presettlement Savannas and Prairies in Southern Wisconsin," *Proceedings of the North American Conference on Savannas and Barrens* (Chicago, 1994), 97–102.

6. The term *presettlement* refers to the period just prior to European settlement and land clearing (i.e., the early 1800s). Given that what is now Wisconsin was "settled" soon after glacial retreat, the term is, strictly speaking, a misnomer.

7. For example, we may know that the amount of oak pollen being deposited at a site changed at a particular time, but we cannot know which oak species were involved.

8. For example, T. Webb III, "Eastern North America," in B. Huntley and T. Webb III, eds., *Vegetation History* (New York, 1988), 386–414; and T. Webb III, E. J. Cushing, and H. E. Wright, Jr., "Holocene Changes in the Vegetation of the Midwest," in H. E. Wright, Jr., ed., *Late Quaternary Environments of the U.S.*, vol. 2, *The Holocene* (Minneapolis, 1983). All dates are in radiocarbon years before present.

9. See Curtis, *Vegetation of Wisconsin*, 15–17.

10. Shade and fire or drought tolerance vary from species to species but are almost always inversely related. Red oak, for example, is the most shade tolerant of all the oaks found in the state, but is also the least able to survive fire and drought. Similarly, white pine is the most shade tolerant and the least fire and drought tolerant of the pines.

11. Curtis, *Vegetation of Wisconsin*, 35–37. R. A. Bryson, "Air Masses, Streamlines, and the Boreal Forest," *Geographical Bulletin* 8 (1966): 228–269.

12. The predominant species in the conifer-hardwood forests studied by Curtis are sugar maple, hemlock, basswood, and yellow birch. Beech is an important component of forests in the counties along Lake Michigan. White, red, and jack pine are the dominant species in the pine forests, though Hill's and red oak, trembling and large-toothed aspen, red maple, and white birch are also common. Conifer swamps are dominated by tamarack, black spruce, and white cedar, while black ash and yellow birch characterize the hardwood swamps.

13. The southern Wisconsin oak savannas were characterized by widely spaced trees—primarily bur, black, and white oak, with shagbark hickory and Hill's and swamp white oak being less common—with an herbaceous understory. Oak woods composed primarily of black and white oaks along with bur, red, and Hill's oaks, black cherry, and shagbark hickory were relatively rare prior to European settlement but are now common because of fire suppression.

Curtis found the southern hardwood forests, like their northern counterparts, to be dominated by sugar maple, but with basswood and slippery elm along with red oak, ironwood, and American elm as the codominants and secondary species. Also as in the north, beech is an important element in the hardwood forests along Lake Michigan. Southern lowland forests are predominantly composed of silver maple, black willow, cottonwood, river birch, and American elm.

14. J. C. Bernabo and T. Webb III, "Changing Patterns in the Holocene Pollen Record of Northeastern North America: A Mapped Summary," *Quaternary Research* 8 (1977): 4.

15. Following Curtis, *Vegetation of Wisconsin* (p. 445), this was taken as the

sum of maple, hickory, basswood, elm, beech, ironwood, and hemlock pollen. Elm pollen makes up most of this sum, though hemlock pollen is abundant in northern sites after 3,500 years B.P.

16. Critical values for these indices were determined by comparing values from the presettlement pollen spectra and the mapped distribution of the pre-settlement vegetation types published in Curtis, *The Vegetation of Wisconsin*. The critical value for the ecotone index (EI = ln[pine/oak]) is −0.31, with values >−0.31 representing northern sites, values <−0.31 representing southern sites, and values of 0.23–0.27 representing sites within the tension zone. For the disturbance index (DI = ln[hardwoods/(pine + oak)]) the critical value is −0.6: values <−0.6 represent disturbed sites.

The approximate ecotone position was mapped from the DI values for each site. In cases where sites were very near the ecotone, the DI results were checked against the pollen profile for the site and, if needed, adjusted accordingly. The disturbance index values are mapped as Theissen polygons and should not be interpreted literally but, rather, as diagrams.

The pollen record for the Driftless Area is problematic for this type of analysis and was not included.

17. M. G. Winkler et al., "Middle Holocene Dry Period in the Northern Midwestern United States: Lake Levels and Pollen Stratigraphy," *Quaternary Research* 25 (1986): 246. See also P. J. Bartlein, T. Webb III, and E. Fleri, "Holocene Climate Change in the Northern Midwest: Pollen-derived Estimates," *Quaternary Research* 22 (1984): 361–374.

18. For an overview of the different interpretations of the mid-Holocene dry period, see Winkler et al., "Middle Holocene Dry Period"; and R. G. Baker, L. J. Maher, C. J. Chumbley, and K. L. Van Zant, "Patterns of Holocene Environmental Change in the Midwestern United States," *Quaternary Research* 37 (1992): 379–389.

19. R. G. Baker, L. J. Maher, C. J. Chumbley, and K. L. Van Zant, "Patterns of Holocene Environmental Change in the Midwestern United States," *Quaternary Research* 37 (1992): 379–389. Winkler et al., "Middle Holocene Dry Period," using data from Lake Mendota (Dane Co.), suggest that the period began around 6,500 years B.P., but this date is probably pushed back by the presence of old carbonate. Devils Lake, a soft-water lake with better dating control, and other sites show the change occurring at 5,500 years B.P.

20. The regional changes that began at 3,500 years B.P. are more evident in the pollen profiles for individual sites than in Figure 6.3. This is an artifact of the crude level of generalization used in the analysis.

21. M. B. Davis, "Quaternary History and the Stability of Forest Communities," in D. C. West, H. H. Shugart, and D. B. Botkin, eds., *Forest Succession Concepts and Applications* (New York, 1981), 132–153; Webb, "Eastern North America."

22. Changes in Kotiranta Lake before 9,000 years B.P. and since 3,000 years B.P. involve spruce pollen and thus are not reflected in the mapped summaries. Birch pollen was not included in the analysis because of the wide range of climatic and disturbance conditions that its species tolerate. Pollen diagrams,

however, reveal changes in birch percentages throughout the northern part of the state at ca. 7,000 and 3,500 years B.P. If these changes are taken into account, only the 3,500-year period between these two dates can be considered as stable.

23.   T. R. Vale, "Clearcut Logging, Vegetation Dynamics, and Human Wisdom," *Geographical Review* 78 (1988): 375–386.

24.   Cf. J. B. Stoltman, "The Archaic Tradition," *Wisconsin Archeologist* 6, nos. 3–4 (1986): 207–238.

25.   Clements' first published his ideas in 1916 and restated his theory more concisely in "Nature and Structure of the Climax," *Journal of Ecology* 24 (1936): 252–284.

26.   For example, the "airplane analogy" of species extinction: if you begin pulling rivets out of a flying airplane, you will be able to do so for a while, but eventually the plane will fall apart and crash. Clements' work was not the only theory available or in use, but it formed the dominant paradigm in ecological thought for most of this century. It persisted as a prominent theme in college textbooks into the 1980s and was the primary model for many nature writers who popularized this view of human-environment interactions among the membership of environmental organizations and the general public.

27.   H. A. Gleason, "The Individualistic Concept of the Plant Association," *Bulletin of the Torrey Botanical Club* 53 (1926): 7–26.

28.   The former view has been promoted by P. A. and H. R. Delcourt, "Vegetation Maps for Eastern North America: 40,000 years B.P. to the Present," in R. C. Romans, ed., *Geobotany II* (New York, 1981), 123–165. The latter perspective is documented in Webb, "Eastern North America."

29.   For example, see N. L. Christensen, "Landscape History and Ecological Change," *Journal of Forest History* 33 (1989): 116–124. D. L. Botkin discusses current directions in ecology as well as the balance of nature as it has played out in ecological theory and management, as well as popular thinking in *Discordant Harmonies: A New Ecology for the Twenty-first Century* (Oxford, 1990).

30.   The adaptations of individual species are continually fine tuned and altered by evolutionary processes, but the short periods of time in which Wisconsin's vegetation communities have existed are not long enough for true evolution to have occurred.

7

Geography, Wisconsin, and the Upland Sandpiper

Robin P. White

Geography's intellectual scope is broad: It is the study of place, of relationships through time and space, and of interactions between human activities and the natural environment. These themes, whether pursued individually or in some combination, appear in a rich variety of contexts, but one situation, the study of birds, blends all three. Wisconsin's birds, for example, have strong links to the characteristics of places; their numbers, distributions, and behavior patterns change through time and across landscapes; they are influenced by human activities. To investigate our birds, or a particular bird, is to study a part of the land and life of the state.

My formal, academic fascination with the upland sandpiper (*Bartramia longicauda*) began with a suggestion from my father. During the several years that the bird was the focus of my research, however, my interest took me to grasslands on quiet spring mornings in familiar Wisconsin, as well as on hikes on the distant Argentine pampas. Those experiences convinced me of the richness of a geographic perspective on the bird, its association with place, its spatial and temporal dynamics, and its connections with people.

Figure 7.1. Upland sandpipers breed on open grasslands (Photograph by the author)

Upland Sandpipers and Place

Upland sandpipers are associated with Wisconsin's short grassy landscapes (Figure 7.1). In his *Sand County Almanac*, Aldo Leopold described the upland sandpiper as a testament of spring: "When dandelions have set the mark of May on Wisconsin pastures, it is time to listen for the final proof of spring. Sit down on a tussock, cock your ears at the sky, dial out the bedlam of meadowlarks and redwings and soon you may hear it: the flight-song of the upland plover, just now back from the Argentine."[1]

The upland sandpiper is a shorebird found in open grassy habitats, including prairies, pastures, and sparsely timbered pine and oak barrens. Its use of upland and inland habitats and its active and nervous behavior patterns have caused many people to call this species a plover. While its habitat and behavior are similar to those of some birds in the plover family, in 1973 the American Ornithologists' Union classified this species as a sandpiper and replaced its former common names, including grass plover, prairie plover, and most notably, the upland plover.[2] Whether called sandpiper or plover, though, the rolling trill of its mating song and its graceful flight have made this bird symbolic of spring on Wisconsin grasslands.

Ecology

Adult upland sandpipers stand about 30 centimeters high (about 1 foot) and have brown, streaked feathers without bright colors or conspicuous marks. They have small heads and long, thin necks. Their bills are short and their legs yellowish. While these characteristics may not stir up a memorable image, one of the bird's most distinctive behavior patterns helps to identify positively the upland sandpiper in the field. It often lands on top of fenceposts or utility poles with its wings held high over its head for several seconds before folding them down over its back.

The upland sandpiper's song also helps to identify this bird. Arthur Cleveland Bent, working for the Smithsonian Institution in 1929, used field observations to describe the bird's melodious whistle.[3] The whistle begins with notes sounding like water gurgling from a large bottle followed by a loud *whip-whee-ee-you* compared with the long, drawn-out whistled cry of a hawk.

Upland sandpipers arrive in Wisconsin in early April. For about 6 weeks, from the end of April to early June, male upland sandpipers perform an elaborate flight display to attract females. This display was observed in detail by Irvin Ailes on the Buena Vista Marsh in southwestern Portage County.[4] The display begins with an upward flight from the ground or a fencepost. When reaching approximately 50 meters (164 feet) above the ground, the displaying bird begins to whistle. While performing the whistle, the bird glides on outstretched wings, stretching its neck forward during the introductory notes.

A flutter stroke, described by Irven Buss and Arthur Hawkins in their study of upland sandpipers at Faville Grove Wildlife Area in Jefferson County, is used at the beginning and end of the flight display.[5] The flutter stroke is a rapid movement of the wings, scarcely rising above the horizontal. This wingbeat has led observers to describe the upland sandpiper as a bird that flies on the tips of its wings.

During the flight display, the bird performs a whistle at about 2–3 -minute intervals, with the complete display lasting up to 15 minutes. After the initial flutter stroke, upward flight, and whistle, the displaying bird flies in a large circle, constantly gaining altitude to a maximum of about 400 meters (1,315 feet). Once at the maximum height, the bird quickly descends to the original takeoff point. Upon landing on the ground or a post, the bird momentarily holds its wings high over its head and, then uttering a last long whistle, folds them into their resting position.

In addition to the flight display, upland sandpiper courtship includes a series of displays performed on the ground. The tail-up display begins with the male upland sandpiper quickly approaching the female while calling softly.[6] He breaks into a chasing pace with his head held high, tail cocked above his back, and gular (or throat) pouch puffed out, uttering a low throaty rattle.[7] When the female is not sexually receptive, she runs off a short distance from the advancing male and continues feeding. When the female moves away, the displaying male stops immediately, gives a low whistle, and resumes feeding. If the female is sexually receptive, she allows the male to approach and mount her from behind. The bird's cloacae then touch, transferring sperm into the female, which fertilizes her eggs.

While observing displaying upland sandpipers in pastures and abandoned agricultural fields in Green County, I found the extended gular pouches particularly striking. The inflated pouches are used to communicate various messages, including aggressive behavior. They give the male bird a distinct, distended white throat not typically noticed while watching this species from a distance or in flight.

Once successful copulation has occurred and the female's eggs are fertilized, the female selects a well-hidden nest site on the ground to lay her eggs. Typical clutch size is four eggs with incubation, shared by both sexes, lasting about 3 weeks. As soon as the eggs hatch, young upland sandpiper chicks are ready to walk, following the adults in search of insects.

Upland Sandpipers, Temporal Change, and Human Activity

Temporal change, or change through recorded history of the upland sandpiper, has been both good and bad for the species. Generally, the entire population has experienced fluctuations in numbers and in range size. It is difficult to separate a discussion of these changes in relatively recent times from a discussion of human activities. While some changes over time, both historical and recent, can be attributed to nonhuman influences, other changes—particularly those critical to the bird's numbers today—are more easily assigned to human interference. Explanations for these changes vary from specific causes to more elusive speculations about the impacts of potential factors.

The upland sandpiper's breeding range extends across North America from the east coast, west to eastern Oregon and Washington, south to Oklahoma and northern Texas, and north into the Yukon Territory and Alaska. In the late nineteenth and early twentieth centuries, market hunting on the breeding grounds greatly depressed upland sandpiper numbers. Arthur Cleveland Bent, in his work for the Smithsonian Institution, described hunting this bird as a "real sporting proposition with the chances much in the bird's favor."[8] Bent described the bird as wary, not coming to decoys or a gunner's whistle, and not easily caught by hunting dogs. Despite this wariness, Bent also reported tales of special refrigerator cars sent out to prairie regions where parties of gunners would ship plovers and curlews back to Chicago markets.

While boxcars full of upland sandpipers shipped to market for food may be an exaggeration, field observers during this period consistently reported lower numbers than had been seen previously. In 1918, legislation was enacted to implement the Migratory Bird Treaty, an agreement signed in 1916 to protect migratory birds in the United States and Canada. A recovery of upland sandpipers, following adoption of this treaty, was documented until at least 1940.

In 1966, the United States Fish and Wildlife Service initiated an annual survey of breeding birds across the United States and Canada. This Breeding Bird Survey (abbreviated BBS) has helped determine the abundance of upland sandpipers on 70 census routes in Wisconsin. Each year the 39.4 kilometers (24.5 mile) road routes are run by automobile with 3-minute stops every 0.8 kilometers (0.5 mile), for a total of 50 stops. Each route is started 30 minutes before sunrise and takes about 4 hours to complete. Wisconsin ornithologist Sam Robbins divided the state and its BBS routes into eight biogeographical regions for easy portrayal of patterns of species abundances (Figure 7.2).

BBS data for upland sandpipers on the 70 Wisconsin routes for 28 years (1966–1993) show several notable patterns (Figure 7.2). Maximum numbers of upland sandpipers have been counted on the three BBS routes in the Lake Michigan Lowland. Upland sandpipers are counted in higher numbers on these three routes than on any of the other Wisconsin routes. The survey route in southern Door County has consistently had more upland sandpipers than any other Wisconsin route for most years. Wisconsin's highest BBS count for any year was 46 upland sandpipers on this Door County route in 1977.

Upland sandpipers are not nearly as abundant on the remaining BBS

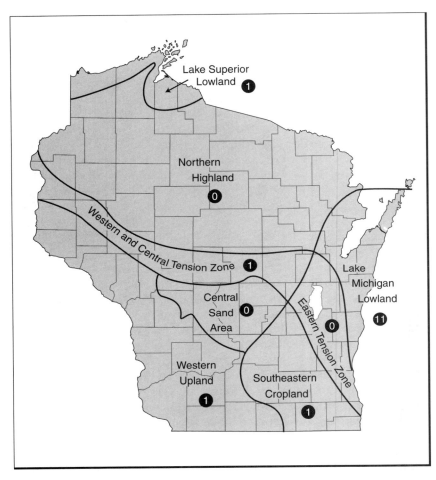

Figure 7.2. Among the biogeographic regions of Wisconsin, upland sandpipers are most common in the Lake Michigan Lowland. The numbers indicate the average number of upland sandpipers counted on Breeding Bird Survey routes from 1966 to 1993.

routes and in the other Wisconsin biogeographic regions. The fewest birds are found in the Eastern Tension Zone and Central Sand Area. In general, there appears to be a downward trend in numbers of upland sandpipers counted in Wisconsin for this 28-year period.[9] The birds are not declining, however, throughout their entire breeding range. BBS data for upland sandpipers across North America showed a statistically significant population increase from 1966 to 1994 for the survey region of the United States and Canada as a whole, as well as for several individual states and physiographic regions in the central part of the continent.[10]

Data collected under another system, initiated by The Nature Conservancy (TNC) in 1974, also can be used to examine the status of the upland sandpiper on its breeding range. This system ranks species according to number of occurrences, typically defined as the number of breeding locations, in a state or province. In Wisconsin, upland sandpipers are assigned an S2 rank, meaning the species is imperiled, with approximately 6–20 occurrences in the state.[11]

The Nature Conservancy's ranking system indicates that upland sandpipers are most abundant in central North America, especially in Nebraska, South Dakota, and Saskatchewan (Figure 7.3). Moving east and west, away from the central part of the breeding range, numbers decline, and the species may be given an endangered, threatened, or special-concern status along with protection under state law. In general, zoologists in eastern and western states and provinces believe upland sandpiper numbers are steady or declining but not increasing. Often, however, aside from BBS data, good data are not available to make conclusive statements about population trends.[12] Data from Illinois' Spring Bird Count may be an exception. Analyses of Spring Bird Count data for upland sandpipers in Illinois (1-day bird counts conducted on the first Saturday of May, from 1972 through 1993) suggest a population decline of approximately 63 percent over this 21-year period.[13]

While numbers of upland sandpipers generally are low in eastern and western North America, some states, such as Idaho, classify the species as peripheral. A peripheral species is one which, although low in numbers, is not a high priority for tracking, because the state is not a significant portion of its breeding range. The continued existence of nesting upland sandpipers in east or west coast states may depend on the presence of healthy populations in the core of its range.[14]

An increase or decrease in numbers of upland sandpipers can be caused by many factors. Although the upland sandpiper may be increasing in some parts of its breeding range, this chapter explores factors that may explain its decline in Wisconsin. Potential factors leading to its decline include predation, loss of habitat, damage by certain land use practices, and bad weather.

Coyotes, foxes, raccoons, other small mammals, crows, and birds of prey are potential predators of upland sandpiper eggs and chicks. Several studies have shown that upland sandpipers vigorously protect their eggs, nests, and flightless chicks. Field studies generally report nest success (number of nests with at least one hatched egg expressed

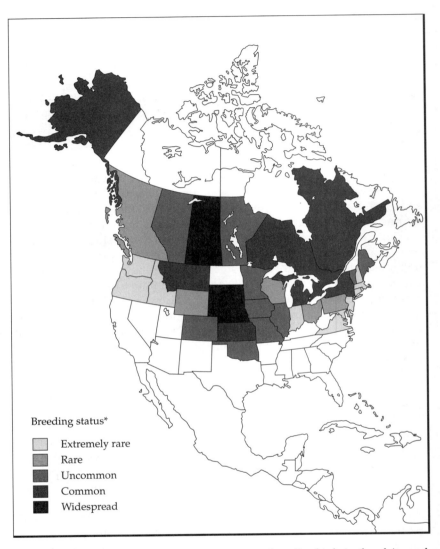

Breeding status*

- ☐ Extremely rare
- ☐ Rare
- ☐ Uncommon
- ☐ Common
- ☐ Widespread

Figure 7.3. Upland sandpipers are most common as breeding birds in the plains and prairies of the midcontinent, as shown here with shadings based on The Nature Conservancy's ranking system. * Unshaded states and provinces indicate that no rank has been assigned, updated rank is not available, or area is a non-breeding area for upland sandpipers.

as a percentage of total nests found) and egg success (number of hatched eggs expressed as a percentage of total eggs found) as greater than 50 percent, with values as high as 100 percent.[15]

In a study of grassland birds on the Kansas prairies, Bowen showed that nesting success was higher for upland sandpipers than for other ground-nesting birds (59 percent for upland sandpipers compared with 27 percent for other ground-nesters). He attributed this higher nesting success to the upland sandpipers' tendency to nest in groups and to defend their nests aggressively against potential predators.

Upland sandpiper adults use several methods of chick defense. One method is guarding their young and giving warning calls when birds of prey fly overhead. The young can then freeze low to the ground and become camouflaged against vegetative cover. A second call may be given as a signal for feeding to continue.[16]

Another method of chick defense used by upland sandpipers is mobbing, a type of social attack in which birds join together in aggressive behavior against a potential predator. Upland sandpipers have mobbed coyotes to distract predators from their chicks, whereas other grassland birds in the same study avoided the coyotes at the possible expense of chick loss.[17] In a study of grassland birds on a military base in Massachusetts, I was mobbed by at least five adult upland sandpipers while trying to capture and place leg bands on their chicks.[18]

Vigorous protection of nests, eggs, and chicks might suggest that loss from predators may be less of a threat to upland sandpiper reproductive success than to that of other ground-nesting species. While some studies seem to report the consistently higher success of nests and eggs of upland sandpipers, others suggest a low survival rate of their chicks.[19]

Determining chick success is more difficult than determining the success of nests or eggs. Although not easily, nests can be found by systematically searching probable nesting fields with dragging devices such as weighted ropes or chains. In contrast, young flightless chicks camouflaged against short grasses and shrubs can be much more difficult to find. Further, chicks are nearly impossible to observe for long periods of time and to follow from hatching to fledging.

Habitat loss on the breeding grounds may contribute to declines in upland sandpiper populations. In Wisconsin, I found upland sandpiper habitat best characterized by large open fields of grassy vegetation and flat topography. However, few large, continuous grassland habitats remain in Wisconsin. Many native grasslands have become fragmented and now are too small to attract nesting pairs and to provide sufficient space for feeding and brood-rearing. David Sample and

Randy Hoffman, of the Wisconsin Department of Natural Resources, reported that the average dry or dry-mesic prairie remnant in Wisconsin is only about 3.6 hectares (9 acres); estimates for minimum habitat size required for upland sandpipers is 10–100 hectares (25–250 acres).[20]

In several other states, where this species has endangered, threatened, or special-concern status, former large open grassland areas have been replaced with urban developments, agricultural fields, and stands of trees. Many of these upland sandpiper populations are now confined to military reserves or airports, the only remaining open-air grasslands in the state.[21]

Controversy may surround the maintenance of upland sandpiper habitat. When habitat loss is attributed to urbanization or development of incompatible agriculture (such as row crops without ground cover), the debate may include suggestions to limit development or to find alternative sites that will not eliminate open grasslands. When habitat loss is attributed to reforestation, the debate involves different questions. For example, should areas cleared for agriculture during European settlement and now returning to forest be maintained as grasslands? How much are people willing to pay for open grassland habitat management such as burning, clearing, and mowing? Answers to these questions require value judgments beyond the scientific determination of actual numbers of birds and survival rates.

In addition to habitat loss, both land use practices and weather conditions in otherwise apparently acceptable, open, flat grassland areas might limit upland sandpiper populations. The upland sandpiper's diet of almost 100 percent insects suggests that insecticides could be detrimental to this bird's survival. Pesticide use has been correlated with eggshell thinning in piscivorous marine birds and in birds of prey, and with mortality in songbirds. To date, however, there is no documentation of pesticide damage to upland sandpipers on their breeding grounds.

Various agricultural practices may affect upland sandpiper survival. Some activities, such as heavy grazing, may discourage adults from nesting and thus reduce nest density.[22] Other practices, such as early spring plowing, mowing or haying, and manure spreading, may destroy nests and eggs as well as discourage upland sandpiper nesting.[23] Cultivation in favored brood-rearing fields could be especially detrimental to upland sandpiper chicks.[24]

Weather conditions during the breeding season may play a role in upland sandpiper survival. Shorebirds may be susceptible to extreme weather conditions and corresponding food shortages, especially at higher latitudes. Upland sandpipers breeding in the Yukon Territory and Alaska may be more at risk to extreme weather conditions on their breeding grounds than are birds breeding in Wisconsin. Several researchers have described losses of large numbers of shorebirds as they arrive on their breeding grounds in the high Arctic.[25]

Although adult upland sandpipers breeding in Wisconsin may not be exposed to extreme weather conditions, flightless, downy chicks may be vulnerable to inclement summer weather such as heavy thunderstorms. Periods of heavy rain or strong winds that coincide with the first few days of life may limit not only food but also the ability to feed successfully, leading to starvation. In his study of upland sandpipers on the Buena Vista Marsh, Irvin Ailes attributed adverse weather at critical times, presumably heavy storms soon after hatching, to upland sandpiper chick mortality.[26]

In sum, upland sandpipers in Wisconsin and throughout North America have changed through time in numbers and distribution. The population has recovered from overhunting by humans and is no longer threatened by this problem, at least on the breeding grounds. The species appears to be doing well in the central part of its range but is uncommon in most eastern and western states. Problems the species may face now while in North America include predation by other animals, especially of the chicks; habitat loss to urbanization, agricultural development, and reforestation; land use practices such as early mowing and haying, as well as grazing; and adverse weather.

Upland Sandpipers, Spatial Change, and Human Activities

As with temporal changes, spatial changes in the upland sandpiper's annual activity involve the effects of both human and nonhuman activities. Spatial change in the upland sandpiper population includes movements from the breeding grounds in Wisconsin and other parts of North America, through Central and South America on migration, to the wintering grounds in Suriname and southern South America (Figure 7.4). Along the migratory route of approximately 20,000 kilometers (12,400 miles) roundtrip, and on the wintering grounds, some changes may be attributed to nonhuman influences, while others are more easily identified as responses to human activities.

Migration

The birds leave Wisconsin in late July or early August, traveling through Central America or across the Caribbean to South America

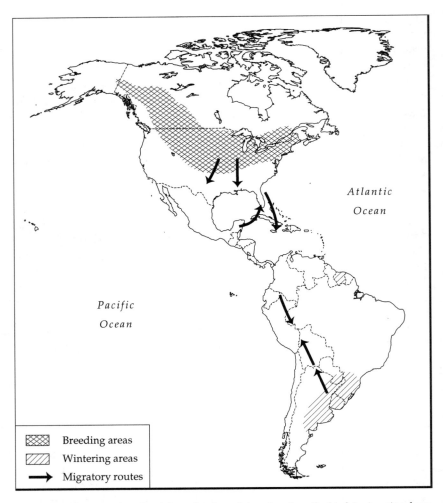

Figure 7.4. The annual cycle of the upland sandpiper involves the birds' migrating from breeding areas in North America to wintering grounds in South America.

(Figure 7.4). In spring, upland sandpipers fly north mainly through Central America. Some may migrate from the Yucatan Peninsula to Cuba and Florida and up the Atlantic coast.

The length of the migration period appears to extend over 3 or 4 months in fall and spring. Records of museum specimens and field observations are scattered throughout South America from August through May. While some birds may undertake a fairly rapid flight between breeding and wintering grounds, there appear to be birds who linger; migratory stopovers may be lengthy, and birds may spend as little as 2 months on the actual wintering grounds. This is apparent from records of birds in Panama and Peru in November, and in Venezuela in November and December. Because year-round observations of upland sandpipers in these countries have not been documented, birds present in these locations in November and December are assumed to be migrants.[27]

Staging areas are sites along the migration route where birds stop to feed. The exposure to predation, to change in habitat or land use, or to adverse weather while using these areas could be critical to upland sandpiper survival. Important staging areas for migrating upland sandpipers have not been carefully documented, and thus the extent of mortality due to these factors has not been quantified.

While hunting of shorebirds for food or sport is no longer a serious threat to populations in North America, south of the United States, hunting may be a critical but largely undocumented problem. The United States Fish and Wildlife Service's Bird Banding Laboratory reported bands of many shorebirds returned from South America as coming from birds taken by local residents for food. Herbert Beck published brief notes about the status of upland sandpipers in Lancaster, Pennsylvania. He cited several ornithologists who observed upland sandpipers being shot along their migratory route.[28] No recent study, however, has determined actual hunting pressures. Similarly, extent of mortality along the upland sandpiper's migratory route attributed to urban expansion, newly developed agricultural lands, and reforestation, as well as to certain land use practices and adverse weather, has not been documented.

Certain aspects of migration might make this a particularly critical time, especially for molting and first-year birds. Molting, or the shedding of old feathers and growth of new ones, occurs in upland sandpiper adults between October and January.[29] Young birds, however, may begin molting as early as August. These first-year birds may suspend their molt as migration approaches, completing the process once they reach the wintering grounds. They may need extra food for growth of new feathers or to compensate for reduced insulation. Thus, young birds, migrating while in partial molt, may need to feed longer each day and may exhibit reduced alertness to predators. Additionally, juvenile birds may be more vulnerable to mammalian predators because of inexperience and greater tendency to roost in more exposed places.[30]

Wintering Grounds

Upland sandpipers winter in southern South America and less commonly in Suriname.[31] The birds do not build nests, lay eggs, or raise young on the wintering grounds. Thus, behavior patterns associated with reproduction, such as the elaborate aerial breeding displays, are not performed here. The lack of conspicuous behavior makes it difficult to find, observe, and count wintering upland sandpipers. In addition, accurate information on mortality of wintering upland sandpipers is practically nonexistent. Wintering birds are scattered over relatively large expanses, and all birds that die are most likely not found, probably having been eaten by scavengers, such as vultures or foxes, soon after death.[32]

Extent of upland sandpiper loss to predators, other than humans, on the wintering grounds is difficult to determine. Potential predators include small mammals, vultures, and birds of prey. Because of the upland sandpiper's wary behavior, it might be difficult for these predators to take an adult of this species. However, juvenile birds that roost in more vulnerable places may be easier prey.

Despite the birds' elusive quality on the winter grounds, changes in historical distribution have been noted by several ornithologists in Argentina. A decline on the pampas south of Buenos Aires was reported in the early 1900s and again, more recently, in the 1960s.[33] As late as 1967, a prominent Argentine ornithologist kept secret the location of several upland sandpipers to protect them from hunters.[34] Landholdings are large in some parts of the Argentine pampas, and legal protection may be difficult to enforce.

The extent and effects of habitat loss on the wintering grounds also are difficult to assess. The amount of wintering habitat in southern temperate areas is small relative to the area used in North America during breeding.[35] Smaller land area for the same number of birds would suggest higher densities on the wintering grounds. To date, however, areas of high densities of upland sandpipers have not been reported.

Effects of pesticides on wintering upland sandpipers are unknown. Insecticides were sprayed in Argentina in the 1940s to eradicate locusts, and several observers reported the birds ate large quantities of these insects.[36] However, no cases of debilitating effects of pesticides on this species on the wintering grounds have been documented.

Wary and mobile juveniles and adults are likely not as vulnerable to mowing of agricultural fields on the wintering grounds. No information is available on the effects of grazing or adverse weather on wintering upland sandpipers.

Any of several factors—hunting, predation, habitat loss, certain land use practices, and adverse weather—could account for reduced survival of this species on the wintering grounds. An important basic question to address, however, is, how well are wintering upland sandpipers observed and counted? Reliance upon periodic observations of birdwatchers or notes from ornithologists, without systematic census techniques or surveys, leads to questions regarding blank spaces on a map: do they represent no birds, or birds but no observation? Very few of the birding hot spots in Argentina—Iguassu Falls, Barlioche, Peninsula Valdes—include grassland habitat. Very few of the potential wintering sites in other South American countries, such as Uruguay, southern Brazil, Paraguay, and Bolivia, are birded intensely and systematically in potential upland sandpiper winter habitat.

Conclusion

Upland sandpipers are reported in the highest numbers in several states and provinces in the center of their breeding range. In other parts of North America, including Wisconsin, the species may be stable but low in number, or decreasing. Along the migratory route and on their wintering grounds, numbers of birds seem to be low but are generally undocumented. Potential limiting factors include predation, habitat loss, land use practices, and adverse weather (Table 7.1). Identification of specific factors limiting upland sandpiper survival and reproductive success requires time and effort devoted to field observation, systematic data collection, and year-round study.

Upland sandpipers are an established part of Wisconsin's geography. Birders associated this species with specific places in the state—short grass, open-air habitats such as pastures in Door County or barrens in Burnett County. Within Wisconsin, the abundance of upland sandpipers has changed through time, with the population experiencing declines as well as increases. But the well-being of this "Wisconsin" bird involves what happens far beyond the state borders, particularly in northeastern and southern South America. The species' ability to survive, our objective skill in eliminating deleterious factors, and our emotional desire to encourage the bird will determine whether the upland sandpiper remains a testament to the glory of the Wisconsin spring.

Table 7.1 Factors affecting upland sandpiper survival[a]

| Potential problem | Breeding grounds | | Migration | Wintering grounds |
	Nests, eggs, chicks	Juveniles, adults	Juveniles, adults	Juveniles, adults
Predation				
Human	no	no	?	?
Other	?	?	?	?
Habitat loss				
Urban development	yes	yes	?	?
Agricultural development	yes	yes	?	?
Reforestation	yes	yes	?	?
Land use practices				
Pesticide application	?	?	?	?
Grazing	?	yes	?	?
Mowing or haying	yes	yes	?	?
Adverse weather	yes	?	?	?

Key
Yes: Is a problem affecting survival
No: Is not a problem affecting survival
?: Effects uncertain; need documentation
[a] Problems may affect survival or reproductive success directly (for example, predators killing chick) or indirectly (for example, decreased nest density due to grazing).

Notes

1. A. Leopold, *A Sand County Almanac* (New York, 1966), 37.

2. American Ornithologists' Union, *Checklist of North American Birds*, 6th ed. (Lawrence, Kans., 1983).

3. A. C. Bent, "Life Histories of North American Shorebirds," *U.S. National Museum Bulletin* 146 (1929): 55–69.

4. I. W. Ailes, "Ecology of the Upland Sandpiper in Central Wisconsin: Behavior of the Upland Sandpiper in Central Wisconsin," M.S. thesis, University of Wisconsin-Stevens Point, 1976.

5. I. O. Buss, and A. S. Hawkins, "The Upland Plover at Faville Grove, Wisconsin," *Wilson Bulletin* 51 (1939): 202–220. In the same article, Buss and Hawkins describe the Faville Grove Wildlife Area as "a 2,400 acre tract composed of 10 farms, situated in Jefferson County" (202).

6. E. H. Miller, "Communication in Breeding Shorebirds," *Behavior of Marine Animals* 5 (1984): 169–241.

7. I. W. Ailes, "Ecology of the Upland Sandpiper in Central Wisconsin: Behavior of the Upland Sandpiper in Central Wisconsin," M.S. thesis, University of Wisconsin–Stevens Point, 1976, 19–35.

8. Bent, "Life Histories of North American Shorebirds," 64.

9. BBS data for 1966–1994 continued to show a significant population de-

cline for the upland sandpiper in Wisconsin. J. R. Sauer, S. Schwartz, B. G. Peterjohn, and J. E. Hines, "The North American Breeding Bird Survey Home Page," Version 94.3, Patuxent Wildlife Research Center, Laurel, Maryland.

10. Sauer et al., "The North American Breeding Bird Survey Home Page."

11. This ranking system was developed by The Nature Conservancy in 1974 using biodiversity element ranks. The state and provincial ranks range from 1 to 5, the higher numbers indicating greater abundance. Additional TNC ranks identify other information on status, such as exotic, accidental, or wintering species. Not all states and provinces have adopted this system or assigned ranks to all species.

12. A survey was conducted in August and September of 1994. Several rankings were updated in June 1996. Zoologists, ornithologists, and data managers in states and provinces with breeding records of upland sandpipers were contacted to verify status and to comment on population trends. Results are mapped in Figure 7.3.

13. Unpublished data from Jim Herkert, wildlife biologist with the Illinois Endangered Species Protection Board. The population index of Illinois' Spring Bird Count data is the number of upland sandpipers recorded during a 1-day count with attempts to adjust for the fact that not all counties are censused each year and for an increase in observer effort. These data suggest a significant 4.6 percent per year population decline in Illinois between 1972 and 1993 (approximately a 63 percent decline over the entire period).

14. Washington Department of Wildlife, "Washington State Recovery Plan for the Upland Sandpiper (*Bartramia longicauda*)," typescript, Nongame Program, Wildlife Management Division, Olympia, 1993, 41 pp.

15. *Egg success reported by the following:*
Buhnerkempe and Westemeier, 1988: 91 percent (29 of 32 eggs hatched), Illinois (from J. E. Buhnerkempe and R. L. Westemeier, "Breeding Biology and Habitat of Upland Sandpipers on Prairie-chicken Sanctuaries in Illinois," *Transactions of the Illinois Academy of Science* 81 [1988]: 153–162).

Buss and Hawkins, 1939: 97 percent (101 of 104 eggs hatched), Faville Grove, Wisconsin (from I. O. Buss and A. S. Hawkins, "The Upland Plover at Faville Grove, Wisconsin," *Wilson Bulletin* 51 [1939]: 202–220).

Higgins and Kirsch, 1975: 94 percent (377 of 400 eggs hatched), North Dakota (from K. F. Higgins and L. M. Kirsch, "Some Aspects of the Breeding Biology of the Upland Sandpipers in North Dakota," *Wilson Bulletin* 87 [1975]: 96–102).

Nesting success reported by the following:
Buhnerkempe and Westemeier, 1988: 50 percent (12 of 24 nests successful), Illinois (from Buhnerkempe and Westemeier, 1988).

Buss and Hawkins, 1939: 66 percent (31 of 47 nests successful), Faville Grove, Wisconsin (from Buss and Hawkins, 1939).

Kaiser, 1979: 80 percent (26 of 33 nests successful), South Dakota (from P. H. Kaiser, "Upland Sandpiper Nesting in Southeastern South Dakota," *Proceedings of the South Dakota Academy of Science* 58 [1979]: 59–68).

Kirsch and Higgins, 1976: 65 percent (116 of 178 nests successful), North Dakota (from L. M. Kirsch and K. F. Higgins, "Upland Sandpiper Nesting and Management in North Dakota," *Wildlife Society Bulletin* 4 [1976]: 16–20).

Lokemoen and Duebert, 1974: 100 percent (12 of 12 nests successful), South Dakota (from J. T. Lokemoen and Duebert, "Summer Birds for a South Dakota Prairie," *South Dakota Conservation Digest* 41 [1974]: 18–21).

Oetting and Cassel, 1971: 100 percent (13 of 13 nests successful), North Dakota (from R. B. Oetting and J. R. Cassell, "Waterfowl Nesting on Interstate Highway Right-of-Way in North Dakota," *Journal of Wildlife Management* 35 [1971]: 774–781).

16. I. O. Buss, "The Upland Plover in Southwestern Yukon Territory," *Arctic* 4 (1951): 204–213.

17. D. E. Bowen, Jr., "Coloniality, Reproductive Success, and Habitat Interactions in Upland Sandpipers (*Bartramia longicauda*)," Ph.D. diss., Kansas State University, Manhattan, 1976.

18. R. P. White and S. M. Melvin, "Rare Grassland Birds and Management Recommendations for Camp Edwards/Otis Air National Guard Base," unpublished report, Massachusetts Natural Heritage Program, Boston, 1985, 29 pp.

19. Buss, "The Upland Plover in Southwestern Yukon Territory"; I. W. Ailes, "Breeding Biology and Habitat Use of the Upland Sandpiper in Central Wisconsin," *Passenger Pigeon* 42 (1980): 53–63.

20. D. W. Sample and R. M. Hoffman, "Birds of Dry-Mesic and Dry Prairies in Wisconsin," *Passenger Pigeon* 51 (1989): 195–208. F. B. Samson, "Island Biogeography and the Conservation of Nongame Birds," *Transactions of the North American Wildlife and Natural Resources Conference* 45 (1980): 245–251.

21. D. R. Osborne and A. T. Peterson, "Decline of the Upland Sandpiper (*Bartramia longicauda*) in Ohio: An Endangered Species," *Ohio Journal of Science* 84 (1984): 8–10; White and Melvin, "Rare Grassland Birds and Management Recommendations for Camp Edwards/Otis Air National Guard Base."

22. B. S. Bowen and A. D. Kruse, "Effects of Grazing on Nesting by Upland Sandpipers in Southcentral North Dakota," *Journal of Wildlife Management* 57 (1993): 291–301.

23. Buss and Hawkins, "The Upland Plover at Faville Grove, Wisconsin."

24. Ailes, "Breeding Biology and Habitat Use of the Upland Sandpiper in Central Wisconsin."

25. P. R. Evans and M. W. Pienkowski, "Population Dynamics of Shorebirds," *Behavior of Marine Animals* 5 (1984): 83–123.

26. Evans and Pienkowski, "Population Dynamics of Shorebirds"; Ailes, "Breeding Biology and Habitat Use."

27. R. P. White, "Wintering Grounds and Migration Patterns of the Upland Sandpiper," *American Birds* 42 (1988): 1247–1253.

28. S. E. Senner and M. A. Howe, "Conservation of Nearctic Shorebirds," *Behavior of Marine Animals* 5 (1984): 379–421; H. H. Beck, "Status of the Upland Plover in Lancaster County, Pennsylvania," *Auk* 73 (1956): 135–136.

29. A. J. Prater, J. H. Marchant, and J. Vuorinen, *Guide to the Identification*

and Ageing of Holarctic Waders, British Trust for Ornithology, Field Guide 17 (Beech Grove, Tring, 1977).

30.   Evans and Pienkowski, "Population Dynamics of Shorebirds."

31.   R. Haverschmidt, "The Migration and Wintering of the Upland Plover in Surinam," *Wilson Bulletin* 78 (1966): 319–320.

32.   Evans and Pienkowski, "Population Dynamics of Shorebirds."

33.   R. Dabbene, "Notas sobre los chorlos de Norte America que inviernan en la Republica Argentina." *El Hornero* 2 (1920): 99–128; J. J. Hudson, *Birds of La Plata,* vol. (London, 1920); A. Wetmore, *Our Migrant Shorebirds in Southern South America,* United States Department of Agriculture Technical Bulletin No. 26 (Washington, D.C., 1927); C. C. Olrog, "Observaciones sobre aves migratorias del hemisferio norte," *El Hornero* 10 (1967): 292–298.

34.   Olrog, "Observaciones sobre aves migratorias del hemisferio norte."

35.   J. P. Myers, "The Pampas Shorebird Community: Interactions between Breeding and Nonbreeding Members," in J. A. Keast and E. S. Morton, eds., *Migrant Birds in the Neotropics: Ecology, Behavior, and Conservation* (Washington, D.C., 1979), 37–49.

36.   W. B. Barrows, "Birds of the Lower Uruguay," *Auk* 1 (1884): 315; H. Durnford, "Notes of the Birds of the Province of Buenos Ayres," *Ibis* (1877): 166–203; Dabbene, "Notas sobre los chorlos de Norte America que inviernan en la Republica Argentina."

PART TWO

SETTLEMENT PROCESSES AND CULTURAL PATTERNS

Wisconsin is a place whose inhabitants have always traced their origins to somewhere else. This was true of the very earliest occupiers of the land. It was true of the steady succession of Native American tribes driven westward by European colonial expansion, who entered the region to displace its earlier occupants. And it certainly was true of the great waves of Euro-American settlers who came in the nineteenth and early twentieth centuries to take the land, build towns and cities, and thoroughly organize the economy and life of nearly every corner of the state. The process has continued to the present, as new groups—including substantial numbers of blacks, Asians, and people of Central and South American origins—have been added to the mix.

All these groups came for a purpose: to establish new lives in a place far from their original homes. Each has had its own distinctive set of migration, settlement, and adaptation experiences. Each has labored and created, established communities and neighborhoods, implanted its culture and traditions. All have helped to create the patterns of ethnic and cultural diversity that so strongly characterize the state.

The chapters in this section address the geographic processes and patterns of human settlement in Wisconsin. The first chapters delimit rather broadly the peoples and cultures that first settled the state and the regional variations in their distribution. Others describe the processes by which settlements and towns of various kinds were created across the state or document the emergence in certain places of distinctive cultural identities and built landscapes. A number of the chapters deal with specific groups and the places they settled, detailing the ways in which such forces as religion, kinship, Old World traditions, and economic advancement worked to establish a particular sense of place and identity in Wisconsin's immigrant communities. Wisconsin was and in many ways still is a vast patchwork quilt of ethnic-group settlement, and the chapters that deal with specific groups offer but a small sample of the many stories that could be told. They all, however, point to one of the most important facts about Wisconsin's settlement geography—the wide variation in experiences that took place among different groups in different places.

8

The Euro-American Settlement of Wisconsin, 1830–1920

Robert C. Ostergren

Wisconsin lies deep in the North American heartland. As such, the course of its early development was a product of the great nineteenth-century American effort to convert a vast and remote interior wilderness into a settled and economically productive landscape of farms, cities, and industries. Wisconsin experienced all the events and forces that made up that effort: the mining and timber exploitations, the westering American pioneer and the immigrant land seeker, the drive to bind resource productivity along systems of rail and water transportation to industrial urban processing centers, the peopling of those urban centers by successive waves of immigrant labor and the city-ward relocations of those who quit the land. The aim of this chapter is to see these processes as historical geography, that is, to see the formative period of settlement and development in both its temporal and spatial dimensions. Three themes are explored: the advance of Euro-American settlement, the emergence of an urban system, and the creation of the distinctive ethnic mosaic that best describes the cultural geography of Wisconsin's population.

The Advance of Settlement

The Euro-American settlement of Wisconsin began in earnest during the 1830s. This is not to say that Europeans and Americans were absent

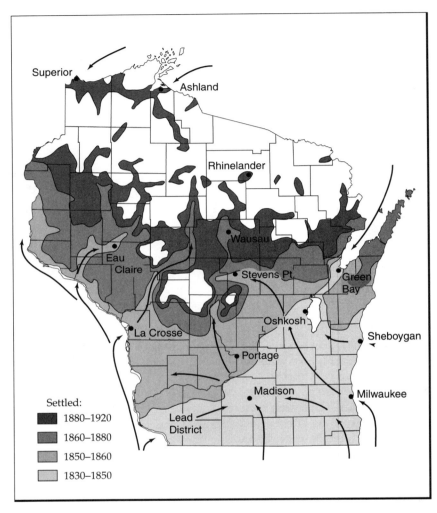

Figure 8.1. The advance of settlement in Wisconsin

prior to that date. In fact, their early presence was considerable. The French were active in organizing the fur trade as early as the mid-1600s and remained engaged in the region for more than a century. The British replaced the French during the latter part of the 1700s and were in turn followed by the Americans, who consolidated their influence in the region by establishing forts at strategic locations along the Wisconsin-Fox River line (Fort Crawford, 1816; Fort Winnebago, 1826;

and Fort Howard, 1816). The long period of fur trade and exploration, however, was a time of external exploitation and territorial aggrandizement rather than a time of serious settlement.

So too, in a fashion, were the early stages of the lead mining frontier that opened in the southwestern corner of Wisconsin during the 1820s. The thousands who swarmed into the lead district during the middle years of that decade were attracted primarily by the prospect of quick wealth. Their presence represented an opportunistic outlier of settlement, thrown out well ahead of an advancing but still distant agricultural frontier.[1] In time the lead mining settlements became agricultural settlements as well. And thus the lead mining district also became Wisconsin's first agricultural frontier, but its development as a mixed mining and agricultural region dates from the late 1820s at the very earliest and was further delayed by the Black Hawk War (1831–1832).[2]

Only after the start of the 1830s did the necessary prerequisities for a full-scale settlement boom in Wisconsin begin to be met. These included a series of treaties (1829–1848) that extinguished the rights of Native Americans to the land; the completion of the federal land survey, which progressed rapidly after 1833; and the opening of land offices at Mineral Point and Green Bay in 1834 and at Milwaukee in 1839. The ensuing advance of the settlement frontier across Wisconsin lasted until around 1920 (Figure 8.1).[3] It moved forward in a series of sweeping rushes and sharp penetrations that brought permanent agricultural settlement, as well as an emerging system of transportation, towns, and commerce, to an ever-increasing portion of the state. In the early decades, the advance was rapid; huge chunks of territory were taken in a relatively short period of time. Later advances took much more time and produced progressively smaller and more marginal additions to the ecumene.

The settlement advance may be divided temporally into four distinct phases, the time of each having been influenced by major national trends and developments. The first advance coincided with a nationwide surge of westward migration that extended from the mid-1830s through the decade of the 1840s. The second boom, which introduced massive numbers of recently arrived European immigrants, began in the early 1850s and continued until the outbreak of the Civil War in 1861. The third began with the resumption of westward movement and the renewal of European mass immigration to America after the Civil War and ended with the financial panic of 1873. The final wave began with improved economic conditions in the early 1880s. This was the most prolonged and discontinuous advance. It ended for all practical purposes around 1920, although a few efforts at colonization persisted into the 1920s. As outlined below, each of these phases had its own spatial or geographic characteristics.

During the first settlement phase (1830s and 1840s) the frontier spread outward initially from two widely separated footholds. One was the area around the old lead mining settlements in the southwest. The other was the extensive area of mixed forest and prairie landscapes in the southeast, which in the early 1830s had begun to receive advance elements of the westward-moving pioneering activity then following the shore of Lake Michigan and the Rock River valley northward out of Illinois. Both advances were slow at first because of the extensive land speculation activity that took place during the mid-1830s and the financial panic that followed in 1837. The late 1830s, however, witnessed heavy influxes of westward-moving Americans, and in the 1840s the stream of westward-moving Americans was joined by the first waves of land-hungry immigrants from western and northern Europe. Together old-stock Americans and Europeans settled extensively, eventually uniting the independent advances from the southwest and southeast and extending settlement over much of southern Wisconsin. By the close of this phase around 1850, the frontier in Wisconsin had reached a line running southwest to northeast that followed, first, the Wisconsin River valley upstream as far as Portage and, then, the upper Fox River valley to the west shore of Lake Winnebago and the lower Fox River valley to Green Bay before sagging back toward the shore of Lake Michigan near Manitowoc.

Roughly a third of the state, including much of the best potential land for agricultural settlement, was consumed in this first great surge. The occupation of the land was most intense in the southeastern part of the state, where settlers were attracted to the fertile soils and readily available fuels and building supplies of the relatively open, mixed forest and prairie landscapes. The agricultural economy of the period depended on the cultivation of wheat, the great American frontier crop. Wheat was ideally suited to the soils of the region and to the deficits of labor and capital that are associated with frontier settlement. Its cultivation as a staple crop resulted in high yields, and Wisconsin became one of the leading wheat-producing regions in the United States during the 1840s and 1850s.[4] Indeed, the transition as one moves northward along Lake Michigan to a more heavily wooded and less hospitable environment for wheat cultivation accounts for the sagging and relatively sparsely settled 1850 frontier line in the northeast. Populations in the heavily dissected southwestern part of the state, where good arable was restricted to the valley bottoms and toplands, were also more scattered than in the south and east, except in the immediate vicinities of the old lead mining settlements.

The renewal of the settlement advance in the early 1850s resulted in the forward extension of the frontier in a number of specific areas. These advances were heavily dependent on immigrant farmers from countries in northern and western Europe, who arrived in record numbers at Wisconsin's Lake Michigan ports and Mississippi landings before moving on to file for land offered at new land offices opened at La Crosse, Stevens Point, Hudson, and Superior. This time the advance of the frontier was geographically more selective, following key river valleys and natural routeways into an environment that was generally more heavily forested and less fertile than that which was settled before 1850. For those who traveled to Wisconsin via the Great Lakes or the new railroad lines then extending westward to Chicago and beyond, the major opportunities for settlement occurred at selected points all along the central and eastern stretches of the frontier line.

One axis of new settlement thrust northward along the Wisconsin River above Portage. Another advanced northwestward from the upper Fox River valley and the western shores of Lake Winnebago across the Central Plain through Waupaca to Stevens Point. There was also considerable settlement activity along both sides of the lower Fox River valley and on both shores of Green Bay. Along the Lake Michigan shore, settlement moved steadily northward to the base of the Door Peninsula. The highly focused nature of this process was facilitated by the construction of railroads within the state that linked the lake ports with key distribution points on the frontier, such as Fond du Lac (1854), Portage (1856), and Berlin (1857). Farther to the west, the river landings at the mouths of Wisconsin tributaries to the Mississippi, such as the Wisconsin, La Crosse, Black, Trempeleau, Chippewa, and St. Croix, became jumping off points for long fingers of settlement that followed these rivers into the more rugged sections of the Western Upland. Between 1850 and 1860 most of the area to the south of a line between La Crosse and Tomah was settled, while deep but widely separated salients were extended for long distances along most of the major river valleys emptying into the Mississippi between La Crosse and Prescott.

By the time the Civil War broke out and temporarily diverted national resources and attention from westward settlement, Wisconsin had a population of more than three-quarters of a million and had become one of the country's most rapidly developing western states. Wheat was still king. During the 1850s the cultivation of the crop intensified and spread far beyond the southeastern counties where it was originally concentrated. A fortuitous combination of good weather, fertile new land, improved transport, and steady markets made the decade a generally prosperous one for Wisconsin's wheat farmers, although some signs of a move away from overspecialization in wheat were already apparent in some of the southeastern counties. The economy was also highly dependent on another abundantly produced resource: timber. Well ahead of advancing settlement was a timber frontier of logging camps and milling sites. This was a rather fluid and impermanent economic frontier, single-mindedly bent on one heavily extractive activity, but it was also a precursor to more permanent settlement, for it served eventually to clear land and to open transportation lines into the more difficult-to-develop agricultural areas that still remained beyond the edge of settlement.

Much like the wave that preceded it, the third wave of settlement played itself out in about the space of a single decade. It was fueled primarily by a fresh surge of European immigration to America that followed the cessation of hostilities between the Union and the Confederacy in 1865. The wave persisted until just after the great American financial panic of 1873. On the frontier of settlement, the effects were twofold. Part of this wave's energy was absorbed by a "filling-in process" in which many of the areas bypassed by the penetrating fingers of the 1850s settlement now received their quota of land seekers. Included in this filling-in process were large portions of the Western Upland's more rugged areas that lie between the valleys of the St. Croix, Chippewa, and Black rivers, as well as the heretofore neglected Door Peninsula in the east. Another part of its energy was directed toward the creation of new spearheads of settlement along the upper reaches of many of the river valleys in the Central Plain region and the southern margins of the Northern Highlands, which until now had been the exclusive domain of the timber barons.

Open to settlement now for the first time were some of the cutover lands previously covered by the great white pine stands along the upper reaches of the St. Croix, Red Cedar, Chippewa, Black, Wisconsin, and Wolf rivers.[5] By the 1870s, agricultural settlement began to enter these areas, although somewhat tentatively given the greater attraction of fertile prairie lands in neighboring states to the west. The construction of railroad lines to connect key lumber-milling towns across northern Wisconsin during the 1860s and 1870s played an important role. While such construction greatly facilitated the extraction of timber resources, it was also instrumental in advancing new spearheads of settlement, particularly beyond Stevens Point to the upper Wisconsin Valley around Wausau, through Marshfield to the upper Black River country, and into the upper Wolf River country around Shawano.

By 1880 all but the final chapter of Wisconsin's settlement era had been written. The agricultural settlement of the southern half of the state was essentially completed; its economy had passed beyond the frontier stage and begun to diversify. The midwestern wheat frontier had moved on to the western prairies of Minnesota, Iowa, and the Dakotas. In Wisconsin's southern and eastern counties, wheat had been replaced by other cash crops (corn, oats, hops, and hay), dairying, and the production of swine and sheep.[6] The agricultural landscape had taken on a solidly prosperous look, with new frame or brick houses in place of original pioneer dwellings. Artists' idyllic-looking renditions of neat and tidy farmsteads surrounded by fields and pastures of bounteous production graced the pages of the many farmer's atlases and directories that were published at the time. What remained of intensive wheat production in Wisconsin had shifted to the more newly settled districts above Lake Winnebago and to the Western Upland counties above La Crosse, where it too would soon disappear. In northern Wisconsin the agricultural frontier was poised along the southern margins of the cutover for what would be the final advance.

The final phase of settlement began in the early 1880s and proceeded somewhat fitfully, its progress mirroring the ups and downs of the economy into the 1920s. Once again, the arrival of fresh waves of European immigration, more often this time from countries in eastern and southern Europe, was an important factor, but so too was the relocation of people from earlier-settled areas. The major achievement of this final advance was to complete the agricultural settlement of the broad arc of "border counties" (Burnett, Polk, Barron, Chippewa, Clark, Wood, Marathon, Lincoln, Langlade, Shawano, Oconto, and Marinette) that occupy the northern reaches of Wisconsin's Central Plain and the southern margins of the Northern Highland region. This was the last continuous frontier—a northward-moving, crescent-shaped line of advance along a front of relatively infertile soils and cutover forest tracts. Elsewhere, this last settlement period saw the further extension of ribbonlike settlement along the transportation corridors that crossed the Northern Highland, the settlement of isolated pockets here and there in the north, especially around the rapidly growing port towns of Ashland and Superior, and the establishment of a string of mining settlements on the Wisconsin portion of the Penokee-Gogebic Iron Range.

The last frontier in northern Wisconsin was never exclusively agricultural or totally successful. The foremost element in the economy of the region throughout the period continued to be the production of timber. The export of iron ore came next. Agricultural settlement proceeded as an adjunct to and in the wake of the region's primary economic activities. The environmental conditions of northern Wisconsin were marginal to farming and could be successfully exploited only through special adaptations, endless hard work, and a certain amount of good fortune. The last frontier in the north was the most highly publicized of the efforts to settle Wisconsin agriculturally. In the 1890s the state legislature engaged in propaganda campaigns to extoll the agricultural possibilities of northern Wisconsin and to recruit European immigrants to the region.[7] Railroads, lumber companies, and land companies also sought to promote and sell the agricultural potential of the "cutover" to prospective settlers, often featuring the incentives of cheap land prices and the promise of special assistance in getting established. Such campaigns began in the 1890s and continued into the early decades of the next century. Some were successful, but many colonization efforts resulted in only temporary settlement gains. The potential was often overstated by promoters. New settlers were beset with uncooperative weather, declining agricultural prices, and poor yields. As their difficulties mounted, disillusionment set in. By the 1920s, farm abandonment across the region was far advanced. The frontier of settlement had begun to retreat.

The Emerging Urban System

The high pace of early settlement and land improvement in Wisconsin was paralleled by the rapid development of towns and cities. Indeed, a quick glance at Table 8.1, which shows at 20-year intervals the rank and size of urban places in Wisconsin whose population exceeded 5,000 between 1860 and 1920, reveals a steady and steeply upward trend. Over this period, the populations of cities in the upper levels of the urban hierarchy roughly doubled every 20 years. Also notable is the persistent primacy of Wisconsin's number-one city, Milwaukee, which maintained a population edge of more than six to one over its nearest rivals throughout the period. The years between 1860 and 1900 were a formative time for the emerging structure of the urban hierarchy. These were the decades in which the relative ordering of established urban places remained reasonably consistent over time and in which the size of the system was swelled by the largest numbers of cities moving into the upper levels of the hierarchy for the first time. In contrast, the period 1900–1920 was a time of redefinition, when relative place in the hierarchy was most volatile. This last period of growth solidified the shape and structure of an urban system that would persist over much of the twentieth century.[8]

As early as 1860, Wisconsin had an identifiable system of towns and cities. In all, there were 32 towns and cities whose populations exceeded 1,000 people. This system was located primarily in the settled southern and eastern portions of the state and was rapidly tied together by the frenetic railroad-building activity that was undertaken to bring to market the abundant surpluses of the new wheat-producing areas. Indeed, the collection, milling, and shipment of wheat were an early engine of development for Milwaukee, the largest urban center, which established itself in the 1850s and 1860s as one of the country's preeminent milling centers with important markets in the east. The milling and shipment of wheat were the primary locational and functional factors for most other towns and cities as well. This made rail connections an early determinant of growth. Aside from Milwaukee, there were relatively few large places in 1860; most Wisconsin towns were relatively modest in size. Racine, Janesville, Oshkosh, Madison, Fond du Lac, and Watertown were the only other places that could claim at least 5,000 inhabitants. Transportation advantages favored most of them. All except Oshkosh had established themselves as important rail centers by 1860; Oshkosh became important because it was especially well placed to receive shipments of saw timber from the booming pineries of the Wolf River country.

By 1880, this early urban structure had reached a new level of maturation. Towns and cities had become markedly larger and more numerous than had been the case 20 years earlier. At least for the established ones, relative position to one another had become somewhat fixed. Milwaukee (115,587) had nearly tripled in size and proportionately maintained its lead over other population centers in the state. Second-rank urban places (10,000–25,000 in population) continued to include, as before, the cities of Racine, Oshkosh, Fond du Lac, and Madison, although Janesville and Watertown grew more slowly and thus remained in the lowest rank of the hierarchy. Nearly half the cities appearing at the upper levels of the urban hierarchy in 1880, however, were new. The most prominent among the newcomers were the new western rail centers La Crosse and Eau Claire.

The function of cities within the system was now more clearly defined spatially. Lake Michigan port cities like Milwaukee, Racine, Sheboygan, Green Bay, and Kenosha had begun to industrialize (milling, brewing, meat packing, tanning, and the manufacture of clothing and farm implements and machines) on the basis of Great Lakes shipping and rail transportation advantages that linked them to the developing American eastern manufacturing belt. Another locus of manufacturing activity was emerging in the towns and cities of the Lake Winnebago–Fox River valley corridor, although here the industrial base was more focused on the lumber industry and considerably

Table 8.1. Wisconsin's urban hierarchy, 1860–1920

Population	1860	1880	1900	1920
400,000				Milwaukee (457,147)
200,000–399,999			Milwaukee (285,315)	
100,000–199,999		*Milwaukee* (115,587)		
50,000–99,999				Racine (58,593)
25,000–49,999	Milwaukee (45,246)		*Superior* (31,091) Racine La Crosse Oshkosh	Kenosha (40,472) Superior Madison Oshkosh Green Bay Sheboygan La Crosse
10,000–24,999		Racine (16,031) Oshkosh *La Crosse* Fond du Lac Madison *Eau Claire*	Sheboygan (22,962) Madison Green Bay Eau Claire *Marinette* Fond du Lac Appleton Janesville *Ashland* *Wausau* Manitowoc Kenosha *Beloit*	Fond du Lac (23,427) Beloit Eau Claire Appleton Wausau Janesville Manitowoc *West Allis* Marinette Waukesha Stevens Point Ashland

146

5,000–9,999	Racine (7,822)	Janesville (9,018)	Stevens Point (9,524)	Watertown (9,299)
	Janesville	Appleton	Merrill	Chippewa Falls
	Madison	Watertown	Watertown	Antigo
	Oshkosh	Green Bay	Waukesha	Merrill
	Fond du Lac	Sheboygan	Neenah	Beaver Dam
	Watertown	Manitowoc	Baraboo	South Milwaukee
		Kenosha	Menominee	Marshfield
			Oconto	Two Rivers
			Menasha	Wisconsin Rapids
			Portage	Menasha
			Marshfield	Neenah
			Antigo	Cudahy
			Beaver Dam	Rhinelander
			Kaukana	Kaukana
				Portage
				Baraboo
				De Pere
				Menominee
				Stoughton

Note: Cities that appear for the first time after 1860 or move up the hierarchy more than one step in a 20-year period are highlighted in **bold italics**.

147

less diversified. Many of the smaller cities that appear for the first time in 1880 are either Lake Michigan industrial port cities or Winnebago–Fox corridor industrial centers. The rising emphasis on manufactures aimed at a regional or national market set these two groups of urban places apart from other centers in the southern half of the state, like La Crosse and Madison, which offered to a largely agricultural hinterland a variety of government, wholesale, and retail services, along with some locally based processing and manufacturing. A fourth category of urban places included northern towns, which were at this time still relatively small, raw, and almost exclusively reliant on the production and shipment of timber products.

The urban system of 1900 was both an intensification and an enlargement of the 1880 system. All across America, the number of people living in urban places was sharply on the rise between 1880 and 1900. Wisconsin shared in this national shift toward urban life. By 1900 Wisconsin boasted 18 cities with more than 10,000 inhabitants. Nearly 600,000 people, or 29.2 percent of the entire population, lived in an urban place of this size. Twenty years earlier there were only seven cities that large, and their combined populations amounted to only a third as many people and only half as great a proportion of the state's population. Except in the north, agricultural populations had peaked throughout most of the state by 1900. There was an exodus from the countryside to towns and cities. This, coupled with the continued arrival of new immigrant labor in the industrial cities and in the lumbering and mining towns of the north, produced an urban boom.

As always, Milwaukee (285,315) stood alone at the top of the urban hierarchy (now more than nine times as large as its nearest rival), ruling over the other larger eastern Wisconsin industrial centers of Racine, Oshkosh, Sheboygan, Green Bay, and Fond du Lac. The balance of the upper levels of urban hierarchy consisted, as had been the case for some time, of the southern and western regional centers at La Crosse, Madison, and Eau Claire, followed by the smaller industrial centers of Wisconsin's eastern industrial belt, like Appleton, Beloit, Janesville, Kenosha, and Manitowoc. The big exception to this established structure in 1900 was the sudden rise of a sizable number of timber and iron processing and shipping centers like Superior, Ashland, Marinette, and Wausau, representing the newly emergent north. Superior, in particular, grew so fast during the 1880s and 1890s, that its sudden emergence as Wisconsin's second largest city in 1900 is absolutely startling. At lower levels of the hierarchy, there were also many first-time additions that came out of the new north. Everywhere else, smaller trade and service centers were flourishing. In 1900 the urbanized proportion of Wisconsin's population (people living in places of 2,500 or more) reached 38.2 percent, a figure roughly comparable to that of the country as a whole and indicative of the state's high level of settlement maturation.[9]

The first two decades of the twentieth century, however, must be seen as a time of redefinition and crystallization. Key adjustments in the relative order of places were being made during this period. Whereas in previous periods almost every city moved upward more or less in concert with other cities of similar size, this time the pattern was more chaotic. Some cities continued to move up smartly. Others remained in place or even lost ground. Also the number of cities breaking for the first time into the upper levels of the urban hierarchy was only half of what it had been in 1900. As a proportion, new additions to this select group in 1920 made up less than a quarter of the total. In 1900 they made up more than half.

The select cities that established themselves as the growth leaders during this period are those that emerged from the settlement era with the best advantages for growth in the modern era. The most important of these were the more diversified industrial centers of southern and eastern Wisconsin, like Racine, Green Bay, Kenosha, Beloit, and Waukesha. Also notable is the rapid rise of places like West Allis, South Milwaukee, and Cudahy, which contributed, as industrial outliers of Milwaukee, to the emergence of a Greater Milwaukee urban region.[10] On the other hand, the regional centers for the more agriculturally based economies of the southern and western parts of the state became relatively stable in their growth patterns. The exception to this was Madison, which continued to grow rapidly as the state's major institutional center.

Meanwhile the "boom and bust" nature of Wisconsin's northern cities, dependent as they were on highly cyclical extractive industries, was already becoming apparent. Superior and Ashland, both of which had experienced meteoric rises at the end of the nineteenth century, fell back sharply in relative position by 1920. Ashland's decline is particularly striking. The city's population peaked around 1900 at 13,074. By 1920 it had fallen to 11,334 and would continue to slide for most of the century. At the same time other northern cities, such as Rhinelander, Marshfield, Wausau, Wisconsin Rapids, and Chippewa Falls, enjoyed rapid, but ultimately unsustainable, spurts of growth. By 1920, the broad outlines of an urban system that was made up of several regional components, each with its own functional and growth characteristics, were set. This system would remain in place, with only minor modification, until well after World War II.

The Cultural Mosaic

Much has been made of the heavily "ethnic" makeup of Wisconsin's population around the turn of the century. The fact that the proportion of foreign born at that time was among the very highest in the country is a constant leitmotif in just about every statement about the cultural and social foundations of the state. Indeed, if one chooses to speak of the foreign stock—that is, the combined number of people who were born abroad or had parents who were born abroad—more than 7 out of 10 Wisconsin residents in 1900 (71.1 percent) belonged to this group.[11] References also abound to the respective contributions of the individual elements that made up the whole: to the pervasive influence of the large German majority; to the varied contributions of the many others, the Scandinavians, Poles, Bohemians, Finns, Belgians, Welsh, Dutch, Swiss, Luxembourgers, Czechs, Italians, Greeks, Slovenes, Lithuanians, and so on; to the fact that westward-moving North Americans were also a significant and not-to-be-forgotten element in the nineteenth-century peopling of the state. Wisconsin has always been, as one historian has remarked, "something of a living ethnological museum," and the many and sometimes contradictory implications of this condition are a perennial subject for discussion.[12] Our focus, however, is more on questions of geographic process and pattern. By what processes did such a culturally varied population migrate to and settle within the boundaries of the state, and what were the broad geographic patterns produced by these processes?

Movements of large numbers of people are usually not random. They are in response to certain events and opportunities that are connected in space. Often they are highly influenced by specific flows of information about events and opportunities, and take the form of streams of movement that are quite specific in their points of origin, direction, and destination. Most of the migrants who arrived in Wisconsin, whether conscious of it or not, were part of some kind of migration stream. People moved to Wisconsin because they became aware of opportunity there and because certain events or circumstances at home made them receptive to the idea of migrating. Probably few people ventured out without some idea of where they were going and what they might expect to find there, whether from printed words, the appeals of railroad and state immigration agents, or the trusted advice of neighbors, friends, and relatives. Once set in motion the few initial decisions to move had the habit of snowballing into much larger and prolonged sequences of similar decisions involving many people. The process was natural. Those who arrived first sent word back encouraging others to follow. The result was the creation of streams of migration that brought large numbers of people of specific geographic origins to Wisconsin.

Also critical to the process was the element of timing. Migration streams could be as specific in timing as they were in space. They occurred at a point in time when the pull of opportunity in Wisconsin and the push of circumstances at the place of origin were optimally coincidental and lasted only as long as those coincidental conditions persisted. Thus one migration stream could be quickly replaced by another. Timing had much to do with where in Wisconsin a particular stream may have been directed. Land seekers were more or less forced to places where land was currently available; industrial workers, to places where industries were currently hiring. At different times, different streams of migration were responsible for fueling Wisconsin's advancing frontiers or filling the immigrant quarters of her cities. The natural outcome was a high degree of spatial clustering. Although a great deal of mixing did occur, the varied intensity and timing of migration streams were powerful forces in helping to create the patchwork quilt effect so apparent on any map of ethnic groups in the state.

The earliest streams were those of westward-moving Americans, who are generally thought of as having belonged to one of several latitudinal zones of westward expansion emanating from old cultural hearths on the Atlantic seaboard. For example, the initial rush to the lead mines of southwestern Wisconsin in the 1820s was made up of individuals who came up the Mississippi from established mining districts in Missouri. These miners hailed from parts of southern Illinois, Missouri, Kentucky, and Tennessee and belonged to a "midland-upland south" population that had long been moving westward from a string of mid-Atlantic and southern hearths stretching from southern Pennsylvania to the Carolinas. In contrast, the early streams of American settlers who arrived in such large numbers in southern and eastern Wisconsin in the 1830s and 1840s were derived from an old "Yankee" population that originated in southern New England and settled across New York State and westward to the shores of the Great Lakes.

Wisconsin was affected over the course of the nineteenth century by streams of American migrants belonging to three westward migratory traditions. The Driftless Area of southwestern Wisconsin drew much of its American population from a tradition of westward migration that mixed people of Yankee and Middle Atlantic origins. These migrants came from a Yankee-Midland zone that was populated rather equally by people from New York and southern New England, on the one hand, and by people from southern Pennsylvania and Ohio, on the other. It was often the Midlanders who arrived first, as was the case with the early lead miners. Their settlements were then enveloped by those of later-arriving streams of Yankees. In addition to the lead mining areas, an outlier to this zone developed in extreme western Wisconsin along the Mississippi between La Crosse and the upper reaches of Lake Pepin. Another section of Wisconsin was affected primarily by migration streams that came directly from western New England and the New England Yankee–settled districts of western New York State and northeastern Ohio. These Yankee farmers were the dominant element in the early settlement of large parts of southeastern and southern Wisconsin, but they also mixed readily with newly arriving waves of Europeans all along the westward and northward advancing agricultural frontiers until as late as the 1870s. As a consequence, migrants from states like New York, Connecticut, Massachusetts, and Ohio figure prominently in the histories of many of the townships, towns, and cities all across a wide swath of central Wisconsin. The northern portion of the state drew its American element from a third migratory tradition. Streams of migrants who entered this region were derived originally from populations that had been moving westward across the northern Great Lakes region from the St. Lawrence Valley and the northern portions of New England and New York State. They brought with them long traditions of lumbering, mining, and farming in a northern woods environment that were particularly suitable to the conditions of northern Wisconsin.[13]

European immigrants came in a series of waves, each of which had its own set of distinctive streams.[14] The first wave coincided more or less with the latter half of Wisconsin's first settlement surge during the 1830s and 1840s. These early immigrant streams came from places located in the northern and western parts of Europe and were directed into identifiable catchment areas in Wisconsin. The largest number of immigrants were German speaking. They came from the southern and western principalities and states of Germany, as well as from parts of Austria, and settled primarily and rather densely in the counties immediately to the north and west of Milwaukee. Because of their numbers and the intensity with which they established their large farming communities, these early German immigrants became the pervasive element throughout most of east-central Wisconsin and almost entirely dominated the lakeshore counties north of Milwaukee. German-speaking Swiss immigrants settled in several different areas, but most prominently in the form of a colonization effort organized from the Swiss canton of Glarus to Green County. Dutch Protestants colonized lands near Fond du Lac, Sheboygan, and Milwaukee, while streams of immigrants from Norway established a series of large settlements across southern Wisconsin, the key ones being at Jefferson Prairie, Rock Prairie, Muskego, and Koshkonong. Immigrants from the British Isles were also important. The lakeshore counties to the south of Milwaukee received substantial influxes of Irish, as did the city of Milwaukee itself. English and Scottish settlers were found scattered in many places across southern Wisconsin. The Cornish, Irish, and English were all present from early on in the lead mining communities of the southwest, while the Welsh established exclusive rural settlements in Waukesha, eastern Columbia, and western Dodge counties. By 1850 roughly a third of Wisconsin's population was foreign born. Patterns of regional and local ethnic exclusivity were already well established.

Emigration from Europe accelerated during the 1850s. New arrivals in Wisconsin poured into settlements already established by their respective groups, where they either settled permanently or paused briefly before moving to new settlements on the advancing frontier. Many of the early settlements became important staging areas for the extension of migration streams westward. The large Norwegian settlement at Koshkonong, for example, acquired a reputation as the "gateway" for Norwegians heading to Wisconsin's western frontier. Many new settlements on the frontier were in fact "daughter settlements" that could trace direct linkages back to earlier settled areas. The German population was greatly augmented during this period. The large zone of predominantly German settlement in east-central Wisconsin spread westward to encompass portions of the upper and lower Fox River valleys as well as new settlement areas immediately to the west of Lake Winnebago. It was during this period, too, that "German Town" became an important feature of Milwaukee's residential divisions, and Germans began to emerge as the city's dominant immigrant group.[15] New streams of Norwegian immigrants overflowed the older settlements in southern Wisconsin and spread to dozens of new settlements in Columbia and Sauk counties, and especially in the Western Upland counties above La Crosse.

The 1850s brought new groups to Wisconsin as well. Small groups of Dutch Protestant Frieslanders appeared in the west, particularly around La Crosse, while large numbers of Dutch Catholics began to colonize tracts in the lower Fox River valley between Little Chute and Green Bay. Settlements of Czech immigrants were established in several western counties and in an extended zone that cut through Manitowoc and Kewaunee counties. Thousands of Belgian Catholic Walloons arrived to settle east of Green Bay, while groups of Luxembourgers established a string of settlements in Ozaukee and Sheboygan counties. Other new arrivals included Swedes along the shore of Lake Pepin and in the St. Croix Valley border country, and Danes around Racine, Waukesha, and Waupaca.

In essence, the 1850s worked to solidify and extend already established geographic patterns of regional and local ethnic exclusivity. High degrees of segregation between groups continued to result from the strong specificity of migration streams. The makeup of Norwegian settlements, for example, was often based on provincial or local affiliations rather than on national identity; whole communities consisted of people from a particular fjord or valley who often knew one another or at least could feel comfortable with one another because of a commonality of dialect and customs.[16] Local and regional affiliations were important. National identifications were generally not strongly developed among Europeans at this time. No one, for example, had yet established a popularly accepted concept of German nationhood. Germany would not be unified as a single state until the 1870s. People from parts of Bavaria did not necessarily identify with those from Baden or Rhenish Prussia and quite naturally chose to segregate themselves into distinctive communities in Wisconsin. Religion was also a factor. Dutch Protestants kept themselves separate from Dutch Catholics, and German Protestants removed themselves from areas dominated by German Catholics, a division that would become more pronounced after the Civil War with the arrival of much larger numbers of German Protestants. Norwegians even divided themselves over theological disputes within the Lutheran church.[17]

The post–Civil War boom in European immigration that occurred in the late 1860s and early 1870s was a large one. It was also one that exhibited some shifting in the geographic distribution of those emigrating from Europe, who now also came from areas farther north and east on that continent. Many of the new immigrants from Germany had departed from the largely Protestant states in the north and east. Promotional efforts by railroad companies, lumber companies, and the state of Wisconsin were instrumental in directing many of these fresh streams of German immigrants to the forested sand plains of central Wisconsin, particularly to Shawano, Marathon, and Lincoln counties. Associated with this development was the arrival of large contingents of German-speaking Poles near Stevens Point in Portage County. Other streams of migration from Poland were incorporated in the "filling in process," whereby previously ignored areas of the Western Upland were settled. Upper Silesia provided most of the immigrants who founded a string of Polish settlements in western Trempealeau County. The eastward and northward spread of emigration in Europe added much larger numbers of Swedes and Danes to the streams of Scandinavians entering Wisconsin. Norwegians still dominated and continued to build new settlements, particularly in Pierce, St. Croix, Dunn, and Barron counties, but a Danish settlement in Polk County and a score of new Swedish settlements in western Polk and Burnett counties extended the already heavily Scandinavian flavor of western Wisconsin northward. Meanwhile immigrants from Iceland established themselves on the tip of the Door Peninsula.

Large numbers of immigrants from Germany, Austria, Poland, and the Scandinavian countries were also arriving during this period in industrial cities, like Milwaukee, Racine, Sheboygan, and Manitowoc, to engage in urban industrial occupations. More than 35 percent of the population in each of these cities was foreign born (39.9 percent in Milwaukee) in 1880, making their populations more immigrant than the population of the state as a whole (30.8 percent). Continued high levels of immigration to urban areas reflected diminishing agricultural opportunity on the frontier as well as a growing shift in the background of many European immigrants toward higher levels of urban and industrial experience.

The makeup of the final waves of European immigrants to reach Wisconsin was markedly different. The period after 1880 saw sharp changes in the source areas and character of European immigration. Immigration occurring during the period before 1880 is often thought of as "old immigration"—largely agriculturally based and focused on western and northern Europe. The "new immigration" that began to emerge in the 1880s and prevailed until European immigration was restricted in the 1920s, was increasingly fueled by migrants from new source areas in southern and eastern Europe. These new immigrants were drawn either to the mines, woodworking industries, and residual agricultural frontiers of northern Wisconsin or to the immigrant quarters of industrial towns and cities in southern and eastern Wisconsin. The mines, mills, and port cities of the north, for example, drew a diverse mixture of Finns, Poles, Italians, and Slovaks to take their place alongside older American, Canadian, Scandinavian, or German workers. The northern cutover received extensive agricultural colonizations of Polish immigrants throughout the 1880s and 1890s, particularly in the new districts north and west of Green Bay. There were also scattered agricultural settlements of Slovaks, Russians, Italians, Finns, Estonians, and Czechs.

The largest streams, however, brought immigrant workers to the larger industrial cities. Many of the new urban immigrants arrived to take low-paying jobs in tanneries, breweries, and foundries. Others took skilled positions in lighter industries such as furniture making, found niches in the service economy, or opened businesses to serve a growing ethnic neighborhood clientele. By 1900 half of all Poles living in Wisconsin were located in the Milwaukee area, most of them concentrated on the city's South Side. They were also important in Green Bay, Manitowoc, Racine, and Kenosha. Russian Jews began arriving in Milwaukee during the 1880s. Italians, many of them arriving via Chicago, were well established in Milwaukee by the turn of the century. After 1900, Milwaukee and other cities added sizable communities of Hungarians, Greeks, Slovenes, Slovaks, Lithuanians, and Latvians. The major industrial cities with the largest proportion of foreign born in 1920 included Kenosha (31.4 percent), Racine (27.7 percent), Superior (26.9 percent), Sheboygan (26.7 percent), Ashland (24.9 percent), and Milwaukee (24.2 percent). In contrast, much lower proportions of foreign born prevailed in western and outstate centers, like Madison (12.6 percent), Eau Claire (16.9 percent), La Crosse (14.6 percent), and Janesville (12.1 percent).

The geography of immigrant cultures that has resulted from all of this is extremely complex. No sizable part of Wisconsin in 1920 was exclusively the domain of any particular group. Timing and migration streams had conspired to produce a complex patchwork quilt, with large and small patches of cultural homogeneity, interspersed with areas of heavy mixing.[18] Even the most homogeneous larger patches (there was, for example, no shortage of counties in which at least 75 percent of the foreign born were German) had plenty of enclaves settled by people of other origins. Small groups were sometimes found almost nowhere in the state except in half a dozen townships or a few city wards where they constituted an overwhelming majority. Some groups, like the Irish, were more or less ubiquitous, living virtually everywhere in addition to a few areas of heavier concentration.

Despite the complexity, it is possible to present a generalized map of immigrant culture areas within the state (Figure 8.2). The most striking feature on the map is the very large zone of German dominance that extends rather broadly in a northwestward direction from the Milwaukee area. The apparent solidarity of the German zone hides a fairly high degree of local differentiation, not only in the form of numerous non-German enclaves, including the extensive area of Polish settlement in Portage and Marathon counties, but also in the spatial exclusivity of German provincial and religious groups. The city of Milwaukee is part of the German zone, despite its internal ethnic diversity, for it remained the most German of American major cities even after the massive influxes of eastern and southern European groups following the turn of the century.[19] The other large zone with a relatively high

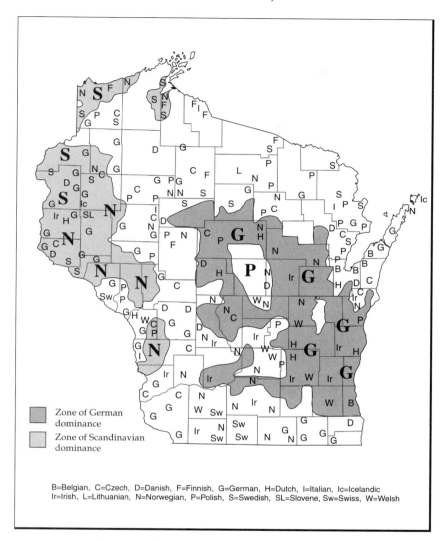

Figure 8.2. The cultural mosaic in Wisconsin, ca. 1920

degree of cultural coherence is the extended, but at times discontinuous, swath of Scandinavian settlement that extends along the western margins of the state from the northern portions of La Crosse County to the shores of Lake Superior. Here too the apparent solidarity of the zone belies the many non-Scandinavian enclaves and the spatial segregation of Norwegian, Swedish, Danish, and Finnish settlements, not to mention the spatial segregation of regional and local groups within these national designations.

What remains beyond these two zones are three areas of heavy mixing. One stretches across the southern two tiers of counties and consists of relatively high densities of old-stock Americans mixed with British and Irish immigrants, punctuated here and there by intensely settled local communities of Swiss, Germans, Norwegians, Danes, Welsh, and Czechs. A second area forms a buffer between the German and Scandinavian zones and is characterized by large areas of sparsely settled and ethnically mixed populations, interspersed with small concentrations of primarily German or Scandinavian settlement. The third is a wide zone of cutover northern lands populated by widely scattered colonies of various origins, but especially Poles, Finns, Norwegians, Swedes, Danes, Czechs, Italians, Lithuanians, and Icelanders.

Germans were the dominant group in the majority of cities (Table 8.2). This was true in 1920, despite the fact that German immigration slackened considerably after the 1890s. The falloff in numbers of new immigrants from Germany and the rise of American-born second and third generations mean that the German element can be somewhat underrepresented when compared with newer groups purely on the basis of number of foreign born. Aside from Milwaukee, Wisconsin's most German cities were places like Appleton, Wausau, Fond du Lac, and Oshkosh, all centers within the predominantly German zone of east-central Wisconsin. Other cities with large German elements included many of the major industrial centers of the southeast and a few outliers like La Crosse and Madison.

Most towns and cities had mixed ethnic populations, within which individual neighborhoods could be quite homogeneous. Milwaukee was known for its distinctive ethnic neighborhoods, as were any number of other large Wisconsin urban places. Vernacular names for places, such as Madison's "Little Italy," Racine's "Kringleville," and Milwaukee's "German Town" or "Polonia," were a commonplace means of conveying what everyone knew to be the dominant ethnic character of some neighborhoods. In some cities no single group held a numeric edge. This was especially true for northern cities like Ashland, Superior, and Marinette, which were home to rather diverse populations, although one might say that the Scandinavian element was larger than any other. It was also true for some southern Wisconsin industrial centers like Beloit, Kenosha, Racine, and West Allis, where the relative size of newer immigrant groups was greater than it was in more German-dominated areas. The exceptions to the above seem to have been Eau Claire, with its large Norwegian majority, and the strikingly Polish city of Stevens Point.

Table 8.2. Percentage distribution of foreign-born major ethnic groups in Wisconsin cities of more than 10,000 in population in 1920

City	1st group	2nd group	3rd group	4th group	5th group
Appleton	*Germans* 62.0	Austrians 5.3	Czechs 3.9		
Ashland	Swedes 19.3	Poles 13.9	Norwegians 13.3	Canadians 12.4	*Germans* 12.0
Beloit	*Germans* 18.8	Norwegians 14.9	Italians 14.3	Greeks 6.5	English 6.4
Eau Claire	Norwegians 47.4	*Germans* 21.7	Canadians 10.9	Swedes 5.5	
Fond du Lac	*Germans* 50.6	Russians 9.0	Greeks 6.1		
Green Bay	*Germans* 23.5	Belgians 22.2	Poles 9.2	Danes 6.2	
Janesville	*Germans* 33.1	Norwegians 11.0	Irish 10.4	English 8.4	
Kenosha	Italians 15.0	*Germans* 14.9	Russians 13.9	Poles 11.9	Lithuanians 5.5
La Crosse	*Germans* 38.4	Norwegians 23.7	Czechs 6.6	Austrians 4.3	Poles 3.8
Madison	*Germans* 24.9	Norwegians 20.3	Italians 10.0	Russians 8.2	English 5.8
Manitowoc	*Germans* 38.3	Poles 19.8	Czechs 8.6	Austrians 8.0	Russians 7.1
Marinette	Swedes 21.4	*Germans* 19.5	Fr. Canadians 13.2	Norwegians 11.1	Canadians 9.1
Milwaukee	*Germans* 36.1	Poles 21.0	Russians 6.5	Austrians 5.4	Czechs 4.1
Oshkosh	*Germans* 50.9	Russians 8.3	Czechs 8.3	Austrians 5.7	Danes 5.1
Racine	Danes 24.4	*Germans* 15.0	Russians 9.8	Czechs 5.7	Poles 5.5
Sheboygan	*Germans* 42.8	Russians 24.3	Austrians 7.6	Yugoslavs 6.5	Dutch 6.4
Stevens Point	Poles 48.4	*Germans* 20.5	Norwegians 6.4	Canadians 5.4	
Superior	Swedes 23.1	Norwegians 18.0	Finns 8.4	*Germans* 4.7	Russians 4.3
Waukesha	*Germans* 39.8	Italians 14.2	English 7.5	Swiss 6.5	Austrians 4.3
Wausau	*Germans* 64.9	Poles 6.4	Russians 4.8	Canadians 4.4	
West Allis	Jugoslavs 25.6	*Germans* 18.5	Poles 14.9	Czechs 6.0	Austrians 5.9

Note: Germans, the predominant group in the state, are highlighted in **bold italics**.

Aftermath: The Seventy-five Years Since

What has been described here is formative. These are the processes and patterns of settlement and development that shaped the quintessential cultural landscape of Wisconsin. Over the last three-quarters of a century that fabric has been modified and stretched in a variety of ways. Factors of environmental marginality have caused the frontier of agricultural settlement in the northern and central parts of the state to recede while powerful economic forces have conspired to lower the density of farm populations nearly everywhere. The move to towns and cities has accelerated, advancing significantly the relative importance of urban dwellers in the state's population. The Greater Milwaukee area still occupies the primate position in the urban hierarchy. At nearly 1.5 million people, it remains roughly four times the size of its nearest rival, which today is Madison.[20] But there has also been some significant reordering of the upper levels of the hierarchy. The highly concentrated, heavier industries of the past have declined in importance relative to newer and lighter industries that are more footloose in their locational requirements. This shift has helped to intensify urbanization throughout southern and eastern Wisconsin in both small and large places, bringing with it a widespread suburban and exurban invasion of the countryside. The Fox River valley corridor from Green Bay to Fond du Lac, in particular, has evolved into an important urbanized region. Regional centers in western Wisconsin have also grown significantly, while places both small and large in the northern parts of the state have struggled to maintain their position.

The effects of time and the arrival of new groups have also modified the old cultural mosaic. Time has moved many of the old immigrant groups into what might be called the twilight of their ethnicity, where remembrance has become a matter of summertime community celebrations, associational memberships, and a genealogical interest in "roots."[21] Especially in the cities, the older groups have been displaced by new arrivals. The African-American population, which was already present in a modest way during the settlement period, began to expand strongly in the 1940s with the northward migration of southern blacks to take wartime industrial jobs, and this expansion continues to the present, primarily as a function of midwestern interurban migration. So, too, has the cultural mix been altered by the more recent arrival of Hispanic and Asian populations. But to an important degree, it is still the legacy of the processes and patterns of migration and settlement in the more distant past that provides the state with its overarching sense of progressive and multicultural identity.

Notes

1. This point is made in G.-H. Smith, "The Populating of Wisconsin," *Geographical Review* (1928): 405. For a sense of the position of Wisconsin relative to the advancing flow of settlement across the American Middle West, see the maps and discussion in R. C. Ostergren, *A Community Transplanted: The Trans-Atlantic Experience of a Swedish Immigrant Settlement in the Upper Middle West, 1835–1915* (Madison, 1988), 10–14. See also J. F. Hart, "The Middle West," *Annals of the Association of American Geographers* 62 (1972): 260–262.

2. J. Schafer, *The Wisconsin Lead Region* (Madison, 1932), 132–133.

3. This generalized map of settlement advance in Wisconsin was produced by combining information from three sources: (1) G.-H. Smith's meticulously drawn dot maps of Wisconsin population, which were completed for the census years 1850, 1860, 1870, 1880, 1900, and 1920, and first published as an inset to "The Populating of Wisconsin," (2) the series of decadal population-density maps compiled by H. Gannett for the Eleventh Census of the United States in 1890, and (3) population data for minor civil divisions published for individual census years by the U.S. Bureau of the Census.

4. The classic history (and geographic treatment) of wheat in Wisconsin is J. G. Thompson, *The Rise and Decline of the Wheat Growing Industry in Wisconsin,* Bulletin of the University of Wisconsin, no. 292 (Madison, 1907).

5. R. F. Fries, *Empire in Pine: The Story of Lumbering in Wisconsin, 1830–1900* (Madison, 1951), 17–21.

6. The transition from wheat farming to other forms of agriculture is well covered in J. Schafer, *A History of Wisconsin Agriculture* (Madison, 1922); and E. E. Lampard, *The Rise of the Dairy Industry in Wisconsin: A Study in Agricultural Change, 1820–1920* (Madison, 1963).

7. The best-known effort was a 192-page guidebook prepared in 1895 by the dean of the College of Agriculture at the University of Wisconsin, W. A. Henry, entitled *Northern Wisconsin: A Handbook for the Homeseeker* (Madison, 1896).

8. A useful historical analysis of Wisconsin's emerging urban system is found in I. Vogeler, *Wisconsin: A Geography* (Boulder, 1986), 147–154. Vogeler sees the basic urban pattern as set by about 1930. See also M. P. Conzen, "Capital Flows and the Developing Urban Hierarchy: State Bank Capital in Wisconsin, 1854–1895," *Economic Geography* 51 (1975): 321–338.

9. This point, about Wisconsin's rural-urban mix beginning around 1900 to approximate the national average, is often made. See, for example, G.-H. Smith, "The Settlement and the Distribution of the Population in Wisconsin," *Transactions of the Wisconsin Academy of Sciences, Arts and Letters* 24 (1929): 90; and R. C. Nesbit and W. F. Thompson, *Wisconsin: A History* (Madison, 1989), 341.

10. Indeed, the city of Milwaukee's proportional size advantage over its nearest rival declined from a ratio of more than nine to one in 1900 to just under eight to one in 1920. Yet, if other components of the Greater Milwaukee area are factored in, there is no decline.

11. According to the 1900 U.S. Census, 1,472,008 of Wisconsin's 2,069,042 people were of foreign stock. Only Minnesota, among midwestern states, could claim a higher proportion.

12. R. N. Current, *Wisconsin: A Bicentennial History* (New York, 1977), 56–62.

13. For a definitive treatment of the latitudinal zonation of American migrant origins in the Middle West see J. C. Hudson, "North American Origins of Middlewestern Frontier Populations," *Annals of the Association of American Geographers* 78 (1988): 395–413.

14. Source materials on the settlement of immigrant groups in Wisconsin are too numerous to list here. Most are group-by-group surveys or are entirely devoted to a single group. None are organized along the "timing–settlement advance scheme" employed here. From a geographer's point of view, however, one of the most comprehensive and useful surveys is found in the "Settlement" section of *Cultural Resource Management in Wisconsin*, vol. 1: *A Manual for Historic Properties* (Madison, 1986).

15. K. N. Conzen, *Immigrant Milwaukee, 1836–1860: Accommodation and Community in a Frontier City* (Cambridge, Mass., 1976), 143–148.

16. P. A. Munch, "Segregation and Assimilation of Norwegian Settlements in Wisconsin," *Norwegian-American Studies and Records* 18 (1954): 102–140.

17. A. M. Legreid and D. Ward, "Religious Schism and the Development of Rural Immigrant Communities: Norwegian Lutherans in Western Wisconsin, 1880–1905," *Upper Midwest History* 2 (1982): 13–29.

18. See, for example, G. W. Hill's map, "The People of Wisconsin According to Ethnic Stocks," which appears as a frontispiece in *Wisconsin's Changing Population*, Bulletin of the University of Wisconsin, ser. no. 2642 (Madison, 1942).

19. See the discussion of urban-immigrant destinations in D. Ward, "Population Growth, Migration, and Urbanization, 1860–1920," in R. D. Mitchell and P. A. Groves, eds., *North America: The Historical Geography of a Changing Continent* (Totowa, N.J., 1987), 310–312.

20. The 1990 census population of the Milwaukee metropolitan area was 1,432,149. Madison's was 367,085.

21. The term *twilight of ethnicity*, which is a particularly apt way of describing the present situation of old European immigrant groups, comes from R. Alba, "The Twilight of Ethnicity among Americans of European Ancestry: The Case of the Italians," *Ethnic and Racial Studies* 8 (1985): 134–158. See also S. D. Hoelscher and R. C. Ostergren, "Old European Homelands in the American Middle West," *Journal of Cultural Geography* 13 (1993): 87–106.

9

The European Settling and Transformation of the Upper Mississippi Valley Lead Region

Michael P. Conzen

Lead replaced furs as the leading stimulus to European activity in the upper Mississippi Valley during the early nineteenth century. Moreover, lead achieved what furs could not: it changed the human and ecological history of southwestern Wisconsin and nearby portions of Illinois and Iowa forever. Its extraction from shallow, accessible deposits drew 10,000 people within six years and helped to skew the frontier of mass colonization well to the northwest of the main continental "margin" of American-settlement advance at the time. And in doing so, lead triggered the rapid political development of what ultimately became the state of Wisconsin in a way matched by none of the region's other natural resources.

This mineral rush, however, was unlike most others in American history, for it focused on a district well endowed with other possibilities. Most metallic deposits in the United States before and after were found in mountainous and remote places offering few if any alternate means of livelihood when the ores gave out. The Upper Mississippi Valley Lead Region,[1] by contrast, occupied a strategic position in the transition zone between forest and prairie, lay astride the continent's largest navigable river within reach of major markets, and offered not one but eventually two minerals near the land surface—a surface equally hospitable to intensive agriculture. As a result, southwest

Wisconsin boasts a long and varied history of European development without the cycle of collapse and total abandonment, ghost towns, and grotesquely ruptured landscapes typical of later mineral frontiers. A price was paid for the winning of lead, zinc, wheat, and corn from the region, of course, but through all the vicissitudes a sustainable future was never in doubt. The emergence of lead mining and its economic and social results, the spatial patterns they etched upon the land, and the significance these have for the present character of the region form the core of the following exploration of the processes by which this region gained its geographical identity.[2]

The Ending of the Ancient Order

Human occupation of the Upper Mississippi Valley Lead Region already had a long history by the time Euro-Americans arrived on the scene, as did the knowledge of and exploitation of the lead deposits. Paleo-Indians arrived in the present area of Wisconsin as nomadic hunters from the south or southwest about 9500 B.C., when the area had already developed as a mixed zone of hardwood forests with grassland corridors on the upland margins separating the incised valleys, and contained a mixture of small game and some large animals (Figure 9.1). In the millennia that followed, increasingly sedentary peoples occupied the region in very small numbers, leaving their mark in burial sites, small horticultural villages, effigy mounds, and scattered collections of rock art. By A.D. 1100, the region supported about 3,000 inhabitants, balancing hunting and agriculture in various proportions.[3]

In the centuries that followed, human settlements in the area continued their slow evolution until affected by catastrophic population collapse sometime between the fourteenth and sixteenth centuries.[4] Whether from disease or environmental degradation, this change occurred before sustained European contact. What links historical Indian groups of the seventeenth and eighteenth centuries had with the prehistoric peoples in the region remains unclear, though direct connections between the ancient Oneota and the modern-era Winnebago seem to be emerging from recent archaeological and historical research.[5] Local Indian communities then encountered further disruptive changes, first through contact with intruding French missionaries and traders, whose only interest in Indian products lay in animal furs, and then through the combined pressures of displacement, warfare, and societal breakup precipitated by British and later American political and commercial interests (Figure 9.1). By 1810, the fur trade had virtually exhausted the region and moved north and west.[6]

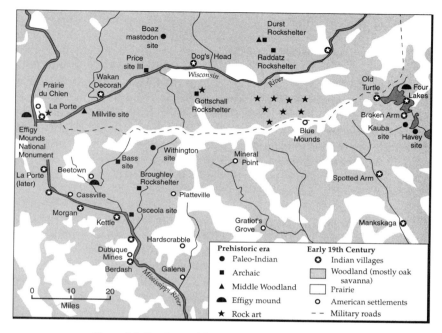

Figure 9.1. Southwest Wisconsin before the lead rush

Lead in minute quantities was widely collected, fashioned, and traded among the Indians of eastern North America, beginning in Late Archaic times. In nodule form, it had value as a magical charm, and ground into powder it served for face paint and as ceremonial powder to be thrown at the sun. Lead from the upper Mississippi Valley traveled to numerous settlements along the middle Mississippi and lower Illinois rivers, and as far away as Poverty Point, Louisiana, and Ontario from the eleventh to the fourteenth century.[7] Presumably, it was gathered from widely dispersed surface deposits, though was concentrated near the principal rivers, especially the Mississippi, because these were the main corridors of Indian settlement and movement.

Aboriginal knowledge of lead in the region survived until the seventeenth century, when French explorers arrived. Radisson and Groseilliers learned of upper Mississippi Valley lead in 1658 from the Sioux, and mines in the region are shown on Hennepin's map of 1688.[8] A few years later, Nicholas Perrot is believed to have established a trading post near the Miami at what is now East Dubuque and begun the mining of lead there, which Le Sueur visited in 1700 with a party of French miners who briefly extracted lead. By 1743 Le Guis noted about 20 functioning mines around a camp on the Fever River, from which lead was carried by packhorse to Kaskaskia, it having value for making ammunition by then.[9] More stability appeared in the person of Julien Dubuque, a recent Prairie du Chien settler who from 1788 to his death in 1810 founded a major lead mining and smelting settlement, the "Mines of Spain," on the west bank of the Mississippi where the modern city which bears his name now stands (Figure 9.1). Dubuque's monopoly of the lead trade did not outlive him, and local Indian groups accounted for what production occurred through the 1820s, when Americans and Europeans began arriving in significant numbers to mine lead in the Fever River area.[10]

A century and a half of French presence in the upper Mississippi Valley had barely altered aboriginal settlement geography or the corridors of migration and exchange. "The French came as exploiters, not as settlers, and as a result their settlements were temporary and unsubstantial in character," notes a local historical geographer, Glenn T. Trewartha.[11] These consisted mainly of a few trading and military posts, forming a locally important string from Fort Le Sueur in the heart of the fur-rich upper Mississippi country south to Fort St. Nicholas, near the confluence of the Mississippi and Wisconsin rivers, which guarded the Fox-Wisconsin portage route to the Great Lakes and Montreal, the great colonial fur entrepôt 1,000 miles away. The site of Fort St. Nicholas, known from about 1740 onward as Prairie du Chien, has seen the longest continuous European occupation of any in the general vicinity of the lead mines (Figure 9.1).[12] Strung out along a broad river terrace on the east bank of the Mississippi River, the settlement was laid out by the French in agricultural long-lots, which have given the town as it urbanized an enduringly different character from others in the region. This quality has been enhanced by the survival of quite a number of the old French log dwellings and early American Fur Company buildings built of stone.[13] Following the War of 1812 the town and its military functions passed to the Americans, and a new stronghold, Fort Crawford, was built in 1816 to prevent a British return to the region, as well as to police the Indian trade. It anchored the western terminus of the strategic Military Road, constructed 1835–1837, which followed the Military Ridge heading east across the high ground of the lead region and then northeastward by way of Fort Winnebago (Portage) ultimately to Fort Howard at Green Bay.[14]

The Lure of Lead

The Military Road was built in response to the scare introduced by Indian revolts against the growing pressure of white incursions into the region, most notably the shock produced by the Black Hawk War of 1832. Several Indian mines were in operation in the Fever and Apple river valleys just south of the Wisconsin border around 1820, their proprietors hostile to white interest in the lead deposits. Trading lead became lucrative because the market for it was expanding. The practice of painting frame buildings had become widespread in America by the early nineteenth century, which greatly raised the demand for lead, as did the need for bullets, pipes, lead sheeting, and printer's type.[15] Most important, however, the Upper Mississippi Valley Lead Region promised to become America's first major domestic source of this strategic mineral, and thereby lessen if not eliminate dependence on British importation.[16]

Enterprising Americans from the Ohio Valley corridor and St. Louis became keenly interested in the region. A clutch of prospectors from Kentucky appeared between 1819 and 1822, led by Colonel James Johnson, who, despite the lack of a required federal government license, opened a mine in the vicinity of what would become the town of Galena, Illinois. In 1822 these men were granted leases, and mining by Americans began in earnest, with a workforce that at first included some slaves.[17] By June 1825, 69 permits in the region had been issued, a figure that rose to 4,253 four years later.[18] The total population is estimated to have shot up between 1822 and 1828 from 20 to 10,000 inhabitants.[19] The Upper Mississippi Valley Lead Region vied with and then quickly overtook the Missouri lead mines in total production, averaging 5,000 tons of pig lead per annum by the late 1820s and 11,390 tons by 1836, compared with a slow rise to an average of 3,000 tons per annum for Missouri by the mid-1830s.[20] For 25 years, from the early 1820s to the late 1840s, the Upper Mississippi Valley Lead Region boomed on the basis of lead mining.

Wisconsin, noted John Davidson, a local historian, might be regarded as the first state settled by steamboat, and indeed the early migrant routes overwhelmingly favored the Ohio and Mississippi rivers, drawing people from Kentucky, Tennessee, Ohio, Missouri, and southern Illinois (Figure 9.2).[21] This gave the lead region an immediate southern flavor, which left its mark on the landscape through southern forms of vernacular architecture and local government organization. But as time passed, source regions and travel routes shifted northward, drawing in settlers from New England and the upper Midwest,

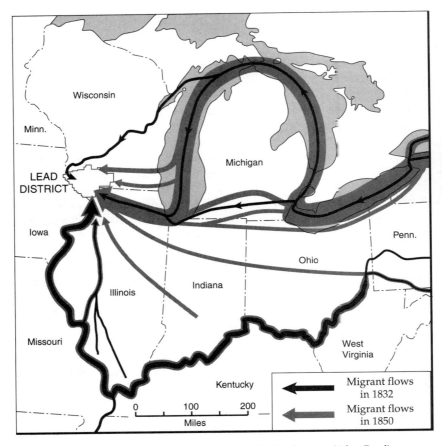

Figure 9.2. Changing migration routes to the lead region (After Read)

as well as European immigrants, especially from western Cornwall in southwest England. These experienced hard-rock miners and their families, dislodged by plunging tin and copper prices in Europe, came in force to the Upper Mississippi Valley Lead Region from the early 1830s on, and proved the perfect agents to guide the transformation of the lead workings from shallow "diggings" to shaft mines. "As they came to this country they were unlettered, shrewd, industrious, and skillful," observed Davidson. "They would go to mines that had been abandoned, and would make them pay."[22]

There emerged from the 1820s to the 1840s a clear geography of lead extraction, processing, and export in southwest Wisconsin, and this spatial pattern permanently influenced the economic and social development of the region as a whole.[23] The pattern was extensive and superficially chaotic, but structured like the root-and-trunk system of a tree: the mines provided nourishment, which, like sap, was suitably concentrated at various root joints, passed along to the main trunk, and sent for distribution along distant branches. This dendritic pattern echoed the topography and drainage of the land over which it was spread.

Lead was deposited in crevices and bedding planes of a Paleozoic magnesian limestone, 100–275 feet thick, which forms the subsurface rock of much of the uplands and higher portions of the river valleys of the Driftless Area. These deposits form mineral veins and clusters distributed widely over the region, often very close to the land surface. As a result, lead mining quickly assumed a highly scattered pattern as prospectors fanned out and combed the area for rich workings. The main zone of activity, however, developed within a polygon centrally placed within the region as a whole, and anchored in the corners by Galena, Potosi, Highland, Dodgeville, and Gratiot's Grove, with major inner foci at Mineral Point, Platteville, and Benton–New Diggings (Figure 9.3). Few localities which proved ultimately productive escaped early notice, and expansion occurred locally around earlier activity, especially as shallow workings played out or hit the water table. At first, workings, or "leads," were little more than surface pits—hence the general term *diggings.* Only as veins were followed to greater depths were shafts sunk, and the identity of the region shifted to "mines." If miners made promising discoveries, they came "in companies proportioned to the reputation the new diggings have acquired," wrote James T. Hodge in 1842, "and in a month's time a little village of log cabins has sprung up in the midst of what was just before wild woods or an uninhabited prairie."[24] Rules of the mines permitted stakeholders only a few square rods of ground (40 by 20 rods) for their operations, permitting later comers to work nearby if they desired. Even after mature development the districts of active mining probably covered less than 8 percent of the total region.[25]

The dispersed mining pattern dictated a network of smelting furnaces only slightly less scattered, where the ore was reduced to pig lead. The furnaces required proximity to dependable supplies of ore, timber for the logs that heated the furnace, and waterpower to drive the bellows that provided the "blast." The smelters were quite distinct from the miners: they had some capital, and their origins were different. If the miners were often Cornishmen, the smelters were Yorkshiremen, responsible for introducing an efficient furnace to the lead

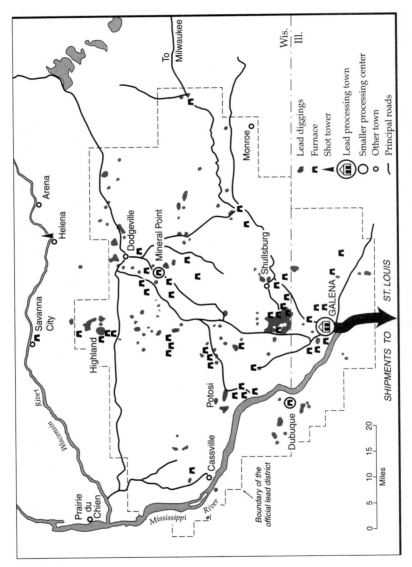

Figure 9.3. Geographical structure of the lead extraction system in 1839

Figure 9.4. Dodgeville furnace (Photo: M. Conzen)

region in the 1830s, a type long used in the Pennine hills of the north of England.[26] A surviving example is preserved in relict form in northeast Dodgeville (Figure 9.4). Production rose dramatically until the mid-1840s, when the district was producing more than 23,000 tons of lead annually, after which levels fell back considerably during the 1850s and 1860s.[27]

The pigs were transported by ox-cart generally southward and downslope toward Galena, which, with several of Wisconsin's river valleys leading to it (Figure 9.3), served as the gateway to the lead region. Though not directly on the Mississippi River, Galena lay only 1.5 miles up the Galena River, and for 40 years functioned as a major Mississippi River port. Mineral Point, an early center of mining, served as an inland collecting point. Galena soon dominated as the entrepôt for the whole region, exporting the lead in riverboats south to St.

Louis and New Orleans (whence by sea to New York) or up the Ohio River. Lesser quantities left the region from Dubuque, or were moved overland by wagon to Milwaukee, Racine, and Chicago. Galena's position in this system of flows encouraged a small manufacturing sector consisting of white lead works, foundries, and similar firms. Entrepreneurs at Helena, on the Wisconsin River, sought to establish a shipping point there for exports via the Great Lakes, and a shot tower at the village flourished briefly. This cumbersome but profitable—and essentially riverine—system lost its spatial simplicity when Galena's small river began to silt up in the 1850s, thanks to disturbances of the soil in the watershed caused by mining, tree cutting, and agricultural activity, and trade passed to Dubuque. Railroads beginning in the 1850s would further modify the regional pattern of mineral shipments.

The Economic Impact of Lead

Lead far outpaced all other natural resources as a fount of wealth in the region before the Civil War, with the possible exception of agriculture. By the 1820s the fur trade was unimportant. Timber was relatively sparse and quickly diminished in the cause of smelting ores. Lead mining and agriculture together generated opportunities in related businesses, both commercial and manufactural, thus drawing in additional people and capital. It is difficult, of course, to calculate even the most general measures of relative regional economic stimulation for the period, but selected statistics are suggestive. Between 1821 and 1865 the mines are said to have pumped at least $40 million in direct cash into the region's economy.[28] By contrast, a total of 1,444,053 acres of land had been bought as farms in the region by 1860, representing value added of $17.7 million.[29] The mines stimulated an early immigration of 10,000 people within six years of mine permits being issued, though some of these migrants doubtless moved directly into farming, since the miners provided an immediate market for foodstuffs.

Mitigating these beneficial conditions, however, the chancey nature of mineral prospecting and development bred instability in the capital markets and encouraged wild speculation. Furthermore, the same siren call that brought men to mine in the region could also take them off to the next discovery, and did so.

Less directly but more permanently, however, the lead trade accelerated the development of the region's infrastructure, its water transportation facilities, its roads and carrier services, and its towns. Lead mining brought steamboats to the area in large numbers and led to landing and storage facilities and interest in navigation improvements far beyond what agriculture alone would have wrought. The first steamboat docked at Galena in 1821, and by 1828 there were 99 arrivals during the year, and 350 arrivals by 1837, about one a day.[30] Such accessibility greatly reduced the frontier isolation of townspeople and farmers in the vicinity. The lead trade also created a network of primitive roads in the region, which farmers were free to utilize (Figure 9.3), and many of them became permanent elements of the later township road pattern of the area. Perhaps most dramatically, lead mining created towns more rapidly and made them grow larger than agriculture alone would have done. The mining itself was physically dispersed, but it concentrated wealth in the hands of tradesmen, who lived and supplied services in towns, and also in the hands of enterprising industrialists, who were also largely urban. By 1860, aside from Dubuque (14,000) and Galena (13,500), the lead region boasted half a dozen towns containing between 1,500 and 5,000 inhabitants.[31]

The Social Impact of Lead

Along with the Missouri mines, the Upper Mississippi Valley Lead Region served as the cradle of all later mining societies that sprang up in the American West. It was here that the social consequences of a serious mineral "rush" were first felt, and here that various experiments in government coordination of enterprise were tried. The restless, ravenous, and raucous nature of fortune-hunters attracted to a mining frontier tinged the whole early mining history of the district, and just as the underlying rocks were riddled with lead flats and crevices, so were society and politics shot through with swindles, bribery, and corruption. Fights over the location and relocation of seats of government, charters for internal improvements, land scams, and stunning appropriations of the public interest for private gain of all kinds enmeshed the leaders of the region, good and bad.[32] It was here that the practice of leasing, rather than selling, public lands containing mineral deposits was first tried, in order to provide continuous income for the national government. But this approach was found highly unpopular and unworkable, and was abandoned and never attempted again in any western mining arena.[33]

Without lead, southwest Wisconsin would not have acquired its Cornish population, and they epitomize the region's legacy of folk communities shaped by lead. While eastern Americans, Irish, Welsh, Germans, and Swiss came to the district in numbers and occupations proportionately little different from other areas where they were significant, the Cornish came there specifically to mine, and their

Figure 9.5. Cornish cottages on Shake Rag Street, Mineral Point (Photo: M. Conzen)

concentrations were highest in the richest parts of the lead fields.[34] Theirs is a classic case of individually based chain migration from a clear initial source area (mainly Camborne Parish, Cornwall). Though not clannish, the Cornish settled densely enough—half the population of towns like Mineral Point, Dodgeville, and Hazel Green hailed from Cornwall—that they formed coherent folk communities with ease and studded the landscape with their modest limestone dwellings (Figure 9.5). Their Methodist orientation, as with the Welsh, ensured a strong showing for that church in the region, and sprinkled chapels across the rural landscape and in the towns. Some distinctive foodways and folk music survived.[35] Similar foreign communities developed in the region as a whole, though on a smaller and spatially more limited scale, as in the case of the Welsh in Iowa County and the Swiss in Green County.[36]

Another enduring social impact of lead is the stimulus it gave to missionary activity in the cause of civilizing the raw communities it spawned. In the shadow of the large American and Cornish Protestant presence, many Irish, Germans, and others of Catholic faith were drawn to a region in which there was no minister active in the early years. This circumstance brought Samuel Mazzuchelli, a gifted Italian Dominican friar, to Galena in 1835 to continue a remarkable career in the Middle West as a founder of parishes, schools, and an order of nuns; as a designer and builder of churches; as an architect of rectories, market halls and court houses, and an all-round participant in the politics of the region for almost three decades (Figure 9.6). Mazzuchelli was not alone in seeking to regenerate the region—Methodist and Presbyterian circuit riders and others criss-crossed the district too—but his many talents and vast energy made him uniquely effective, and many of his architectural creations remain as witnesses in the landscape.[37]

The Environmental Impact of Lead

Beside the economic and social impacts of lead mining, there were also environmental consequences. Capable of sustaining agriculture, the region suffered the early withdrawal of large quantities of timber for smelting purposes, timber which otherwise would have gone only to buildings and fences or remained to anchor the soils upon which it stood. The depletion of timber and its substitution with costly imports was greatly accelerated by mining. In addition, the diggings churned up substantial areas of generally smooth and undulating ground and rendered them useless for agriculture. "In every direction within scope of the eye," wrote a chronicler, "heaps of mineral refuse blackened and disfigured the verdant hillsides, and the clank of the windlass made merry music to the accompanying sounds of the crowbar, pick and drill."[38] Such disturbance, together with the leaching it produced, permanently altered the soil cover and polluted the streams of the district.[39] The surface character and wide distribution of the deposits exacerbated this problem, for in discovering mineral veins that would repay exploitation, miners also dug up large areas barren of ore. "Certain townships and parts of townships," wrote the historian Joseph Schafer in 1932, "are so honey-combed with mines as to place farming at a physical disadvantage."[40] And Trewartha adds, "Even down to the present time one can observe areas of pitted mining terrain which have remained out of cultivation."[41]

A second impact was that deforestation and soil disturbance (both from mining and farming) accelerated the rates of valley-floor sedimentation, leading in some instances to more local flooding. Evidence from the Galena and Big Platte rivers shows shifts in sediment buildup from a presettlement rate of 0.02 centimeter per year to 0.3–5.0 centimeter per year during the period from colonization to the 1930s.[42] The most dramatic result was the rapid silting up of the lower Galena River during the 1840s and 1850s, stranding steamboats and thereby dooming Galena's status as a major Mississippi River port.[43]

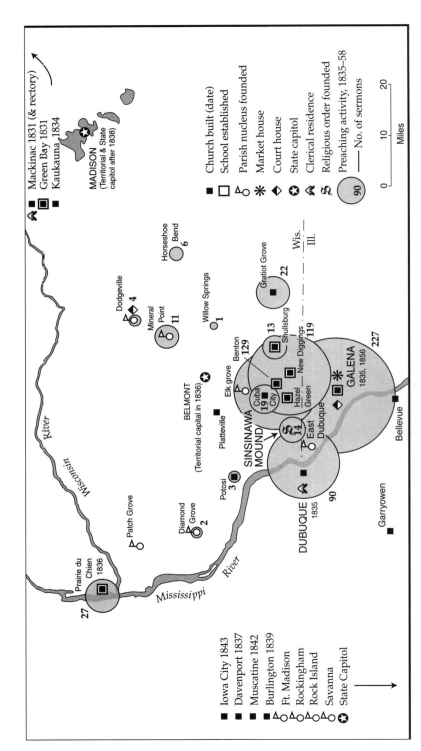

Figure 9.6. Mazzuchelli's social impact in the region, 1831–1860

Third, lead mining possibly affected the health and life-span of its practitioners. Little is known about the risks to the upper Mississippi Valley populations in the past, when medical knowledge and industrial practices were less advanced, but hazards from toxic lead buildup in humans have long been recognized.[44] Workmen in the smelters were more likely to have been affected than the miners, because they breathed fumes and dust in the plant. Lead from mine waste also shows up as a trace metal in the stream sediments of the district today, and perhaps has affected drinking water, though whether the concentrations have been high enough to affect people and grazing animals in this area is unclear.[45]

The Turn to Farming and Zinc

Viewed down the corridors of time, mining is usually a cyclic phenomenon, and this was certainly true in the lead region. Even where reserves are ample, the costs of extraction rise with increasing inaccessibility. Unless market prices also rise or technological innovations reduce the retrieval costs, exploitation of any one field is likely at some point to decline. Competition from more profitable fields can quickly hasten the process. This was the case in the upper Mississippi Valley, but the region's special advantage lay in the farming alternative. The land had supported light forms of agriculture for centuries, and when the Euro-Americans came, the plow was ready to follow the pick and shovel. While early government mining regulations aimed to discourage all but garden-scale farming, the high prices of imported foodstuffs soon stimulated local production. Statistics about farms before 1850 are scarce, but by 1844 Grant County, according to one account, had as many farmers as miners: 600 of each, together with 100 smelters, 60 merchants, 32 millers, 16 tavern-keepers, 15 grocers, 17 lawyers, and 16 physicians.[46] Farming was hindered in the early years by higher investment costs and lower returns compared with mining, by poor transportation, and by limited knowledge of agricultural possibilities. But soon these drawbacks disappeared, and farming attracted settlers on its own terms.[47] In addition, lead mining fell back in importance during the 1850s, as the easily worked deposits played out and gold fever propelled miners to California. At 12,227 short tons, lead production in 1860 was but half of what it had been at its peak in 1845.[48]

As the decades wore on, the miners who did not move away turned to farming, undoubtedly encouraged by the success of their neighbors who had settled the region for that purpose. Thus many Cornish, whose pioneer migrants had come as proud miners, utterly changed what in many cases were generations-old family traditions and began tilling the soil, and in so doing lost the interest in and the skills for mining. This trajectory is typified by the Trewartha family.[49]

Peter Trewartha, born in Cornwall in 1818, came to the lead region in his late twenties sometime between 1846 and 1849 with his wife, Elizabeth, and two daughters and settled in Hazel Green, where he worked as a miner. The 1850 census recorded that he could neither read nor write. He escaped the census-takers' notice 10 years later, but by 1870 a great change had occurred: he was listed as a 52-year-old farmer with a comfortable farm of about 130 acres (which he valued at $3,000) just south of the village, and with about $900 in personal property. By then his family had greatly expanded, doubtless benefiting from the local schools which had been organized by the time the Trewarthas arrived.[50] None of the sons went into mining, however, though the industry was still functioning. In the 1870s Peter died, and by 1880 his wife was recorded as a farmer, with five children and a grandchild on the farm. As the children matured they left the house and started their own families or moved away, and Elizabeth was left living by herself. Frank, the third son, in 1887 married a local woman, Eva, and by the early 1890s was running a meat market in the village and farming as well.[51] Soon a daughter, Jane, was born to them, and she was followed in 1896 by a brother, Glenn. Glenn Trewartha finished high school, went on to college at the State Normal School at Platteville, and then built a 44-year career as a distinguished professor of geography at the University of Wisconsin in Madison.[52] Thus in the tapestry of this family's first three generations in America are woven several fundamental strands of Cornish social history in the district: central involvement in lead mining, a key turn to farming, better schooling, entry into small-town trading, access to higher education, and then the addition of professional pursuits—all within the frame of 60 years.

Farming and Agricultural Specialization

In the secure period following the Black Hawk War of 1832, farming advanced, shaped by the terrain and soils of the various parts of the region. William O. Blanchard recognized two topographic zones in southwestern Wisconsin, the north-facing escarpment, essentially north of the Military Ridge, and the backslope to the south, including the Galena district of Illinois. Within these broad zones, Joseph Schafer noted three agricultural land types: the mineral districts proper, the nonmineral lands within the mining realm, and the steep, eroded slopelands beyond its margins.[53] Farms developed most easily in the second zone, the nonmineral lands of the backslope, their number spreading into mineral localities where diggings did not interfere with the surface. The steep valleys at the margins developed last, despite being closer to river transportation. Corn far outweighed oats and wheat in the economy of the district, which never fully participated in the wheat craze that swept through other parts of Wisconsin, partly because of the more varied food market of the mining population, ranging from cornmeal to pork.[54] Prairie farms became noticeably more productive during the 1850s, especially in the vicinity of the new Mineral Point Railroad, built in 1857 to link the town with the Illinois Central Railroad, which gave access to Chicago and points south.

Once established, these general patterns continued right through the Civil War period and the rest of the century, with production best on the nonmineral prairie lands within the heart of the district and least impressive on the fringe lands with the steepest slopes.[55] What did change, however, was the level of interest in dairying. As wheat yields across Wisconsin slumped during the 1870s, many localities experimented with new specialities, creating a broad new pattern of agricultural diversity.[56] Southwestern Wisconsin moved vigorously into dairying, in common with other parts of the state but specializing in certain types of cheese making. One consequence of this shift was that farms in the poorer fringe districts of the region were able to narrow the gap between their farm income and that of their prairie neighbors, because the valley farms were well suited to grazing by dairy cattle.[57] Little more than a domestic appendage in earlier days (like poultry), milk production became serious business for lead region farmers during the 1880s and 1890s. Several factors seem locally important in this development: the transfer of New York cheese making by migrants from that state, a campaign by state and local officials to promote dairying, special railroad rates to ship Wisconsin cheese to the East Coast, and the settlement in compact communities of Swiss farmers in southwestern Wisconsin, especially throughout Green County.

Cheese making did not occur in all dairy districts of the state; some stressed butter or fluid milk for urban markets. Cheese production is best revealed through the distribution of cheese factories, since they were placed at numerous crossroads and hamlets, the local central places where farmers delivered their milk (Figure 9.7). The first cheese factory in the state was established in 1870 by Niklaus Gerber, a Swiss immigrant, near Monticello in Green County. Loyal Durand has identified five specialized cheese-making regions in Wisconsin by the end of the century, two of which—the Swiss and Limburger cheese area and the American cheese area—developed in the lead region broadly

Figure 9.7. Roadside cheese factory west of New Glarus (Photo: M. Conzen)

defined (Figure 9.8a), and an extension centered on Jo Daviess County in Illinois.[58] The Swiss and Limburger cheese area developed because the Swiss settlers were preadapted to undertake cheese manufacture, though they had participated in the wheat boom as enthusiastically as their Yankee neighbors. However, a process of landholding concentration through the departure of non-Swiss from the vicinity of the New Glarus colony between 1860 and 1920 helped solidify the Swiss community in Green County, and they turned to cheese in the 1870s with alacrity. Swiss cheese factories peaked around 1905, with 213 in Green County alone.[59] The American cheese area that emerged in parts of Iowa, Lafayette, Grant, Richland, and Sauk counties owed its origin to Ohio and eastern settlers and contained about 150 factories at the turn of the century.[60] These regions remained strong islands of specialization well into the middle of the twentieth century.

Southwestern Wisconsin's development of a favorable farming economy over the first century of American occupance came at a considerable ecological price that the pioneers' children and grandchildren had to pay—serious soil erosion on their farms. Not only did mining contribute to soil loss and downstream sedimentation, but so also did the sometimes negligent and uninformed agricultural prac-

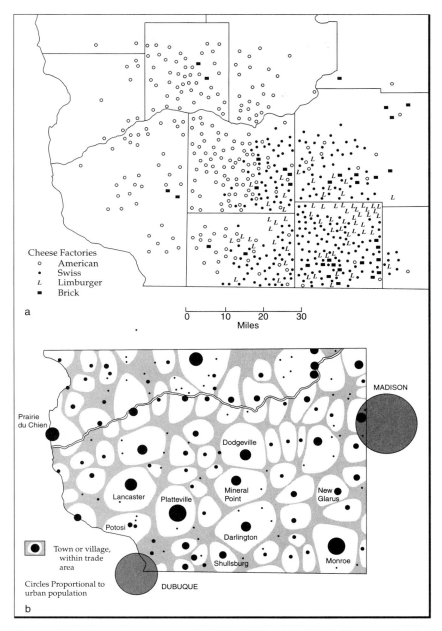

Figure 9.8. (*a*) Cheese factory types, 1915 (After Hibbard, Benjamin Horace, and Asher Hobson, *Markets and Prices of Wisconsin Cheese.* Madison, 1915); (*b*) the central place system, 1950 (After Brush, John E. The hierarchy of central places in southwestern Wisconsin. *The Geographical Review,* 43:3 [July 1953] pp. 380–402).

tices of a farming class that included many with but little accumulated farming experience in their families, a situation made worse by the deep valleys of the region. Soil conservation became a cause célèbre among a new breed of environmentalist, most notably Aldo Leopold, whose local studies of the problem and public campaigns for solutions led ultimately to such innovations as contour plowing, a farming technique particularly suited to and widely adopted on the hilly lands of the lead region.[61]

Zinc Galvanizes a Mining Renaissance

Lead mining continued its decline during and after the Civil War, but did not disappear from the region, though it became highly sensitive to price fluctuations and competition from the Missouri-Kansas fields, and later from Colorado and Utah. Within the region it became geographically ossified, fixed at familiar sites where capital-intensive shaft mining at greater depths than before had become the rule. During the 1860s and 1870s, mining and smelting became concentrated in fewer hands at fewer locations, and quickly made room for a new activity — the recovery of zinc (known locally as drybone). As the lead harvest dwindled, the zinc harvest swelled and provided a remarkable reprise for a mining region that had seen farms sprout in its midst and most of its skilled workforce disperse.

A particular stimulus to zinc development, but by no means the only one, was the activity of two German chemical engineers, Frederick W. Matthiessen and Edward C. Hegeler.[62] They came to America from the Royal Saxon School of Mines in Freiberg, Germany, in the late 1850s, looking for ways to develop zinc manufacture. Zinc-refining technology was advancing rapidly in Europe, and, besides its use in brass making, many new industrial applications were being discovered, but in the United States the industry was underdeveloped. Around Mineral Point, Matthiessen found the zinc ore he needed among the tailings left over from lead mining, and realized that here was a long-term source of supply. He contracted for shipments to a smelter he and Hegeler built at LaSalle, Illinois, the most northerly point where coal was available. The recently opened Mineral Point Railroad, which connected with the Illinois Central line to La Salle, could bring the zinc ore to meet the coal. The La Salle plant became nationally prominent in American zinc manufacture, and provided a steady market for Wisconsin ore. In 1863, a rival smelter was constructed at Mineral Point, but it soon foundered because special rates to bring in coal could not then be secured from the railroad.[63] A more lasting operation, the Mineral Point Zinc Company, opened in 1882 and survived into the late 1920s.

The increasing economic diversity of the lead region continued to attract railroads, both for the minerals and the specialized agriculture, particularly dairying. Between 1868 and 1882, the Chicago and Northwestern, the Chicago, Milwaukee, and St. Paul, and the Illinois Central railroad companies built additional feeder lines into the district, their very alignments reflecting the intense competition between the terminal cities of Milwaukee and Chicago for the trade of the territory. Some of them were narrow-gauge, to negotiate the steep valley inclines. The region boasted a dense pattern of railroads by the late 1880s in comparison with much flatter, richer farming areas. After a severe lull in the late 1880s and early 1890s, mining advanced again, and a little railroad was even built as late as 1905 to connect the rich zinc fields of Highland, north of the Military Ridge, with the smelter at Mineral Point. Zinc and lead mining experienced one last, glorious surge between 1896 and 1920, after which price declines sucked away the vitality of the remaining mines. As mining fared, so did some of the railroads; line closures began early, with the first abandonment occurring on the Woodman to Fennimore route in northern Grant County in 1926.[64]

A Hierarchy of Local Central Places

Towns in the region—urban places, that is—came with the first Europeans, even if at first small in size. They formed the control points within far-flung networks of long-distance and local exchange, communication, and local production. During the French occupation, Prairie du Chien existed to serve as a frontier trading post in a gossamer network that stretched thousands of miles to Montreal, and then through Le Havre to Paris, France. During the early American mining period, Prairie du Chien fell into regional eclipse, while Galena emerged as the undisputed entrepôt of a completely new system based on lead mining and grew to 13,500 souls by 1860.[65] Small inland towns of between 500 and 1,000 inhabitants, such as Mineral Point, Dodgeville, Shullsburg, Potosi, and Platteville, developed to concentrate the flow of exports from their localities to Galena and to offer frequent services to the surrounding miners' camps and villages for which Galena was too distant. As mining became a fixture and farmers settled the remainder of the land, some of these towns prospered accordingly and grew to new levels, generally between, 2,500 and 3,000, while smaller new towns emerged in the purely farming districts, such as Lancaster,

Darlington, Monroe, and New Glarus, some aided by county seat status (Figure 9.8b).[66]

Within only a few decades, the network of towns assumed a general spatial and hierarchic arrangement that would then endure for a century. Only rarely did towns and cities lose rank, as Galena did to Dubuque, when steamboats could no longer reach it and the lumber trade blossomed on the Mississippi River to Dubuque's advantage. By 1880 most towns had found their particular "niche" within the larger pattern, and residents' dreams of radical growth yielded to sober acceptance of their towns' limited role within the regional system. This pattern was measured and mapped by John E. Brush for 1950 and little had changed: of the 11 places he classified as "towns" in the area south of the Wisconsin River (but including Prairie du Chien), 8 already had such status by 1880, and 6 of them had emerged as early as 1860 (Figure 9.8b).[67] Once admitted to the rank, towns held their own; only the old lead center of Shullsburg had slipped to the functional level of a village. Notwithstanding significant changes in the region's economic profile and layout, the urban pattern was defined rather early and proved essentially stable. Even the humble hamlet, so typical a feature of the hilly terrain of the region, survived remarkably intact well into the modern period.[68]

The character of the lead region's towns reflects their deep mining roots, the often dramatic topography within which they are set, the native building materials, and the ethno-regional landscape preferences of past residents, as well as their debt to general American town-planning practices. Prairie du Chien, as a cultural throwback, echoes yet its French and early American heritage as a river town, if ever so softly. A glorified street village in basic plan, with long-lot property lines still evident in the town pattern, there survive still a few modest buildings from the fur trade era in both the upper and lower village.

Towns built by lead mining, however, form a second group with their own singularities, and differ quite sharply from the typical farming centers of the upper Middle West. The lead towns had informal street plans, shaped to the hilly and winding valley sites into which they were initially crammed by miners unversed and uninterested in town-planning theory (Figure 9.9). They began as elemental linear street villages, but they were built up with stone- and brick-lined "High Streets" fairly shouting their mixture of American southern and British west-country building roots in the galleried river-bluff houses, Greek revival public buildings, late-federalist warehouses, Cornish and Welsh stone cottages, steepled churches, Italianate shop fronts, and the colonnaded mansions and grand villas of merchants, riverboat

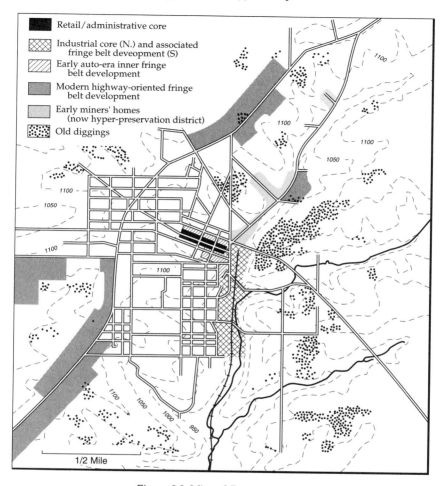

Figure 9.9. Mineral Point town plan

captains, and smelting magnates that occupied treed streets and bluff-top aeries.[69] The jewel in the crown is Galena, but the grouping must surely include Mineral Point and Dodgeville, too.[70] Even Shullsburg fits this type, though on a much smaller scale, and is graced with a handsome Mazzuchelli-inspired church, as can be found also in such quintessential mining villages as New Diggings and Benton. And the other small fry mimic their more urban cousins with the most basic of linear but irregular patterns—Beetown, British Hollow, Potosi, Hazel Green, and Fair Play.

A third group of towns, the farm trade centers laid out on simple grid plans and found in the flatter agricultural districts, do link the region to western American urban norms. Whether sporting street grids with courthouse squares suggestive of civic spirit or central railroad reserves that pay homage to business, these towns display the more regular, open-ended, mechanistic townscape of the prairies.[71] But even here, the undulating country, the distinctive building materials, the occasional small-town brewery or cheese factory, and the Cornish, Germans, and Swiss who live there to this day—these factors still infuse these towns with a southwest Wisconsin character not found in Indiana or western Iowa.

Geographical Heritage of the Lead Region

What has the European wrought in the Upper Mississippi Valley Lead Region during the last three centuries? The outcome has been a fundamental transformation of the ecology, settlement, and economy of this stretch of the continent. It is a mature portion of a larger entity, the American Middle West, and it sits awkwardly at the margins of most boundaries, physical or cultural, that scientists and scholars employ in attempting to mark off subdivisions within this amorphous realm. The lead region occupies the southern portion of the hilly unglaciated Driftless Area of the upper Mississippi Valley, and it nestles within that broad transition zone between forest and prairie. It does not obviously belong to either the upper or the central Middle West (though certainly not the lower);[72] and it fits uncomfortably within the overlap zone of the midwestern Dairy and Corn belts, displaying strong features of both. Few regionalists with national schemes venture to pronounce it distinctive in any way, reflecting perhaps the difficulty of the challenge.[73]

However, one might suggest that the crucial early settlement processes examined here shaped a human geography for the Upper Mississippi Valley Lead Region that has endured, becoming only more deeply etched with time, rather than less, composing an intricate pattern of things generic and things unique. The generic features—the level of economic development, the standards of internal and external accessibility, the infrastructure of modern life—stem from its midwestern character, produced by a modernizing, aggressive, at times progressive, practical, self-absorbed, and nonideological culture.[74] The unique features result from the particular *mixture* of elements—peoples, resources, terrain, and cultural imperatives—concocted here, that is, *in situ*, and from some of the idiosyncrasies consequent upon that mixture.[75] One can no more account for the presence here of Frank Lloyd Wright's Taliesin than one can account for Father Wernerus' Dickeyville Grotto, or explain the abundance of appealing vernacular architecture as opposed to the dearth of major literary talent.[76]

What can be said, however, is that there developed a distinct geographic symbiosis of local resources and settling peoples that produced a joint "architecture" in the region not duplicated anywhere else. Perhaps the key feature is that in all its development the region remains remarkably nonmetropolitan but, moreover, set within a surrounding rim of transit corridors between metropolitan centers peripheral to it. The region has no obvious "capital" city today. Galena lies at its margin and, as it has always done, in another state, and without the lead trade or Mississippi River frontage the town is but a shriveled, albeit museum-grade, little burg. Dubuque is also peripheral and lies on the far bank of the continent's largest river to most of the lead region, as well as under different state jurisdiction with its own traditions. Madison would be a fine regional capital, were it not already the state capital for a broader territory, a consequence of Belmont (in the heart of the lead district) losing its role as capital very early in a northward-expanding state.

This nonmetropolitan island today, however, falls well within the commuting zones of weary, recreation-seeking metropolitans from Chicago, Milwaukee, Madison, Dubuque, and even farther afield, and a new geographic symbiosis is emerging that commodifies history in places like Mineral Point and the landscape of the region's rolling hills and valleys (Figure 9.10).[77] Indeed, the process of historical celebration is well rooted in the historic preservation and ethnic revival movements of such places as Mineral Point and New Glarus, and can be seen advancing with alacrity elsewhere, such as Mount Horeb.[78] And given the rising importance that congenial rural environments have as attractions for technical personnel in the location of new high-tech industries, the region may well benefit from a new wave of permanent settlers, outfitted not with pick axes but with modems. They will be plugged into the delights of small-town and rural life for the sake of the inner and outer child, and within easy reach of the metropolitan, face-to-face, cultural sophistication of intellectual centers, such as Madison, at needed intervals, courtesy of the Wisconsin Highway Department.[79] As an added bonus, many of the railroad rights-of-way, now all abandoned, are being fashioned into popular bike and hiking trails. History is priceless, and the future of the Upper Mississippi Valley Lead Region will likely be bound up with an approach to environ-

Figure 9.10. Hill and dale landscape west of Monticello (Artist: Frank S. Moulton)

mental conservation that pays close attention to the surviving cultural resources. Already, the forces are being marshaled to make the most of this rare heritage.

Notes

1. Historical accuracy should dictate that the region be called the Upper Mississippi Valley Lead and Zinc Region, since mining there experienced a second wind through zinc extraction, but for brevity's sake the name used will recognize only the first metal mined.

2. The general development of the Upper Mississippi Valley Lead Region is rich in conventional literature, marred only by the rigid compartmentalization of thought encouraged by state boundaries. Too often, the complete picture requires tedious assembly of separate and often not-quite-comparable studies focused on the Wisconsin, Illinois, and Iowa components of the region. The best single historical account is J. Schafer, *The Wisconsin Lead Region* (Madison, 1932), although a fine, brief overview is available in R. C. Nesbit and W. F. Thompson, *Wisconsin: A History* (Madison, 1989), chap. 8: "The Wisconsin Lead Region," 106–117. The only monographic, geographical treatment is W. O. Blanchard, *The Geography of Southwestern Wisconsin* (Madison, 1924), which provides some historical perspective. The most comprehensive historico-geographical study is D. R. Fatzinger, "Historical Geography of Lead and Zinc Mining in Southwest Wisconsin, 1820–1920: A Century of Change," Ph.D. diss., Michigan State University, 1971. More specialized historico-geographical treatments covering the lead region include M. J. Read, "A Population Study of the Driftless Hill Land during the Pioneer Period, 1832–1860," Ph.D. diss., University of Wisconsin, 1941; R. R. Polk, "A Geographical Analysis of Population Change in the Hill Land of Western Wisconsin, 1870–

1950," Ph.D. diss., University of Wisconsin, 1964; and C. Rosen, "Images of Wisconsin's Settlement Frontier," Ph.D. diss., University of Wisconsin–Milwaukee, 1990.

3. The broad sweep of Indian settlement history in the region is treated in C. I. Mason, *Introduction to Wisconsin Indians: Prehistory to Statehood* (Salem, Wis., 1988); and W. Green, J. B. Stoltman, and A. B. Kehoe, eds., *Introduction to Wisconsin Archaeology* (Madison, 1986). On village and horticultural sites, see J. E. Freeman, "The Millville Site, a Middle Woodland Village in Grant County," *Wisconsin Archeologist* 50 (1969): 37–87; C. Arzigian, "The Emergence of Horticultural Economies in Southwestern Wisconsin," in W. F. Keegan, ed., *Emergent Horticultural Economies of the Eastern Woodlands* (Carbondale, Ill., 1987), 217–242; and J. P. Gallagher, "Prehistoric Field Systems in the Upper Midwest," in W. I. Woods, ed., *Late Prehistoric Agriculture: Observations from the Midwest* (Springfield, Ill., 1992), 95–135. On effigy mounds, see W. M. Hurley, *An Analysis of Effigy Mound Complexes in Wisconsin* (Ann Arbor, Mich., 1975). On rock art, see R. J. Salzer, "Preliminary Report on the Gottschall Site (47 Ia 80)," *Wisconsin Archeologist* 68 (1987): 419–472; and D. Lowe, "Rock Art Survey of the Blue Mounds Creek and Mill Creek Drainages in Iowa and Dane Counties, Wisconsin," *Wisconsin Archeologist* 68 (1987): 341–375. For a splendid overview of historic Indian settlement history, see also R. E. Bieder, *Native American Communities in Wisconsin, 1600–1960: A Study of Tradition and Change* (Madison, 1995).

4. W. Green, "Examining Protohistoric Depopulation in the Upper Midwest," *Wisconsin Archeologist* 74 (1993): 290–323.

5. R. J. Mason, "Oneota and Winnebago Ethnogenesis: An Overview," *Wisconsin Archeologist* 74 (1993): 400–421.

6. B. H. Hibbard, "Indian Agriculture in Southern Wisconsin," *Proceedings of the State Historical Society of Wisconsin* 53 (1905): 145–155; J. C. Kay, "Wisconsin Indian Hunting Patterns, 1634–1836," *Annals of the Association of American Geographers* 69 (1979): 402–418.

7. J. A. Walthall, *Galena and Aboriginal Trade in Eastern North America* (Springfield, Ill., 1981), 2, 37, and 43.

8. L. R. Abbott, "Frontier Lead Mining in the Upper Mississippi Valley," *Journal of the Iowa Archeological Society* 35 (1988): 1–15.

9. T. A. Rickard, *A History of American Mining* (New York, 1932), 150 and 156.

10. J. C. Kay, "Indian Responses to a Mining Frontier," in W. W. Savage and S. I. Thompson, eds., *The Frontier: Comparative Studies,* vol. 2 (Norman, 1979), 193–203; M. Meeker, "Early History of the Lead Region of Wisconsin," *Wisconsin Historical Collections,* vol. 6 (Madison, 1872), 271–296. As lead mining was taken up by Americans, the Fever River, to improve its reputation, became known as the Galena River.

11. G. T. Trewartha, "French Settlement in the Driftless Hill Land," *Annals of the Association of American Geographers* 28 (1938): 182.

12. G. T. Trewartha, "The Prairie du Chien Terrace: Geography of a Confluence Site," *Annals of the Association of American Geographers* 22 (1932): 119–158.

13. M. A. DeJulio, "Prairie du Chien and the Rediscovery of Its French

Log Houses," in M. O. Roark, ed., *French and Germans in the Mississippi Valley: Landscape and Cultural Traditions* (Cape Girardeau, Mo., 1988), 98–110.

14.   R. D. Durbin and E. Durbin, "Wisconsin's Old Military Road: Its Genesis and Construction," *Wisconsin Magazine of History* 68, 1 (1984): 3–42.

15.   D. R. Fatzinger, "Mineral, Bone, and Jack and the Mining Industry of Southwest Wisconsin," paper presented to the Mid-America Historical Geography Association, Galena, Ill., 1981, 7. For more detail, see Fatzinger, "Historical Geography of Lead and Zinc Mining."

16.   For statistics on imports and exports of lead and lead products to and from the United States, 1791–1861, see Schafer, *Lead Region*, 253–255.

17.   R. G. Thwaites, "Notes on Early Lead Mining in the Fever (or Galena) River Region," *Wisconsin Historical Collections*, vol. 13 (Madison, 1895), 287–290. For details on the mineral land leasing system, see J. E. Wright, *The Galena Lead District: Federal Policy and Practice, 1824–1847* (Madison, 1966). A total of 59 slaves were recorded in the 1830 census for the upper Mississippi Valley. Read, "Population Study," 33.

18.   A. E. Smith, *The History of Wisconsin*, vol. 1: *From Exploration to Statehood* (Madison, 1973), 183.

19.   G. T. Trewartha, "A Second Epoch of Destructive Occupance in the Driftless Hill Land," *Annals of the Association of American Geographers* 30 (1940): 126.

20.   J. D. Whitney, *The Metallic Wealth of the United States, Described and Compared with That of Other Countries* (Philadelphia, 1854), 421.

21.   J. N. Davidson, "Some Distinctive Characteristics of the History of Our Lead Region," *Proceedings of the State Historical Society of Wisconsin* 46 (1899): 185; Read, "Population Study," 30 and 37.

22.   Davidson, "Some Distinctive Characteristics," 191. A measured, scholarly treatment of the Cornish in the Upper Mississippi Valley Lead Region is offered in J. Rowe, *The Hard Rock Men: Cornish Immigrants and the North American Mining Frontier* (Liverpool, 1974), 37–61. For the global context of Cornish emigration, see G. Burke, "The Cornish Diaspora of the Nineteenth Century," in S. Marks and P. Richardson, eds., *International Labour Migration: Historical Perspectives* (London, 1984), 57–75.

23.   The technical aspects of lead mining are well described in such sources as D. D. Owen, *Report of a Geological Exploration of Part of Iowa, Wisconsin, and Illinois . . .* , 28 Cong., 1 Sess., 1844, S. Doc. no. 407, serial no. 437. For a fine, modern-era summary, see H. F. Bain, *Zinc and Lead Deposits of the Upper Mississippi Valley* (Washington, D.C., 1906), which also contains a detailed bibliography of previous writings on the lead and zinc resources and technical mining history of the district. A good national discussion of the industry is contained in W. R. Ingalls, *Lead and Zinc in the United States: Comprising an Economic History of the Mining and Smelting of the Metals and the Conditions Which Have Affected the Development of the Industries* (New York, 1908). The regional structure of the Lead Region as a production system in its heyday is portrayed in the detailed reconstruction derived from Owen's manuscript master map (never

published), presented in H. R. Friis, "The David Dale Owen Map of Southwestern Wisconsin," *Prologue: The Journal of the National Archives* 1 (1969): 9–28.

24.   J. T. Hodge, "On the Wisconsin and Missouri Lead Region," *American Journal of Science and Arts,* o.s. 43 (1842): 41.

25.   Trewartha, "Second Epoch of Destructive Occupance," 135.

26.   H. R. Davidson, "The Yorkshire Dales Smelting Mill in the Mississippi River Valley Lead Fields: A Study in Adoption-Diffusion in the Mid-Nineteenth Century," *Pioneer America Society Transactions* 14 (1991): 37–46. Comparison of this furnace can be made with the simpler log furnace as sketched in Abbott, "Frontier Lead Mining," 5–7; and Hodge, "Wisconsin and Missouri Lead Region," 45–49. Concerning the proprietors, see D. Morris, *The Dalesmen of the Mississippi River* (York, England, 1989).

27.   For the historical geography of lead production, see S. L. Schubring, "A Statistical Study of Lead and Zinc Mining in Wisconsin," *Transactions of the Wisconsin Academy of Sciences, Arts and Letters* 22 (1926): 9–98. For the Illinois district around Galena, see B. H. Schockel, "The Settlement and Development of Jo Daviess County, Illinois," in A. C. Trowbridge and E. W. Shaw, eds., *The Geology and Geography of the Galena and Elizabeth Quadrangle* (Urbana, 1916), 173–228.

28.   *Hunt's Merchants Magazine* 40 (1857): 244. Miners sold land only for cash, and thereby attracted much gold specie into the region, which lessened the impact of the 1837 and 1857 financial panics there. Thwaites, "Notes on Early Lead Mining," 179.

29.   This dollar amount was calculated on the basis of the total value of farms in 1860 at $19.5 million (as reported in the census), less $1.8 million estimated as the acquisition price of all agricultural land in the region to that time from the government and intermediaries. This estimate assumes an average raw land price of $1.25 per acre (the official government figure), making due allowance for the possible balance between such land as was acquired without cash through preemption and such land as was obtained at higher prices from speculators. Clearly, such assumptions are at best crude.

30.   Schockel, "Settlement and Development of Jo Daviess County," 215.

31.   The river town portion of this trading system is carefully studied in T. R. Mahoney, "Urban History in a Regional Context: River Towns on the Upper Mississippi, 1840–1860," *Journal of American History* 72 (1985): 318–339; and his *River Towns in the Great West: The Structure of Provincial Urbanization in the American Midwest, 1820–1870* (New York, 1990). For the wealth created by manufacturing, see M. Walsh, *The Manufacturing Frontier: Pioneer Industry in Antebellum Wisconsin, 1830–1860* (Madison, 1972), chap. 3: "Grant County: Manufacturing in the Mining Region," 69–97. The character of Wisconsin's chief lead-mining town during the period is chronicled in G. Fiedler, *Mineral Point: A History* (Mineral Point, Wis., 1962).

32.   See, for example, Schafer, *Lead Region,* chaps. 6 and 7; Smith, *History of Wisconsin,* vol. 1, esp. chaps. 7 and 10–13.

33.   Wright, *The Galena Lead District.*

34.   J. Rowe, "Cornish," in Stephan Thernstrom, ed., *Harvard Encyclopedia of American Ethnic Groups* (Cambridge, Mass., 1980), 243–245. B. P. Birch, "From Southwest England to Southwest Wisconsin: Devonshire Hollow, Lafayette County," *Wisconsin Magazine of History* 69 (1985–1986): 129–149.

35.   Notable Cornish foods that survived include the Cornish pasty (a triangular pastry filled most often with potato, but also with meat, turnip, apple, or whatever the cook had on hand), meat pies, saffron cake (filled with candied lemon, raisins, and currants) at Christmastime, scalded or clotted cream, and plum or figgy hoggan. See L. A. Copeland, "The Cornish in Southwest Wisconsin," *Wisconsin Historical Collections*, vol. 14 (Madison, 1898), 327–328. See also H. Hachten, *The Flavor of Wisconsin: An Informal History of Food and Eating in the Badger State* (Madison, 1981), 9 and 199–200. On Cornish folk singing in the region, which lasted at least into the 1940s, see H. B. Peters, *Folk Songs Out of Wisconsin: An Illustrated Compendium of Words and Music* (Madison, 1977), 6, 23, 29, 32, 61, and 66–73.

36.   Copeland, "Cornish in Southwest Wisconsin," 301–334. For the Welsh settlements, see P. G. Davies, *The Welsh in Wisconsin* (Madison, 1982), 16; and, for the Swiss, see F. Hale, *The Swiss in Wisconsin* (Madison, 1984), 9.

37.   M. Quaife, "Reverend Samuel Charles Mazzuchelli," in his *Wisconsin: Its History and Its People, 1634–1924*, vol. 4 (Chicago, 1924), 411–417; J. B. Alderson and J. M. Alderson, *The Man Mazzuchelli, Pioneer Priest* (Madison, 1974).

38.   C. W. Butterfield, "Life in the Diggings," in his *History of Iowa County* (Chicago, 1881), 495.

39.   C. L. Schmidt, "Effects of Jig Tailings on Soils in Southwestern Wisconsin," master's thesis, University of Wisconsin–Madison, 1985, 68–71.

40.   Schafer, *Lead Region*, 131.

41.   Trewartha, "Second Epoch of Destructive Occupance," 135.

42.   J. C. Knox, "Human Impacts on Wisconsin Stream Channels," *Annals of the Association of American Geographers* 67 (1977): 323–342; J. C. Knox, "Historical Valley Floor Sedimentation in the Upper Mississippi Valley," *Annals of the Association of American Geographers* 77 (1987): 224–244; S. W. Trimble, "Modern Stream and Valley Sedimentation in the Driftless Area, Wisconsin, USA," *Proceedings of the Twenty-third International Geographical Congress*, vol. 1: *Geomorphology and Paleogeography* (Moscow, 1976), 228–231.

43.   Schockel, "Settlement and Development of Jo Daviess County," 220.

44.   The historical consequences of lead mining for human health in the upper Mississippi Valley is an unexplored topic, but the general dimensions of the problem are ably covered in N. Castellino, "History of Lead Poisoning and Uses of Lead over the Centuries," in N. Castellino and P. Castellino, eds., *Inorganic Lead Exposure: Metabolism and Intoxication* (Boca Raton, Fla., 1995), 3–11. The retarded industrial and governmental response to the toxic hazard of lead in the United States is examined in W. Graebner, "Private Power, Private Knowledge, and Public Health: Science, Engineering, and Lead Poisoning, 1900–1970," in R. Bayer, ed., *The Health and Safety of Workers: Case Studies in the Politics of Professional Responsibility* (New York, 1988), 15–71.

45.   See Knox, "Historical Valley Floor Sedimentation," 227–229; and, for

a specific study of tailings, see *The Silver Creek Mine Tailings Exposure Study* (Atlanta, Ga., 1988).

46. C. W. Butterfield, *A History of Grant County, Wisconsin* (Chicago, 1881), 489.

47. Blanchard, *Geography of Southwestern Wisconsin*, 66–67. Chandler's map of the lead region in 1829 (see Trewartha, "Second Epoch of Destructive Occupance," 127) already showed seven prominent farms, and there were at least 65 separate, cultivated patches of ground at the time of the government land survey in 1832–1834; Trewartha, "Second Epoch of Destructive Occupance," 140.

48. Schubring, "Statistical Study," Exhibit I. The California gold rush of the early 1850s had a special effect among the miners of the upper Mississippi Valley because, by then, the region boasted a population of experienced miners. Grant County is said to have lost two-thirds of its population, and Mineral Point Township lost 250 people out of a total not more than double that, and 50 people left the little village of Lancaster. See Butterfield, *History of Grant County*, 489–490; and Butterfield, *History of Iowa County*, 489, 497. Numerous migrants returned, most of them disappointed by their experience, but many were able to regain a niche in their home districts. John Coad, a Cornish miner who settled in Mineral Point in 1844 was fairly typical: he left for California in 1852, stuck it out until 1860, and then returned to his wife and family in their Mineral Point stone cottage, where he lived for another 20 years. See S. Ewart, "Cornish Miners in Grass Valley: The Letters of John Coad, 1858–1860," *Pacific Historian* 25 (1981): 38–46.

49. The following family information is drawn from the manuscript schedules of the U.S. censuses from 1850 to 1920, from state and local business directories of Wisconsin, and from local plat maps. The surname Trewartha means, in Cornish, "upper homestead." For a portrait of the vicinity in its early mining days, see R. P. Fay, "Hardscrabble: A Case Study," *Wisconsin Academy Review* 30 (1984): 8–10. I am indebted to Sarita Trewartha, widow of Glenn Trewartha, for checking the statements about the family.

50. Butterfield, *History of Grant County*, 750.

51. Long ago, Peter Coleman drew attention to the high turnover rate of families in Hazel Green and other Grant County villages in this period, but his statistical methods failed to capture and interpret the local movements between village and township, illustrated by the Trewartha family, which betokened perhaps considerable stability. See P. J. Coleman, "Restless Grant County: Americans on the Move," *Wisconsin Magazine of History* 46 (1962): 16–20.

52. R. Hartshorne and J. R. Borchert, "Glenn T. Trewartha, 1896–1984," *Annals of the Association of American Geographers* 78 (1988): 728–735. He made major scholarly contributions to several fields of geography, including, early on, the genetic study of American settlement morphology and the historical geography of the upper Mississippi Valley, reflected in several citations in this chapter. Glenn Trewartha's typification of the three-generation shift from miners to professors is noted with satisfaction in A. L. Rowse, *The Cousin Jacks: The Cornish in America* (New York, 1969), 229.

53. Blanchard, *Geography of Southwestern Wisconsin*, 70–71; Schafer, *Lead Region*, 131.

54. S. Hartnett, "Harvesting the New Land: A Geographical Appraisal of the Wheat Frontier in the Upper Midwest, 1835–1885," master's thesis, University of Wisconsin–Madison, 1981. Grant and Iowa counties, however, did not join the specialized midwestern "Corn Belt" until the 1880s, and Iowa, Green, and Dane counties not until the 1950s, according to J. C. Hudson, *Making the Corn Belt: A Geographical History of Middle-Western Agriculture* (Bloomington, 1994), 11 and 153.

55. O. G. Libby et al., "An Economic and Social Study of the Lead Region in Iowa, Illinois, and Wisconsin," *Transactions of the Wisconsin Academy of Sciences, Arts and Letters* 13 (1901): 188–281.

56. T. R. McKay and D. E. Kmetz, eds., *Agricultural Diversity in Wisconsin* (Madison, 1987).

57. Schafer, *Lead Region*, 181. See also G. T. Trewartha, "The Dairy Industry of Wisconsin as a Geographic Adjustment," *Bulletin of the Geographical Society of Philadelphia* 23 (1925): 93–119.

58. L. Durand, Jr., "The Cheese Manufacturing Regions of Wisconsin, 1850–1950," *Transactions of the Wisconsin Academy of Sciences, Arts and Letters* 42 (1953): 109–130; G. T. Trewartha, "The Green County, Wisconsin, Foreign Cheese Industry," *Economic Geography* 2 (1926): 292–308; and L. Durand, Jr., "Cheese Region of Northwestern Illinois," *Economic Geography* 22, 1 (1946): 24–37.

59. D. Brunnschweiler, "Tradition and Environment as Counter-Influences in the History of New Glarus," in L. Schelbert, ed., *New Glarus, 1845–1970: The Making of a Swiss American Town* (Glarus, Switz., 1970), 160–189. This is a summary in English of his impressive, larger work, *New Glarus: Gründung, Entwicklung und heutiger Zustand eine Schweitzerkolonie im amerikanischen Mittelwesten* (Zürich, 1954).

60. Durand, "Cheese Manufacturing Regions," 119; B. H. Hibbard and A. Hobson, *Markets and Prices of Wisconsin Cheese* (Madison, 1915), 5–8, and 16.

61. A. Leopold, *The Erosion Problem of Steep Farms in Southwestern Wisconsin* (r.p., 1943). On the spread of contour plowing, see H. E. Johansen, "Diffusion of Strip Cropping in Southwestern Wisconsin," *Annals of the Association of American Geographers* 61 (1971): 671–683.

62. M. E. Lenzi, "Zinc Comes to LaSalle and Peru: A Historical Geography of the Matthiessen and Hegeler Zinc Company and the Midwestern Zinc Industry," in M. P. Conzen, G. M. Richard, and C. A. Zimring, eds., *The Industrial Revolution in the Upper Illinois Valley* (Chicago, 1993), 119–134.

63. Butterfield, *History of Iowa County*, 700–701.

64. Schubring, "Statistical Study," 50–72; Fatzinger, "Mineral, Bone, and Jack," 20–28; G. and R. M. Crawford, eds., *Memoirs of Iowa County Wisconsin* (n.p., 1913), 199. On the railroad pattern, see D. J. Lanz, *Railroads of Southern and Southwestern Wisconsin: Development to Decline*, 3d ed. (Monroe, Wis., 1986); and P. S. Nadler, "The History of the Mineral Point & Northern Railway Company," *Wisconsin Magazine of History* 38 (1954–1955): 3–6, 47–50, and 95–105.

65. Mahoney, *River Towns,* 109, 183 and 275.

66. Cash and locally made new capital traditionally enjoyed high repute among miners and ex-miners, and this cash-based prosperity can be seen reflected in the local origins of most bank capitalization in the region. See M. P. Conzen, "Capital Flows and the Developing Urban Hierarchy: State Bank Capital in Wisconsin, 1854–1895," *Economic Geography* 51 (1975): 321–338, especially fig. 2, 328.

67. For 1860 "town" designations, see Read, "Population Study," 217; the 1880 count was based on populations above 1,000; 1950 designations from J. E. Brush, "The Hierarchy of Central Places in Southwestern Wisconsin," *Geographical Review* 43 (1953): 389.

68. Compare the material in G. T. Trewartha, "The Unincorporated Hamlet: One Element of the American Settlement Fabric," *Annals of the Association of American Geographers* 33 (1943): 32–81, which is largely focused on southwest Wisconsin circa 1940, with D. R. Every, "Changes in the Geography of Hamlets in South-Western Wisconsin: 1937–1964," master's thesis, University of Wisconsin–Madison, 1973.

69. The urban morphology of the Lead Region has still to be written, but hints of the pattern are contained in G. T. Trewartha, "The Earliest Map of Galena, Illinois," *Wisconsin Magazine of History* 23 (1939): 40–43; G. H. Krausse, "Historic Galena: A Study of Urban Change and Development in a Midwestern Mining Community," *Bulletin of the Illinois Geographical Society* 13 (1971): 3–19; M. N. Rowe, "Factors Influencing Early Town Planning in Southwestern Wisconsin: A Morphogenetic Analysis," master's thesis, University of Wisconsin–Madison, 1985; and C. S. Zellie, "Nineteenth-Century Townsite Plans in Southern Wisconsin: A Typology and Analysis," master's thesis, University of Wisconsin–Madison, 1989.

70. See, for example, C. H. Johnson, *The Building of Galena: An Architectural Legacy,* 2d ed. (Galena, Ill., 1977); A. Jenkin and F. Humberstone, Jr., *The Homes of Mineral Point* (Mineral Point, Wis., 1976); M. Taylor, *An Intensive Architectural and Historical Survey of Lancaster, Potosi/Tennyson, Dodgeville, Belmont, Gratiot* (Platteville, Wis., 1985).

71. Edward Price has distinguished various types of American courthouse plans, of which at least two are represented locally: the Shelbyville-type square (Monroe, Lancaster, Wiota) and the Harrisonburg-type square (Darlington). On courthouse square morphology, see E. T. Price, "The Central Courthouse Square in the American County Seat," *Geographical Review* 58 (1968): 29–60. The region contains a few examples of railroad towns (e.g., Boscobel and Cuba City), but not many; it already had most of the townsites needed before the railroad entered. For railroad town morphology, see J. C. Hudson, "Towns of the Western Railroads," *Great Plains Quarterly* 2 (1982): 41–54.

72. For a historian's acceptance of the confusion, see S. E. Gray, "The Upper Midwest," in M. K. Cayton et al., eds., *Encyclopedia of American Social History,* vol. 2 (New York, 1993), 989.

73. For the two most widely discussed regionalization schemes, see

W. Zelinsky, *The Cultural Geography of the United States,* 2d ed. (Englewood Cliffs, N.J., 1992), fig. 4.3; and R. D. Gastil, *Cultural Regions of the United States* (Seattle, 1975), map 31, 205, on which Gastil portrays boundaries that include the lead region with an elongated "German-Swiss District" extending from Milwaukee to northeastern Nebraska.

74. For a larger list of regional social traits, see J. F. Hart, "The Middle West," *Annals of the Association of American Geographers* 62 (1972): 258–282. An attempt to portray the regional mind-set is given in T. T. McAvoy, "What Is the Midwestern Mind?" in T. T. McAvoy, ed., *The Midwest: Myth or Reality? A Symposium* (Notre Dame, Ind., 1961), 53–72. For an additional theme, "the cult of the average," see G. Hutton, *Midwest at Noon* (Chicago, 1947), 246–264. See also J. R. Shortridge, *The Middle West: Its Meaning in American Culture* (Lawrence, 1989).

75. There are no generally accepted generalizations about the lead region in the cultural-geographical terms broached here. For some useful insights for the state of Wisconsin as a whole, however, see E. Davis, "Wisconsin Is Different," *Harper's Monthly Magazine* 165 (Oct. 1932): 613–624; and B. Hofmeister, "Wisconsin: Eine kulturgeographische Skizze," *Jahrbuch für Amerikastudien* (1959): 249–283.

76. Hamlin Garland and August Derleth, two notable regional writers, were born north of the lower Wisconsin River, and thus were not native sons of the lead region. Derleth, however, did set several of his novels in the lead region: *Bright Journey* (New York, 1940) and *The House on the Mound* (New York, 1958), set in Prairie du Chien; *The Hills Stand Watch* (New York, 1960), set in Mineral Point; and *The Shadow in the Glass* (New York, 1963), set in Cassville.

77. C. W. Olmstead, "A Place and Point of View: Southwestern Wisconsin," in D. G. Janelle, ed., *Geographical Snapshots of North America* (New York, 1992), 337–341.

78. Regarding Mineral Point, see K. W. Hannaford, "Complexity and Simplification: The Dynamics of Historic Preservation at Mineral Point, Wisconsin," master's thesis, University of Chicago, 1995; on New Glarus, see W. E. Millard, "The Sale of Culture," Ph.D. diss., University of Minnesota, 1969 (which focuses on Galena and Nauvoo, Ill., and New Glarus, Wis.); and S. D. Hoelscher and R. C. Ostergren, "Old European Homelands in the American Middle West," *Journal of Cultural Geography* 13 (1993): 87–106.

79. For some data on what might be termed the "high-tech in the hollows" phenomenon, see J. Burdack, "Kleinstädte und die Renaissance des ländlichen Raums: Das Beispiel Wisconsins," *Mitteilungsblatt des Arbeitskreises USA im Zentralverband der Deutschen Geographen* 7 (1987): 11–35.

10

The Creation of Towns in Wisconsin

John C. Hudson

According to the most recent U.S. Census of Population, there are nearly 600 incorporated places in the state of Wisconsin. Many of these places are incorporated as villages, others as cities, but all of them, large or small, once were (or still are) commonly known as towns. They are not to be confused with townships, or "civil towns"—the 36-square-mile subdivisions of territory formed by the township-and-range land survey system. Towns, as they will be known in this chapter, are agglomerated settlements ranging upward in size from only a few inhabitants to those containing many thousands. Whether classified as urban (at least 2,500 inhabitants) or nonurban, and whether legally incorporated or simply recognized as named localities, Wisconsin's towns have played a.significant role in the state's growth from the beginnings of Euro-American settlement to the present day.

Because towns are part of the landscape, they offer an enduring record of economic and social history, reflecting many years later what was important in the lives of people who once lived there. Most Wisconsin towns, like those elsewhere, were created for the purpose of commanding the trade in primary resources or manufactured goods between one locality and the rest of the economic system of which each town was a part. Thus, Wisconsin's first towns, Green Bay and Prairie du Chien, were created as outposts of the fur trade, but they did not disappear when the fur trade ended 150 years ago. Wausau, Wisconsin Rapids, and Oshkosh began as sawmilling centers, but they became even larger cities after the era of white pine logging on the Wisconsin and Wolf rivers had ended. Burlington and Waukesha were originally the sites of waterpowered gristmills, yet neither waterpower nor wheat production has been important in southeastern Wisconsin for most of the life of these two places.

The locations of many Wisconsin towns, both large and small, were chosen on the basis of technologies and resources that have long since diminished to insignificance. Many other town locations were the work of land speculators who merely hoped that something would come of the sites they chose. A flurry of speculative townsite activity accompanied the opening of southeastern Wisconsin lands to settlement in the 1830s. Later, railroad companies would plat dozens more new towns along their tracks, especially in the northern half of the state, where few towns existed near railroad lines built during the 1870s and later. The supply of new towns has always kept pace with the demand. Unlike the spread of agricultural settlement into unoccupied territory, which was regulated by the federal government through the system of public land laws, town settlement has been almost entirely a private-enterprise affair, largely unregulated, governed only indirectly by the swings in business activity which determined the level of investment in new ventures.

Legacy of the Fur Trade

To speak of the fur trade as a phase in the development of Wisconsin towns is to conflate what was no more than an incidental aspect of that trade with the more deliberate and calculated schemes to create towns in the era of American settlement that was to follow. There was only one important city in the fur trade and that was Montreal. Mackinac was an outpost of Montreal, Green Bay an outpost of Mackinac, and Prairie du Chien, in turn, was an outpost of Green Bay. French habitation in the Green Bay area began when Father Claude Allouez established the St. Francis Xavier mission at De Pere in 1669. The French built a fortification, Fort de la Baie-des-Puants (Puants was the French name for the Winnebago), on the west side of the mouth of the Fox River the next year. Green Bay became the center of a substantial trade in skins and corn, which the native people sold to the French traders (*coureurs de bois*) who passed along the Fox-Wisconsin river route to the Mississippi. In 1757 it was reported that eighty bark canoes, which could carry 4,000 pounds of fur each, were paddled, eight men to the canoe, from Green Bay to Mackinac every year.[1]

Trader Nicholas Perrot built Fort St. Nicolas at the mouth of the Wisconsin River in 1686. Fur traders entered the upper Mississippi Valley for the purpose of reaching the Dakota, who then inhabited what is now the state of Minnesota. Typical of this period was an almost total reliance on water transportation. River junctions, such as the Wisconsin-Mississippi confluence at Prairie du Chien, were strategic locations. Twice a year native people and French traders gathered at the "Prairie les Chiens" (Dog Plains). In October credits were issued to those who were to rendezvous there the following May. Parties of Ojibwa, Menominee, Potawatomi, Winnebago, Sauk, Fox, Kickapoo, Iowa, and Dakota then dispersed in various directions.[2] In May they returned to Prairie du Chien with furs and were met not only by the usual Mackinac-based traders but also by boats loaded with trade goods that had been rowed up the Mississippi River from New Orleans.

Although France surrendered Canada to Great Britain and Louisiana to Spain in 1763, French Canadians remained important participants for the duration of the fur trade. Prairie du Chien was a French community into the early decades of the nineteenth century. When Illinois achieved statehood in 1818, Wisconsin became part of Michigan Territory and was divided into two counties, Brown on the east, with its seat at Green Bay, and Crawford on the west, with a seat of government established at Prairie du Chien. The two French settlements thus became the first established sites of government in Wisconsin.

With United States law finally in force, it became necessary for the land claims of individual residents to be determined according to legal procedures of survey. In 1820 the United States government sent a surveyor, Isaac Lee, to Prairie du Chien for the purpose of establishing property rights. Lee's map shows long, narrow strips of land fronting the river and extending to the bluff line (Figure 10.1).[3] It is an example of the French long-lot system of survey, which is also found near Green Bay and nearly everywhere else the French settled in North America. The French had made no such map of landownership, and neither had the British during their brief period of control. It took the imposition of American law to force this expression of French custom onto the map of Crawford County.

The French had been far more interested in trade than in land. It was the activity associated with particular sites, not who owned the sites, that made places acquire value in the French system of business.

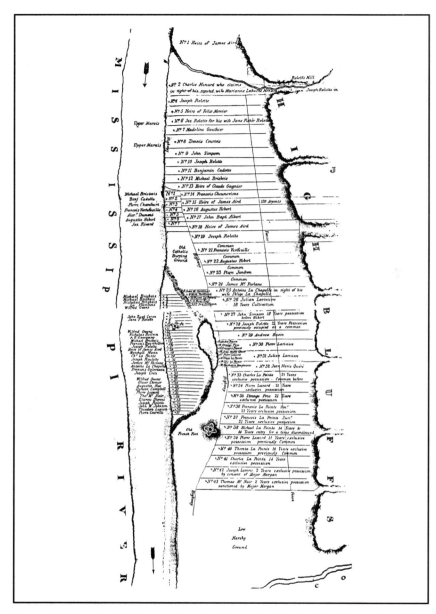

Figure 10.1. Isaac Lee's 1820 map of landholdings at Prairie du Chien (Reproduced from Scanlan, 1937)

French Canadians had received various grants of land in Wisconsin, but they lost the more valuable of these tracts to American entrepreneurs who understood how valuable land would become in the future and contrived ways of relieving the French (and the native peoples) of their lands.

Before 1830 there was not a single village of Green Bay but rather a scatter of habitations, known as Shantytown, on both sides of the Fox River near its mouth. The local population of approximately 500 was three-fourths French Canadian or mixed blood in composition. The role of the French traders was slowly diminishing as the fur trade itself declined. John Jacob Astor's American Fur Company, organized in 1807, held a virtual monopoly during the last years of the trade in the United States. When Astor liquidated his interest in the fur trade at Green Bay in 1834, the traders with whom he had dealt settled their accounts with Astor, to whom many owed money, by giving Astor mortgages on their land. An American trader at Green Bay, Daniel Whitney, acquired land there in a similar fashion.[4] By 1829 Whitney had secured mortgages and powers of attorney from four heirs and the widow of trader Pierre Grignon to their share of the undivided lots on the east side of the Fox River.

Astor and Whitney then foreclosed on the mortgages and used the land for townsites. Whitney's town, Navarino, located on the east bank of the Fox River opposite Fort Howard, was platted in 1829. Astor platted his town, named Astor and situated adjacent to Navarino, in 1835. The two towns were rivals but merged to become the borough of Green Bay in 1838. Astor apparently lost money on the town he had named for himself.

Other Wisconsin towns also have their beginnings in the fur trade period. Solomon Juneau's post at Milwaukee was an outgrowth of the trade at Green Bay. Sheboygan and Manitowoc began as French trading posts; and Fond du Lac, Oshkosh, and Appleton were the sites of French habitations. But unlike the French who remained at Detroit and St. Louis when the fur trade ended, and who profited from the increase in value of their lands as American settlers began to arrive in large numbers, the French at Prairie du Chien and Green Bay grew poorer in the transition. As Louise Phelps Kellogg noted, the "French settlements never were [progressive] in the American sense of the word; planted in a distant region, the residuum of the fur trade, with little contact with the larger world, they tended to a static rather than to a developing condition."[5]

French forts and trading posts were always linked by water routes, the bark canoe being their universal means of transportation. Wisconsin's first artery of commerce, the Fox-Wisconsin route from Green Bay to Prairie du Chien, was optimal only in terms of canoes, portages, and the need for taking the shortest route between the upper Mississippi River and Montreal. Americans avoided canoes and traveled by horseback or wagon whenever possible. Wisconsin's first overland route, the Military Road built from Fort Crawford to Fort Howard via Portage in 1835, mimicked the fur trader's water route, but the Military Road had little economic impact because it was not built on the geographic alignment that Wisconsin's development would take thereafter. The fur-trade era thus left Wisconsin with the nucleus of several important towns, but the trade itself left no impress of economic transactions over a distance that would endure in the era of American settlement that replaced it.

Speculators and Townsites

New Yorkers and New Englanders were prominent among the thousands of migrants who came to Wisconsin in the 1830s. Although many of them may have carried in their heads the image of some pleasant, long-settled village that they hoped to re-create in the West, many also were familiar with the speculator's townsite plat. By 1810 the creation of new towns had become a business in New York State.[6] It was rows of stakes driven into unoccupied ground, rather than a covenant drawn up among settlers bound together in a pioneering venture, that best marks the beginnings of New York towns founded after 1810.

Whenever a substantial amount of land was opened to white settlement there was typically a rush to establish various named sites around the countryside where investments might be made. One rather inexpensive means of doing so was to survey a town into all its streets and building lots, give it a name, and record a map (plat) of the would-be town at the appropriate courthouse. Those engaged in the business were known as townsite speculators, because neither the speculator nor anyone else could guarantee that a prospering town would actually grow on the platted site. These conventions were common and well known to the "Yorkers" who settled southeastern Wisconsin in the 1830s.

Native land titles in southern Wisconsin were extinguished in the 1820s and 1830s. Although the Black Hawk War (1831–1832) discouraged white settlement for a brief time, a boom was underway by the middle 1830s. New arrivals came by lake vessel and headed west in search of land. Sites most often selected for towns were those where power could be developed on a flowing stream, where a stream

Figure 10.2. Waterpowers mark the site of many of Wisconsin's early towns: Blair on the Trempealeau River (Photograph by the author)

could be forded easily, or where an existing trail crossed a stream (Figure 10.2). De Pere, platted in 1835, was the work of the De Pere Hydraulic Company, which built a dam on the Fox River, sold lots near the dam, and developed water mills. Waukesha's town lots were laid out by the developers of waterpowers there as well. Waupaca boasted of having "fifteen good water-powers within one mile of the court house square." East Troy and Burlington grew around mills constructed on Honey Creek, a tributary of southern Wisconsin's Fox River.[7] Every city on the Rock River (Beloit, Janesville, Fort Atkinson, Jefferson, Watertown) represents a site chosen for its waterpower potential.

Because townsite speculation was an activity open to everyone, rivalries sometimes developed around sites where the prospects for growth seemed especially good. Green Bay was only one example where several would-be towns coalesced to form a single, viable one. Janesville, platted in 1837, had two nearby rivals, Rockport and Wisconsin City. Pike River and Southport eventually became Kenosha. Racine was preceded by Port Gilbert and Sage Town. Platteville was originally Platte River; Neenah was Winnebago Rapids; and Ripon was preceded by a short-lived Fourierist colony that called its settlement Ceresco. Buena Vista and Willow River turned into Hudson.[8]

One of the best guesses anyone could have made about Wisconsin's future capital city during the territorial period would have been to choose a site on the Wisconsin River in the middle of the state. In 1835 "Wisconsin City" (a second townsite with that name) was staked out between present-day Merrimac and Prairie du Sac on a site overlooking the Wisconsin River. Wisconsin City had not one but three enormous city squares, named Franklin, LaFayette, and Washington, three market places, and a market street 100 feet wide.[9] But it never amounted to more than a field of wooden stakes; the labels only appeared on paper.

Nor was Madison much more than a field of surveyors' stakes when the capital-location decision was made. Madison was the work of James Duane Doty, a leading figure in early Wisconsin politics and, in the words of Joseph Schaefer, "an inveterate gambler in town sites and water powers."[10] It was Doty, acting as agent for John Jacob Astor, who acquired the French Canadian fur traders' lands at Green Bay through mortgage foreclosure. In addition to interests he had at Green Bay and Prairie du Chien, Doty created Fond du Lac, which at first was little more than a land speculation. He was also involved heavily in the Neenah townsite.

In a single trip from Green Bay to Belmont, Doty and surveyor John V. Suydam staked out three towns along the way. Two were failures: Clifton, at the north end of Lake Winnebago; and Wisconsinapolis, 8 or 10 miles below Portage on the Wisconsin River. The third was Madison.[11] Doty favored Madison and persuaded a majority in the legislature to agree with him, a task made easier by his award of choice, corner lots to some 16 members of the legislature. One supposes that if Wisconsinapolis, or even Wisconsin City, had been Doty's favorite, the state capitol might overlook the Wisconsin River rather than Lake Mendota today.

The 1834–1836 period of town speculation ended abruptly in the Panic of 1837. Wisconsin would not experience a speculative boom in townsites of such magnitude again, but the ensuing decades saw more new towns brought into being in roughly the same manner. Under provisions of an 1824 act of Congress, county governments were allowed to preempt, at a minimum price, two quarter-sections (320 acres) of land for a seat of justice.[12] As one of the few federal land laws designed to promote the growth of towns, county-seat preemptions had the effect of establishing what would become the principal towns of many counties.

Central courthouse squares, typical of the longer-settled portion of the Middle West, were not favored in Wisconsin. With a few exceptions

Figure 10.3. Business blocks constructed of brick and stone still form the common street-scape in Wisconsin towns such as Reedsburg (Photograph by the author)

(such as the busy Green County courthouse square in the center of Monroe), Wisconsin's courthouses were typically located slightly off-center, away from the business heart of the town. The lack of central squares in Wisconsin's towns probably has no single explanation. Like New York, the state from which the largest number of immigrants had come, Wisconsin organized local governments at both the township and the county levels. Township halls were scattered, one per township, around the rural areas of the county. Each was a focus of local activity, which diverted attention from the county seat.

Most Wisconsin towns were laid out around one or more business streets along which the principal urban functions would cluster. Initially a ragged collection of simple frame, even log, structures occupied the main street, but in time the rude structures of frontier life disappeared. As towns prospered, new business buildings replaced the old in a common succession in which frame structures were replaced by stone and brick. Periods of business growth and expansion, such as the 1880s and 1900s, saw the construction of brick business blocks in many established towns (Figure 10.3). The brick business block is by no means unique to Wisconsin, but it is a common feature found on many main streets in the state.

Towns of the Pineries

In 1836 a strip of land along the upper Wisconsin River was taken from the native peoples for the purpose of opening the country to lumbering. Within one year nearly every good sawmill site on the upper Wisconsin had been claimed. Lands in this strip, known as the Cedar Point cession, were offered for sale at the Mineral Point land office and were entered so rapidly that by 1839 all the timber in the upper Wisconsin Valley was in the possession of the makers and vendors of pine boards and shingles.[13] Towns that grew around the sawmills became the region's urban centers.

Rapids and cataracts that mark the Wisconsin River's descent off the hard, Canadian Shield rocks of the northern part of the state were natural sites for waterpowered sawmills. George Stevens, a St. Louis lumberman, came up the Wisconsin River in 1839 and built a mill at Shaw Rapids, the site that became Stevens Point. Little Bull Falls's sawmill, built in 1839, was the nucleus of the subsequent town of Mosinee. Big Bull Falls, the most important drop in the river's course, near the confluence of the Rib and Eau Claire rivers with the Wisconsin, became Wausau, which counted 350 inhabitants in 1847. The 45 sawmills operating on the Wisconsin and its tributaries that year expanded to 170 saws by 1857.[14]

Grand Rapids (Wisconsin Rapids), Plover, and Nekoosa, toward the southern limit of white pine forest, also developed around lumber mills. Point Bause (Nekoosa) actually had the first sawmill on the river, in 1832, although the town of Nekoosa was not platted until 1858. Nekoosa was created by politician and railroad promoter Moses M. Strong of Mineral Point, who formed a stock company to plat Nekoosa, build a dam and mill there, and promote the settlement. Strong had a series of maps and drawings prepared that showed the nonexistent Nekoosa graced by a bluff-top Gothic villa overlooking the river (where a steamboat was shown loading lumber produced at his mill).[15] Strong's advertising probably fooled no one. His fiction merely anticipated the massive forest-based industries that would one day locate at Nekoosa and at other towns along the Wisconsin. The valley produced 30 billion feet of pine lumber before 1900.

Similar developments took place in the Chippewa River basin, although a combination of natural and economic factors restricted the development of mill towns along its course. The Chippewa has few rapids or waterfalls compared with the Wisconsin, the sites of Eau Claire and Chippewa Falls being the principal exceptions. Chippewa Valley white pine was cut by two, often competing, groups of lumber barons. Men who operated mills at Chippewa Falls, Eau Claire, and other scattered sites along the river were rivals of an even more aggressive group of entrepreneurs, of whom Frederick Weyerhaeuser is best known, whose sawmills and lumberyards were located in Winona, Rock Island, Burlington, and St. Louis on the Mississippi River.[16] The two groups eventually found resolution of their competing interests, but the focus on Mississippi River sites, where lumber was most easily prepared for shipment by rail to the western frontier, dampened the prospects for building more mill towns along the Chippewa.

The many falls, dams, and log ponds on the Wisconsin made it impossible to follow the usual practice of river-driving loose logs. White pine logs were rafted down the Wisconsin (and the Chippewa) right over the low dams, which were submerged during spring flooding. But on the Wolf River red-shirted lumberjacks urged a steady flow of loose logs downstream into early summer. Steamboats, which navigated the Wolf as far north as New London, reported that they had to run a gauntlet of pine logs and rafts laden with lumber.[17] Shawano and Winneconne originated as mill towns upstream on the Wolf; Oshkosh and Fond du Lac, on Lake Winnebago, became even more important milling and manufacturing centers for Wolf River white pine.

Fond du Lac held a brief advantage over Oshkosh because it had the same access to timber from the Wolf and because it was also connected to Chicago by railway a few years earlier. But Oshkosh became perhaps the best-known milling and manufacturing center in the state. In 1879 the city was processing more than 100 million feet of lumber annually, producing some 200,000 doors, 600,000 windows, and other wooden goods, including carriages and wagons.[18] Like Oshkosh, Menasha developed wood-product industries in the 1850s. Neenah, which was first a flour-milling center, had four paper mills by 1879. As in eastern Canada and New England, paper manufacturing eventually eclipsed sawmilling, but paper mills occupied roughly the same sites that sawmills had.

Railroads and Towns

Wisconsin's first railroad builders viewed this new mode of transportation as having a function similar to canals. In the 1840s railroads were projected as connections between major rivers and lakes in cases where nature had not provided a navigable stream. Like the early settlers who disembarked a Great Lakes ship and trekked inland in search of opportunity, the first railroads were aimed west from the lake ports of Milwaukee, Racine, Kenosha, and Sheboygan. Some lines were projected toward the southwestern Wisconsin lead district. All sought to connect Lake Michigan with one or more rivers, especially the Mississippi. The need for railroads grew in the public mind, sometimes to alarming proportions. In 1857 the editor of the *Grant County Herald* wrote, "The River and Lake are feeling for each other, and the railroad must unite them even if Sin and Death get the contract."[19]

It soon became clear that railroads would be as important as collectors as they were as connectors. In 1854, the year that the Milwaukee and Mississippi Railroad was completed to Madison via Milton, 14 stations along the line shipped 1.18 million bushels of wheat and 2,500 tons of pork to Milwaukee. Whitewater alone shipped more than 300,000 bushels of wheat and received grain that farmers hauled there by wagon from as far south as McHenry, Illinois. Seven grain warehouses operated at Whitewater in 1854.[20] Of course southern Wisconsin's grain belt was not large, the wheat bonanza did not last long, and the addition of more railroad lines reduced the volume of business handled at each station. But the sudden transformation of Whitewater from a stagecoach stop to a commercial grain market gave people a new perspective on railroads.

What most attracted mid-nineteenth-century entrepreneurs and investors to railroads, however, were the vast sums of money involved as well as the prestige associated with the enterprise. Land grants awarded by the federal government to support railroad construction only added to the appeal. More than 2.8 million acres of federal grant lands—one-twelfth the area of the state—were awarded to Wisconsin railroads.[21] Byron Kilbourn was a failed canal promoter who turned his efforts to railroads in the 1840s. After being discharged as president of the Milwaukee and Mississippi by its board of directors in 1851, Kilbourn organized the La Crosse and Milwaukee Railroad and, with the assistance of Moses Strong, won approval of its charter in the Wisconsin legislature.

Despite lobbying by Kilbourn, Strong, and other railroad schemers, no Wisconsin land grant bill passed the U.S. House and Senate until 1856. In that year Congress mandated that a line be constructed north from either Madison or Columbus to the St. Croix River and then north to Superior. To those who built the railroad came the promise of a land grant of over 1 million acres, in the usual form of odd-numbered square-mile sections within 6 miles of the constructed road; the grant limits were increased to a 10-mile swath west of Tomah.[22] The story of how Strong and Kilbourn acquired the grant by secretly awarding $862,000 in La Crosse and Milwaukee Railroad "corruption bonds" to members of the 1856 Wisconsin legislature, newspaper editors, jurists, the governor, other state officials, and assorted businessmen—simultaneously outmaneuvering the Chicago, St. Paul, and Fond du Lac Railroad, which had an equally nefarious plan to get the grant for their road—forms one of the darker (though amusing) chapters in the state's political history.

Completion of the line to La Crosse in 1858 created a new axis of economic development in the state. By then the objective was not merely to reach the Mississippi River but to cross it and thereby link the new farming districts and bustling towns of Minnesota with the national system. La Crosse, once a trade center of far less importance than Prairie du Chien, was not platted until 1847. Urban real estate there acquired value when it became apparent that La Crosse would become a railroad terminal. Sparta, platted in 1851, was laid out in advance, and in hope, of acquiring the railroad, which arrived in 1857.[23] Portage and Columbus, both established towns, experienced notable increases in population after the railroad came as well.

Although La Crosse was as close to Lake Superior as the La Crosse and Milwaukee Railroad had yet reached, the 1856 land grant bill specified that a little-known, two-year-old town called Superior would become the line's terminus. Native land titles were extinguished at Superior in 1854, and a townsite was platted immediately. Superior grew slowly, even though the syndicate that owned the sprawling 4,000-acre townsite had some well-known members. W. W. Corcoran (donor of the Corcoran Art Gallery in Washington, D.C.) and five United States senators (including Stephen A. Douglas of Illinois, for whom Douglas County is named) were among Superior's founders.[24] Despite the senators' assistance in steering a railroad toward Superior, the town would wait some years for a railroad to arrive (Figure 10.4). But the inevitability of developments there, a site where railroads surely would connect with lake traffic, was enough to entice men of some reputation to invest in what was no more than a marshy estuary at the time.

A second attempt to build a railroad to Lake Superior was more successful. In 1864 Congress bestowed a grant of 837,000 acres to construct what became the Wisconsin Central Railway north from Stevens Point.[25] Attached to the grant was the requirement to build a second line from Portage to Stevens Point via Ripon, an economically pointless zigzag that mainly served political ends. No construction took place until the 1870s, and then, as is generally accepted, it was the prospect of owning 837,000 acres of timber lands, rather than a railroad, that led investors actually to build the Wisconsin Central. In 1874 the obli-

Figure 10.4. Former passenger station of the Great Northern Railway at Superior (originally designated as West Superior by the Great Northern) (Photograph by the author)

gation to build via Ripon was dropped in favor of a direct line from Portage to Stevens Point, which the Wisconsin Central built (and later ripped up) in order to qualify for the land grant.

Colby and Abbotsford, two towns platted by the Wisconsin Central, were named for Boston capitalists Edwin Hale Abbott and Gardner and Charles Colby, who had gained control of the road. Between 1871 and 1877 lines were built from Neenah to Stevens Point and then north to Ashland.[26] The white pine lands accessed by the Wisconsin Central opened a new territory to lumbering which had been remote when logging streams were the only means of transportation. Spencer, Unity, Milan, Athens (Black Creek Falls), Park Falls (Flambeau), and Stratford began as mill towns along the Wisconsin Central and its connections.[27] Railroads built for logging remained in parts of northern Wisconsin into the 1930s (Figure 10.5).

The Milwaukee, Lake Shore, and Western Railway, which was built north along the drainage divide of the Wolf and Wisconsin rivers, also opened new areas to lumbering. In 1882 the line beyond Antigo reached Summit Lake and then Monico. Timber cruisers working west of Monico toward the Wisconsin River that summer met a party of railroad surveyors. Only a single French Canadian trapper was reported to live in the vicinity at the time. The railroad company, whose tracks reached the river at Pelican Rapids in 1883, first named the spot Pelican Station; however, soon thereafter it became Rhinelander, after F. W. Rhinelander, the New York capitalist whose investments brought the town into being.[28] By 1890 Rhinelander was a growing city of 2,658 people.

Attempts to encourage settlement in northern Wisconsin differed from those employed in agricultural areas. It was laborers, not farmers, who were needed in the north. Kent K. Kennan, land commissioner of

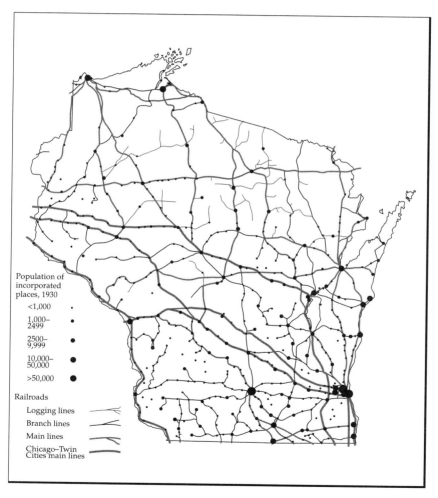

Figure 10.5. Wisconsin towns and railroads in 1930

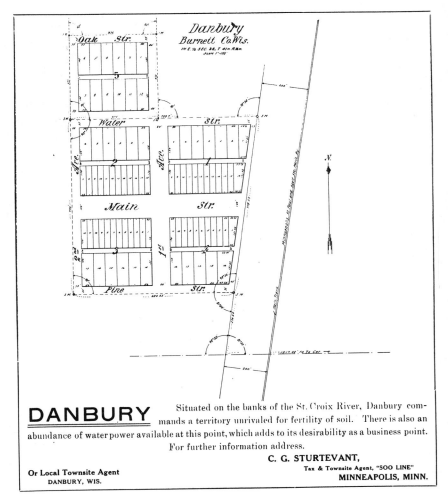

Figure 10.6. Advertising map of Danbury, Wisconsin, prepared by the Soo Line Railroad, ca. 1910 (Courtesy of the Soo Line Railroad)

the Wisconsin Central, was appointed as European agent for the Wisconsin State Board of Immigration in 1880. All the western states had representatives in Europe who competed with one another for immigrants. Among them, Kennan seems to have been notably successful. According to Theodore C. Blegen, Kennan enticed some 5,000 immigrants, "mainly from the forest lands of Bavaria," to settle along the Wisconsin Central line from Stevens Point to Ashland.[29] Good wages in the lumber camps were used as an inducement. The assumption that logging would give way to farming spawned the purchase of land and the building of comfortable homes with the wages earned in lumbering.

Railroads created towns in areas where none had existed, because without towns the railroad had no local business. Grain-growing areas were particularly fertile territory for the creation of towns, not because anyone believed the towns thereby founded would grow into sizable cities but because a regularly spaced series of towns would provide an efficient way to collect grain. Northern Wisconsin lacked agricultural prospects, but it saw many attempt to found new towns in this same manner. Railroad townsite agents used the same approach when booming any and all townsites, regardless of local conditions.

In 1911 the Soo Line Railroad completed its line from St. Paul to Duluth, which had reached north to Frederick, Wisconsin, in 1901. Siren, Webster, and Danbury, plus a half-dozen townsites to the north on the Minnesota side of the state line, were promoted with the usual zeal. Danbury's initial plat, a modest five blocks, suggests that the railroad regarded its own advertising promotion with some skepticism (Figure 10.6). "Danbury commands a territory unrivaled for fertility of soil," they claimed.[30] Waterpower available on the nearby St. Croix River probably added little to Danbury's "desirability as a business point." But Danbury later achieved at least a small amount of recognition (or notoriety), for having set a record low temperature of −54° F for the state in January 1922. Danbury, like its neighbors, was not the success the railroad company had hoped.

Towns founded by railroad companies were focused around the railroad tracks, especially near the train depot. Across from the depot was a hotel where travelers—salesmen, usually—could spend the night. Banks often located at the principal business intersection. Because the 25-foot-wide business lots that lined Main Street were of a single size and shape, so did the initial buildings assume a single size and shape.[31] A small but regular grid of streets, a single business street separating two rows of false-front buildings, and a railroad at the edge of the plat were typical of the railroad towns (Figure 10.7).

Although railroad towns often times were less than spectacular successes, railroads did expand the pattern of town settlement in Wisconsin (Figure 10.5). From the 1850s until the early decades of the twentieth century, railroads were primarily responsible for platting new towns in newly settled territory. Wisconsin's main arteries of commerce—in the railroad era and in the age of superhighways that followed—trended northwestward, toward St. Paul and Minneapolis. As

Figure 10.7. Common structures in Wisconsin's railroad towns included a row of one-story, false-front buildings that could be occupied by various businesses and a hotel near the tracks to accommodate travelers: (*a*) abandoned taverns at Pembine; (*b*) former hotel and restaurant at Sturtevant. (Photographs by the author)

part of a loosely defined northern transcontinental system that crossed Wisconsin, these lines were long-distance corridors more than they were locations for cities.

Secondary railroad main lines had more to do with the pattern of urban growth and resource exploitation in Wisconsin than the Twin Cities lines did. The fortunes of agricultural trade centers in southern and eastern Wisconsin were not closely tied to the railroads, but for those among them that emerged as manufacturing cities the rail network offered all-important access to raw materials as well as to the national market. Railroads replaced rivers as feeder routes for lumber and paper mills in the Fox and Wisconsin valleys.

In 1930 the railroad and town network was complete, even beginning to unravel as highway transportation emerged to challenge it. Lines radiating from Milwaukee dominated most of the state, which helped make Milwaukee the principal wholesaling center. The Driftless Area of southwestern Wisconsin, mostly served by branch lines emanating from Madison, was a region of small, agricultural trade centers. The reverse held true in northern Wisconsin, where most small towns had not incorporated by 1930. The larger milling, shipping, and mining centers there constituted more of a truly urban (rather than small-town) system.

Hamlets and the Driftless Area

Small towns, as a group, are the most numerous of any size class. Individually as well as collectively low in population, small towns are nonetheless dispersed widely over most settled areas of the United States. Explanations of their great frequency tend to emphasize the importance of short-distance travel for frequently performed human activities, such as socializing, worshiping, attending school, or daily trading. For Wisconsin the once-important daily activity of hauling milk to a cheese factory or creamery may be added to the list. When travel was more time-consuming, it was more important to have a local focus of activity.

Wisconsin's small, unincorporated towns, sometimes called hamlets, were described in a 1943 paper by geographer Glenn T. Trewartha; it remains the classic reference on the subject.[32] Trewartha's study, the last in a series of seven articles devoted to what he termed the Driftless Cuestaform Hill Land, focused on the internal structure and function of that region's smallest towns. He defined a hamlet as an unincorporated place which contained fewer than 150 inhabitants but at least four active residence units, six functional units (including houses), and at

Figure 10.8. Gays Mills, one of the prettier town settings in the Driftless Area, exhibits the straight streets, compact business district, and rows of frame houses typical of Wisconsin's small towns (Photograph by the author)

least five buildings. The prevailing view at the time Trewartha wrote, reflecting the influence of German geographer Walter Christaller, was that the rural population has "needs" that are met in small-town service centers. In this view, it is demand that brings towns into existence.

However, Trewartha found that "a relatively large number of hamlets in southwestern Wisconsin did not originate immediately out of the needs of the surrounding countryside." They had been, rather, "conceived, laid out, and platted by promoters who hoped thereby to profit from the sale of lots in the new settlement."[33] Although Trewartha did not make the point, it seems that he understood that hamlets were not much different from other towns; they just had not grown to be very large.

Trewartha regarded the Driftless Area simply as a "convenient laboratory" in which to study hamlets in general, yet there were notable features of the region's town system that favored the existence of the hamlets he studied. The lead industry provided southwestern Wisconsin with the beginnings of an urban system decades before similar developments were found elsewhere in the state. But lead mining did not affect the entire Driftless Area, and, more important, there were few other urban functions at hand that would replace smelting and other mining-related activities once the diggings went into decline. Southwestern Wisconsin is the only region of the state where the majority of incorporated towns either were served by dead-end railroad branch lines or had no railroad at all (Figure 10.5). Largely bypassed by the disruptive effects of the transportation revolution the railroads brought, the Driftless Area's hamlets and other small towns remained as an important part of the settlement system (Figure 10.8).

Initial town plats often bear only a slight resemblance to what actually developed on a site over the years. It is especially difficult to see evidence of deliberate design in many small towns of the Driftless Area where the intersection of two winding roads, following either ridge crests or valley bottoms, defines the center of a town (Figure 10.9). But the haphazard look of the hamlet may have resulted not so much from a lack of initial planning as from the informality that characterized their subsequent, slow-paced growth. Topography and economic history combined to give this area's towns a distinctive look.

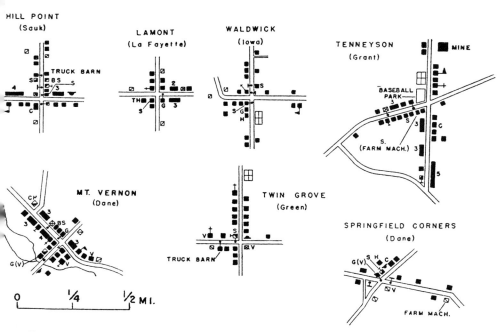

Figure 10.9. Layouts of some Driftless Area hamlets (From Trewartha, 1943, courtesy of the Association of American Geographers)

Conclusion

The founding of new towns in open country was a business, an agency of settlement formation, that operated in Wisconsin for more than three centuries. From the initial period of widespread white settlement on through the railroad era, towns were created by people or corporations that stood to earn money from their investment. When the automobile arrived, railroads stopped founding towns, and, with minor exceptions, the process was not resumed until new suburbs began to appear around the state's major cities in the 1930s.[34]

Most Wisconsin towns have already celebrated their centennials, and some have now reached the sesquicentennial mark. Change has been continuous as each town has simultaneously stimulated and adapted to new developments at scales ranging from local to global. The landscapes of towns both large and small reflect these changes. To know Wisconsin's towns is a good way to know not only the state's geography but also its history.

Notes

1. B. French, ed., *The American Sketch Book: A Collection of Historical Incidents with Descriptions*, vol. 3 (Green Bay, 1876), 26–30; P. L. Scanlan, *Prairie du Chien: French, British, American* (privately printed, 1937), 44; C. E. Heidenreich, plates 37–40, in R. C. Harris, ed., *Historical Atlas of Canada*, vol. 1 (Toronto, 1987).

2. Scanlan, *Prairie du Chien,* 61–65; L. P. Kellogg, *The French Regime in Wisconsin and the Northwest* (Madison, 1925), 405; J. H. Lockwood, "Early Times and Events in Wisconsin," *Annual Report and Collections of the State Historical Society of Wisconsin for the year 1855* — Madison: The Society, 1855–1857 — vol. 2 (1855) 98–196.

3. Lee's map reprinted in Scanlan, *Prairie du Chien,* 186–187, is the source used here.

4. A. E. Smith, "Daniel Whitney: Pioneer Wisconsin Businessman," *Wisconsin Magazine of History* 24 (1941): 283–304; French, *Sketch Book,* 57–71; General A. G. Ellis, "Life and Public Services of James Duane Doty," *Wisconsin Historical Collections,* vol. 5, part 3 (1868), 369–377.

5. Kellogg, *French Regime,* 404.

6. T. C. Bannister, "Early Town Planning in New York State," *Journal of the Society of Architectural Historians* 3 (1943): 36–42.

7. French, *Sketch Book,* 84; *Wisconsin Historical Collections,* vol. 1 (Madison, 1854), 134–135; Rev. J. Lothrop, "A Sketch of the Early History of Kenosha County, Wisconsin, and of the Western Immigration Company," *Wisconsin Historical Collections,* vol. 2 (Madison, 1856), 450–799; *History of Racine and Kenosha Counties, Wisconsin* (Chicago, 1879), 332, 471–472; A. J. Lawson, "New London and Surrounding Country," *Wisconsin Historical Collections,* vol. 2 (1856): 478–

488; J. W. Stewart, "Early History of Green County," *Wisconsin Historical Collections*, vol. 3 (Madison, 1857), 421–426.

8. *History of Rock County and Transactions of the Rock County Agricultural Society and Mechanics Institute* (Janesville, 1856), 160; *Racine and Kenosha Counties*, 289, 356; Gen. C. King, "Early Days in Platteville," *Wisconsin Magazine of History* 6 (1923): 40–42; "Pioneer History of Walworth County," *Proceedings of the State Historical Society of Wisconsin* 6 (1869): 466–468; D. P. Mapes, *History of the City of Ripon* (Milwaukee, 1873), 83–84; J. Schaefer, *The Winnebago-Horicon Basin: A Type Study in Western History* (Madison, 1937), 298; *Proceedings of the State Historical Society of Wisconsin* 3 (1857): 467–468.

9. "Pioneer History of Walworth County," 478. The plat of the stillborn Wisconsin City is shown in J. W. Reps, *The Making of Urban America* (Princeton, 1965), fig. 215.

10. Schaefer, *Winnebago-Horicon Basin*, 258; A. E. Smith, *James Duane Doty: Frontier Promoter* (Madison, 1954).

11. L. C. Draper, "Naming of Madison and Dane County and the Location of the Capital," *Wisconsin Historical Collections* (Madison, 1869–1872), 388–396. Madison's original plat is shown in Reps, *Urban America*, fig. 275.

12. T. Donaldson, *The Public Domain* (Washington, D.C., 1884), 298.

13. Gen. A. G. Ellis, "The 'Upper Wisconsin' Country," *Proceedings of the State Historical Society of Wisconsin* 3 (1857): 435–448.

14. H. J. Klueter and J. J. Lorence, *Woodlot and Ballot Box: Marathon County in the Twentieth Century* (Wausau, 1977), chaps. 1–3; *Proceedings of the State Historical Society of Wisconsin* 8 (1862): 123; E. E. Ladu, *Early and Late Mosinee* (Wausau, 1907).

15. K. W. Duckett, *Frontiersman of Fortune: Moses M. Strong of Mineral Point* (Madison, 1955), 152. Strong also platted the town of Arena in Iowa County.

16. J. N. Vogel, *Great Lakes Lumber on the Great Plains* (Iowa City, 1992), chap. 3; R. F. Fries, *Empire in Pine: The Story of Lumbering in Wisconsin, 1830–1900* (Madison, 1951); and M. Walsh, *The Manufacturing Frontier: Pioneer Industry in Antebellum Wisconsin, 1830–1860* (Madison, 1972), chap. 5.

17. Schaefer, *Winnebago-Horicon Basin*, 270.

18. Schaefer, *Winnebago-Horicon Basin*, 259–270; J. Kuony, "Lusty Days of Lumber," in J. I. Metz, ed., *Prairie, Pines and People: Winnebago County in a New Perspective* (Oshkosh, 1976), 297–319.

19. Quoted in J. S. Griffin, "The Great Lakes in Relation to the Railroad Development of Northern Wisconsin," *Proceedings of the State Historical Society of Wisconsin* 46 (1899): 211–225, quotation on 216.

20. J. A. Leonard, "Sketch of Whitewater," *Proceedings of the State Historical Society of Wisconsin* 3 (1857): 427–434.

21. W. F. Raney, "Building of Wisconsin's First Railroads," *Wisconsin Magazine of History* 19 (1936): 387.

22. Donaldson, *Public Domain*, 801–802; H. I. Deutsch, "Disintegrating Forces in Wisconsin," *Wisconsin Magazine of History* 15 (1932): 391–411.

23. M. McMillan, "Early Settlement of La Crosse and Monroe Counties," *Wisconsin Historical Collections*, vol. 4 (Madison, 1859), 383–393.

24.   F. H. DeGroat, "Henry Sigourney Butler," *Wisconsin Magazine of History* 22 (1939): 222–234; and Griffin, "The Great Lakes in Relation to . . . Northern Wisconsin," 217.

25.   United States, *Session Laws*, 38 Cong., 1 Sess. (1864), chap. 80, 66–68.

26.   *Wisconsin Magazine of History* 15 (1932): 107, 112, 285. Also see A. M. Johnson and B. E. Supple, *Boston Capitalists and Western Railroads* (Cambridge, Mass., 1967).

27.   Klueter and Lorence, *Woodland and Ballot Box*, 23–24.

28.   W. B. Shaw, "Early Days of Rhinelander," *Wisconsin Magazine of History* 6 (1923): 352–354; and a short note, "The Founding of Rhinelander," *Wisconsin Magazine of History* 5 (1922): 416–417.

29.   T. C. Blegen, "The Competition of the Northwestern States for Immigrants," *Wisconsin Magazine of History* 3 (1920): 3–29, quotation on 23. Also see K. A. Everest, "How Wisconsin Came by Its Large German Population," *Wisconsin Historical Collections*, vol. 12 (Madison, 1879), 229–334, especially p. 329.

30.   Copy of Danbury plat and advertising, courtesy John Bergene, Soo Line Railroad Company, Minneapolis.

31.   J. C. Hudson, *Plains Country Towns* (Minneapolis, 1985), chap. 7.

32.   G. T. Trewartha, "The Unincorporated Hamlet: One Element of the American Settlement Fabric," *Annals of the Association of American Geographers* 33 (1943): 32–81. Also see J. E. Brush, "The Hierarchy of Central Places in Southwestern Wisconsin," *Geographical Review* 43 (1953): 380–402.

33.   Trewartha, "Unincorporated Hamlet," 39.

34.   For a fascinating case study of one Wisconsin suburb, launched with goals of social betterment by the Resettlement Administration during the 1930s, see A. R. Alanen and J. A. Eden, *Main Street Ready-Made: The New Deal Community of Greendale, Wisconsin* (Madison, 1987).

11

Lumbering
Wisconsin's Northern Urban Frontier

Randall Rohe

Wisconsin's natural bounty of timber, among the richest of any state, has played a critical role in the state's historical development. Among the many species found in Wisconsin forests, the white pine has had the greatest commercial value. Originally, the state contained over 100 billion board feet of white pine, growing mostly north and east of an irregular line stretching northwest from Milwaukee to near Fond du Lac, Wisconsin Rapids, Sparta, Eau Claire, and Ellsworth. Before the 1890s, lumbering in Wisconsin generally meant white pine lumbering, and the lumbering region became known as the Pinery or Pineries. In large degree lumbering was an urban frontier, and during this period lumbermen established many settlements, ranging in size from the most temporary of logging camps to large cities.[1]

Logging Camps and Sawmill Towns

Logging operations have always been among the most important phases in the production of lumber, and wherever lumbermen carried on their exploitation, they set up temporary settlements. Although some of these settlements had 20 or 30 structures and supported populations of 150, 200, or more, they seldom evolved into permanent

Figure 11.1. Eau Claire, 1870. Eau Claire's location at the junction of the Eau Claire and Chippewa rivers helped it to achieve dominance in the Chippewa River lumber district. By the year of this photograph, Eau Claire had over a dozen sawmills and a population of 4,700. (Courtesy of the State Historical Society of Wisconsin, neg. number WHi[X3]1390)

towns.[2] Sawmills appeared along virtually every stream that could float a sawlog. The mills acted as the focus of early economic activity and, as such, often became the forming agent for larger settlements. As the *Green Bay Advocate* observed in the spring of 1871: "Every stream running in to Green Bay that is large enough to float a sawlog is as sure to breed sawmills along its banks as a swamp is to breed mosquitoes, and that is the way that A.C. Conn & Co. happened to locate their mill near the mouth of the Little Suamico, and most of the town here has grown up around the mill."[3] Especially at or near the mouths of the large rivers that drained the northern part of the state, the establishment of sawmills fixed the location of places like Oshkosh, Eau Claire, La Crosse, Menominee, Oconto, and Peshtigo and spurred their growth into points of importance (Figure 11.1). Elsewhere sawmills brought wholly new places, like Merrill, Mosinee, Wisconsin Rapids, Black River Falls, and Menomonie into existence.

Railroads eventually provided a more efficient means of shipping lumber than the region's rivers and streams. As railroads extended northward, sawmills appeared every few miles, typically forming the

Figure 11.2. Phillips, ca. 1890. Phillip's origin and development into a lumbering center came about because of the building of the Wisconsin Central Railroad, which platted the town and named it for Elijah B. Phillips, its general manager. (Courtesy of the State Historical Society of Wisconsin, neg. number WHi[X3]23308)

nucleus for the development of towns (Figure 11.2). This was particularly true along the Wisconsin Central:

At Manville [Mannville], Medford, Chelsea, Westboro, and other points we had a fine opportunity to observe the great lumbering interests that have grown up along the line of the Wisconsin Central railway since its construction, it having been completed less than five years ago. At these points extensive steam saw mills have been built, and around them little settlements began by hardy pioneers. The piles of pine saw logs and of lumber about these mills is a wonder to behold. On one side acres are covered with logs, piled from ten to twenty feet high, while on the other side vast areas of manufactured lumber greet the eye. The scenes of activity at these settlements are all the more remarkable in that for miles on either hand there is not a house, or a clearing, or a single sign of human habitation.[4]

The building of railroads into the pineries not only brought new towns into existence but revitalized existing ones as well. Before the railroad reached Stevens Point, one scribe described the town as dilapidated, with many dwellings tenantless and unpainted, its streets half grown over with grass, and business languishing. All of that changed after the arrival of the railroad. "Now, [1876] a city of several thousand souls has arisen with wholesale and retail stores—with creditable public buildings, evidencing industry and energy, thrift and prosperity."[5]

Patterns of Milling and Town Development

The distribution of sawmills and lumber towns in Wisconsin changed over the course of the nineteenth century in direct response to a shift in emphasis from water to rail transportation systems and the introduction of new technologies.[6] The first sawmills were practically always built on the banks of some river or stream. The logs came down the river, were held in the boom, and were cut into lumber during the spring and summer using power supplied by the stream. The first requisite of a waterpower mill was, of course, a sufficient head of water to drive the mill. As a consequence, mills were located either on a side channel at the foot of a rapids or waterfall, to which water was directed by a wing dam, or on the stream itself if sufficient head could be obtained by construction of a dam. Driving conditions, which varied from one stream to another, significantly influenced the pattern of urban development. Along the Wisconsin, a difficult river to drive, lumbermen built their sawmills as close as possible to the area of cutting. The resulting scattered nature of lumber production along the Wisconsin gave rise to a number of important lumber centers, including Wausau, Stevens Point, Grand Rapids (Wisconsin Rapids), Merrill, Rhinelander, and Mosinee. On rivers easier to drive, like the Wolf, Black, Chippewa, and St. Croix, sawmilling was more centralized, and Oshkosh, La Crosse, Eau Claire, Chippewa Falls, and Stillwater dominated lumber production in their respective districts.[7] The spatial characteristics of mid-nineteenth-century lumbering activities in Wisconsin show the effect of the limited scale, the ecotechnology, and local market orientation of these early mills.

Early concentrations of mills around Lake Winnebago and the mouths of the Eau Claire, Menominee, and Black rivers are explained by the fact that steam was already becoming a source of power by the mid-nineteenth century. The steam engine was most efficient in large units, and, in contrast with waterpower, it required a costly capital investment. Hence, waterpower, so abundant in the Pineries, came into use at places like Peshtigo, Oconto Falls, Chippewa Falls, St. Croix Falls, and Wausau, while steam-powered mills were concentrated in the towns to the south, such as Fond du Lac, Oshkosh, Eau Claire, and La Crosse. The concentration of steam-powered mills at such points provided economies of scale and encouraged the establishment of related industries such as foundries, logging tool and sawmill machinery manufacturers, and furniture, trunk, and carriage factories.

Although in the 1870s many mills still relied on water transport, those not located on a railroad found it more and more difficult to compete with the mills which could utilize this all-weather route to market. By the 1880s, mills began to disappear from their old locations as the industry moved production facilities closer to the remaining sources of timber. Oshkosh, Eau Claire, La Crosse, and so on, remained major centers of lumber production only because of their good rail connections and their development of diversified wood manufacturing industries. By 1890 all the major clusters of mills and most individual mills were located on a railroad, more often than not in the northernmost portion of the state. After the turn of the century, this trend continued, resulting in fewer but larger mills at points with good railway facilities in the timber regions. By 1905 few mills remained in the old lumber centers like La Crosse, Fond du Lac, Eau Claire, and Oshkosh.

Site and Situation: The Case of Oshkosh

Not fewer than 40 places in northern Wisconsin became important sawmill centers—Marinette, Oconto, Green Bay, Wausau, Stevens Point, Grand Rapids, Merrill, Black River Falls, Eau Claire, Chippewa Falls, La Crosse, Ashland, and others (Figure 11.3). Nearly all grew up at sites which possessed essentially similar characteristics, yet the site and situation of Oshkosh made it one of the foremost cities of this group. No other place in northern Wisconsin ever attained a more sudden celebrity as a manufacturing point than Oshkosh. From a little obscure village in 1852, with three or four sawmills, it arose in only 10 years to the distinction of being one of the greatest lumber manufacturing centers in the Old Northwest.[8]

It would be hard to find a location more clearly endowed by nature to take advantage of forest wealth than Oshkosh. Its site occupies a plateaulike area of nearly eight square miles along the Fox River between Lake Winnebago and Lake Butte des Morts. Its situation at the mouth of the Fox below its junction with the Wolf gave Oshkosh exclusive control over the Wolf and its numerous tributaries, nearly all of which were bordered with pine forests of immense value. The Wolf River pineries were among the finest in the state and the river became famous for the pine it produced. Some townships tributary to the Wolf yielded 125 million feet of pine lumber, and some sections of 640 acres produced 10 million to 12 million feet of pine. Individual trees reached 5–7 feet in diameter and scaled from over 6,000 to over 7,000 feet of lumber.

Lumbermen considered the Wolf River one of the best driving streams in the Great Lakes states. It had relatively few rapids, did not experience the damaging freshets characteristic of the Wisconsin

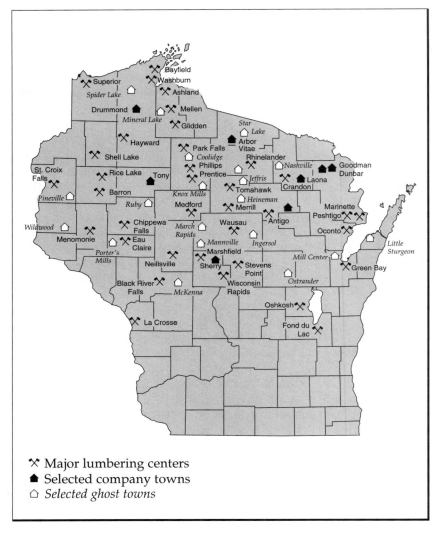

Figure 11.3. Wisconsin lumber towns

River and other streams, and its numerous tributaries generally made the haul a short one. Not until logging reached the upper Wolf and its tributaries did the lumbermen have to undertake substantial improvements, such as dams, to enable log driving. Boom Bay, a large shallow, protected bay in Lake Poygan at the mouth of the Wolf, afforded unequaled boomage capacity. Lake Butte des Morts and the broad, sluggish Fox provided good sites for mill ponds for the sawmills.

In addition to its advantageous position vis-à-vis the Wolf River pineries, Oshkosh was in a position to command excellent transportation links to potential markets. Since the settlement of Wisconsin proceeded from the south and east, Oshkosh was favored with access to the most developed part of the state. Between 1850 and 1875 sailing schooners carried lumber to settlements on Lake Winnebago, in addition to supplying lumber to towns from Fremont on the Wolf to Green Bay on the lower Fox. Occasionally Oshkosh even shipped lumber by raft or scow up the Fox, through the Portage locks, and down the Wisconsin and Mississippi rivers. It was able to supply southern Wisconsin and northern Illinois via the Rock River. Oshkosh's location in the main valley of eastern Wisconsin also assured it an early rail connection. When the Chicago and Northwestern reached Oshkosh in 1859, the city became the first lumber center in Wisconsin with rail transportation.

The Environmental Impact of Wisconsin's Lumber Towns

While the characteristics of its site often greatly influenced the origin and development of a lumber town, its sawmills and other lumber manufacturing plants, in turn, altered the site. Especially during the early years, the mills produced a prodigious amount of waste in the form of sawdust, edgings, slabs, and so on. The circular saws used in most mills until the 1890s cut a wide kerf and nearly buried the mills in their own waste. Lumber companies commonly disposed of their mill refuse by dumping it in nearby streams, filling in lowlands, and using it to pave streets and roads. Marinette continued to pave its side streets with sawdust as late as 1902. At Peshtigo Harbor, the Peshtigo Lumber Company added nearly 10 acres of ground on the west side of the river with slabs and sawdust, and a "desert waste of sawdust" surrounded some of the houses at the harbor. In some lumber towns this sawdust caught fire. After a fire in 1904 at Kimball, the sawdust, which paved the ground between the mills and store, burned for weeks. The thick smoke sometimes caused residents to take to the woods for relief. The old slab yard at Morse burned on and off all during the summer of 1910.[9]

Some of the waste ended up in the rivers on which the mills stood. The waste from the lumber mills created islands in some streams. On the Chippewa River, for instance, a member of a government survey team in the early 1870s reported that "below Eau Claire many of these bars are made up of a very large proportion of slabs, edgings, and saw-dust, the refuse of the saw mills in operation on the river banks."[10] A correspondent of the *Green Bay Advocate,* going up the Oconto River to Oconto by steamboat in 1865, reported: "Each voyage of the plucky little tug which runs from Oconto to Green Bay daily, Providence per-mitting, possesses all the novelty of an experiment. She has to plow a new channel for herself at each trip. After repeatedly "feeling" the accumulated sawdust, a weak spot is at length found and two or three vigorous efforts suffice to carry her through."[11] In streams like the Oconto, Peshtigo, and Menominee, the sawdust and other mill debris accumulated at the river mouths, and the harbors at Oconto, Menomi-nee, and Peshtigo Harbor required periodic dredging to keep them navigable.[12] The Peshtigo Company dumped so much sawdust into the Peshtigo River that it nearly spoiled the fishing between Peshtigo and Peshtigo Harbor. The mill waste also adversely affected fish in Green Bay. The *Green Bay Advocate* claimed in 1877 that the decrease in the fish catch over the last 15 or 20 years had one primary cause—"the fact that the number of our streams emptying into the bay, up which the fish used to run to deposit their spawn, are now occupied by mills, with dams thrown across obstructing the water, and the water itself is so filled with sawdust, bark, etc. that the fish will not run in it."[13] In 1880 the U.S. Fisheries agent at Green Bay reported that sawdust from lumber mills on the Menominee River was clogging fishermen's nets and burying whitefish spawning grounds under as much as eight feet of mill debris. That same year the Wisconsin legislature passed its first law prohibiting the dumping of mill refuse into streams.[14]

Oshkosh has been called the Sawdust City because of the many acres of lowland and marsh along the river that were filled in and re-claimed with sawdust. The lumber mills of Oshkosh were necessarily located on the banks of the Fox River. Stretching for nearly three miles along either side of the Fox to its mouth at Lake Winnebago, there once stood an almost unbroken line of sawmills; lath, shingle, and planing mills; sash, door, and blind factories; and wholesale lumberyards. The river determined the mill line; the mills, in turn, made the banks of the river. The original banks were low and marshy. Some mills had to be built on piles driven into the riverbed. The waste and debris of the mills were then used as fill. Acres were added along the river, narrow-ing it and making its shoreline a regular definite line, especially on the north side. Entire blocks of the city now stand on a foundation of slabs and sawdust.[15]

While sawmill towns never approached the noise and air pollution of the mining towns of the American West, their residents and visitors alike noted the hum of the mills and the smell of fresh-cut pine and wood smoke. Often they regarded such sounds and smells as signs of prosperity. In 1868, for example, the editor of the *Oshkosh City Times* wrote

Standing upon the higher ground of the southern bank, we looked back upon the long line of mills that line the banks of the Fox. The long line of pipes, from which ascend columns of smoke, and about which are clouds of steam—the many vessels towing rafts, and receiving their loading, and the hum of machinery, all conspire to impress the beholder with the industry and wealth that is so manifest in the growth and prosperity of our young city. If any would see the mine from which flow the riches of Oshkosh, let them pass through the "Valley of Saw Dust."[16]

Population

Lumber towns often showed a rapid growth in population. During the decade 1860–1870, the combined population of the lumbering and sawmill centers of Marinette, Oconto, Peshtigo, Wausau, Stevens Point, Grand Rapids (Wisconsin Rapids), Black River Falls, Eau Claire, Chippewa Falls, La Crosse, Green Bay, Oshkosh, and Fond du Lac increased four times as fast as the state as a whole.[17] The population composition of Wisconsin's lumbering towns, however, generally resembled that of the state itself. Most towns had relatively high proportions of foreign born, especially after the 1860s. In detail, they mirrored the regional variations found in the general population. By 1870, for example, places like Marinette and Eau Claire were strongly Scandinavian, and Oshkosh was heavily German, but seldom was any one group completely dominant.[18]

The major lumbering centers of Wisconsin did not display the ethnic segregation characteristic of some similar-sized manufacturing towns in southeastern Wisconsin. That does not mean, however, that some ethnic groups did not dominate certain parts of a lumber town. Oshkosh offers one example. A Volga German community began forming on the West Side of Oshkosh in 1899. The community grew rapidly throughout the early years of this century, continued to practice Volga German customs, and remained cohesive through the end of World War II. The enclave began when the Paine Lumber Company hired four Volga Germans, housed them in apartments on Pearl Street, and encouraged them to invite more Germans from the Volga to come to Oshkosh. Hundreds did and most worked for Paine Lumber and settled on the West Side. A sample of a 1910 directory listed 81 percent of the Volga Germans as employees of the Paine Lumber Company. Between 1899 and 1914, the West Side developed very distinct boundaries and "became like the old homeland, Jagodnaja."[19] In some sawmill towns, street names indicate the early ethnic composition of an area. Eau Claire, for example, had streets named Erin, Germania, and Norsk (now Franklin).[20] On the other hand, one finds surprisingly few examples of ethnic segregation in Wisconsin's lumber "company towns," which usually had quite homogeneous populations with one or two ethnic groups dominating. For example, in Porter Mills (Porterville), a lumber company town near Eau Claire, the foreign born accounted for 56 percent of the population in 1895, and most were Scandinavian (95 percent).[21]

Whatever the ethnicity of a lumber town, its population tended to be young, single, and male. In 1860 males accounted for 79 percent of the adult population of Marinette; most of them were single. In 1870 males accounted for 57 percent of the population in Oconto and for 62 percent in Chippewa Falls. By comparison males accounted for 52 percent of the population of the state as a whole. Even in an established mill village like Centralia (now part of Wisconsin Rapids), two-fifths of the adult male population in 1880 was single, and two-thirds of them lived in mill boarding houses. Hayward, which was almost completely dominated by lumbering, had twice as many males as females in 1885. In time, of course, the gender ratio in the larger, more diversified lumber towns became more balanced. As early as 1870, males accounted for only 51 percent of Oshkosh's population; by 1885, females outnumbered males.[22]

The Morphology of Lumber Towns

In mill towns most of the town proper developed around the sawmills and lumber manufacturers, so it is not surprising that street patterns reflected that fact. The uncoordinated, if convenient, street pattern of Oshkosh, for example, reflected the lumber interests of the city (Figure 11.4). For easy access to mills, long streets were laid in a northwest-southeast direction parallel to the river. Notice the angles in Pearl, High, and Algoma streets, as well as Hancock Street west of Pearl and Church Street east of Algoma; all bend to the sweep of the Fox. Between the factories on High Street, lanes led through the mill properties to the river. Only the streets leading to river bridges interrupted High Street's long extension. The location of the workers' entrance to the factories may well have determined the intersecting streets from the northeast, for the lengths of the blocks between Algoma and High vary. There was a tendency for the workers to

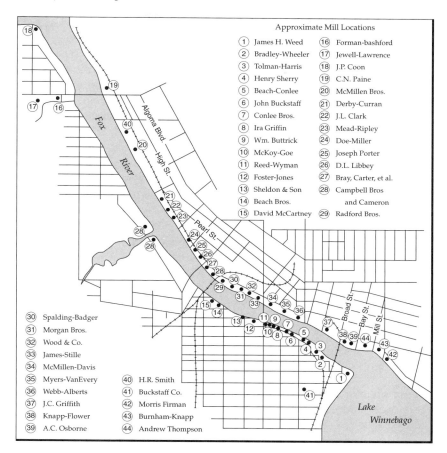

Figure 11.4. Oshkosh sawmills and street patterns

live as near as possible to their work, which often kept them away from home for 11–12 hours a day. In the intervening spaces between factories, and on the northeast side of High Street, were boarding and lodging homes, workers' homes, and small retail establishments. Algoma Street is about 15 feet higher than the river, and here the mill owners built their homes in close proximity to their industrial properties. For many years this street represented the best residential line of the town.[23]

Foust, Vogeler, and de Souza have described Wisconsin's lumber towns as "highly segregated in class and space," primarily featuring "low density housing with a sprinkling of substantial lumber baron homes, or a few high density areas of rooming houses where the workers lived, and a section for saloons and stores in the downtown."[24] While in general outline this simple description may hold true, in detail the lumber towns showed great variety depending on their size, age, specific function, terrain, and time period. Typically, the larger mill centers developed from two or three nuclei. Oshkosh developed from Oshkosh and Algoma; Marinette, from Menekaunee and Marinette; and Eau Claire, from Shawtown, Eau Claire, Eau Claire City (also known as West Eau Claire or Randalltown), and North Eau Claire. Each of the nuclei typically contained saw and shingle mills surrounded by a blacksmith shop, stable, lumberyard, boardinghouse, and perhaps a store. Small frame structures nearby housed the married workers and their families. A description of early Eau Claire noted that it consisted of four struggling villages strung out along the banks of the Eau Claire and Chippewa rivers, "not one could boast more than two streets and the average distance between contiguous houses was some 50 yards."[25]

As in most industrial towns, distinct sections devoted to trade, industry, business, and residences developed over time. Later, the lumber mills attracted related industries that manufactured boilers, sawmill equipment, engines, logging tools, and iron and steel products needed in logging and lumbering. Companies like the Phoenix Manufacturing Company (Eau Claire), the Sanford Tool Company (Oshkosh), and the Prescott Iron Works (Marinette) became part of a lumber town's industrial complex. A commercial district consisting mostly of stores, saloons, and restaurants developed. Usually part of it contained a vice or red light district. Several distinct residential areas typically developed, differentiated on the basis of economic status. Usually within walking distance of their mills and in the most desirable locations, the lumbermen built their mansions in styles popular in the late nineteenth century on very large lots, often surrounded by an iron fence. In Oshkosh it was Algoma Street; in Eau Claire, around Randall Park; and in Marinette, on Riverside and State streets. On the same streets or nearby, the major merchants, bankers, and others of wealth built their homes. Next came the middle class of the lumber towns, the sawmill managers and foremen, the lawyers and doctors, and so on. These neighboring residences were commonly simple, yet comfortable, expressions of the same styles as the lumber barons' mansions. The mill workers lived close to their work, sometimes in company-owned housing. Most workers' homes were of frame, vernacular one-story or one-story-with-attic structures. Churches typically dominated the skyline of the residential area.[26]

Over time, conditions steadily improved. By the 1870s the major lumber towns bore little resemblance to their appearance in the 1850s. Plank and wooden-block pavements replaced dirt and sawdust streets. The business districts now contained unbroken rows of two- and three-story commercial buildings, often of brick in place of the original log and wooden buildings. Gas now lighted the main streets.[27] A correspondent of the *Milwaukee News* who first visited Chippewa Falls in 1860 wrote the following in 1873:

Just thirteen years ago your correspondent first visited this new flourishing little city, after a long and tedious journey through the sand barrens, by stage coach from Wabasha. Chippewa Falls, was at the time one of the extreme frontier settlements of the State, and was known, except in name, to but few persons other than the hardy lumbermen, explorers, and trappers of the northwest; but thirteen years have wrought a marvelous change in the place. Where there was a small village of a few hundred inhabitants, with the pine stumps still standing in the streets, and a dense forest in the immediate background, now stands a compact, well built town of more than 4,000 inhabitants, with handsome three and four story brick business blocks, good school houses and elegant churches, a $112,000 hotel, (the Tremont) which in size and style nearly equals the Plankinton and Newhall, of Milwaukee; the largest saw mill in America, one steam fire engine, a city government in full blast in all its departments, a $60,000 Court House now in course of erection, many new buildings going up, and many other evidences of a driving, thriving, thrifty little city.[28]

It took some lumber towns even less time to achieve a "metropolitan appearance." The North Wisconsin Lumber Company built a sawmill and laid out the town of Hayward in the summer of 1883. Just a year and a half later a correspondent of the *Chippewa County Independent* wrote:

A recent visit to this wonderful burg convinced us that the growth and development of that village has been the most wonderful in the state during the past year. Eighteen months ago there was nothing but the rippling river and lofty pines, now a miniature city has sprung up, and the forest has given way to a hundred houses and a score or more business homes. Two large fine hotels, for one of which, the Clark House, we can say from experience, there is no better, is as good, on the line from Hudson to Ashland. A drug store owned and operated by Clapperton & Co., good looking dry goods and general stores, barber shops, billiard halls, and immense warehouses gives [*sic*] the place quite a metropolitan appearance.[29]

A dozen years later, Otto Christianson, the compiler of a plat book of Sawyer County, described Hayward as a town of 2,000 supplied with all the improvements and conveniences of a city of 10,000.[30]

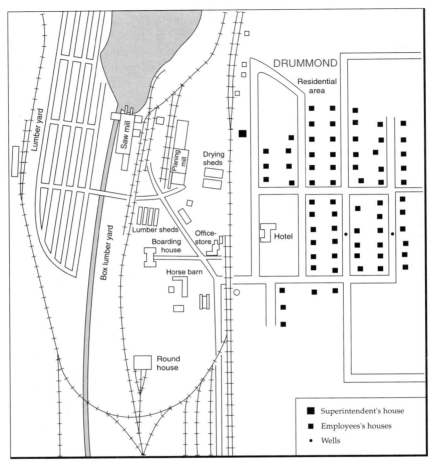

Figure 11.5. Map of Drummond, ca. 1900. This company town of the Rust-Owen Lumber Company displays a layout typical of many lumber company towns.

The Morphology of Company Towns

Not surprisingly, the lumber company towns show the strongest influence of function on form. Since a single lumber company owned the entire site of such a town, it invariably planned and laid out the streets, usually in a grid (Figure 11.5). Almost every task performed in a lumber company town played some part in supplying the mill with logs and converting those logs to lumber. It's not surprising that the sawmill or mills, associated attendant buildings, and the adjacent piles

Figure 11.6. McKenna, ca. 1892. Built by McMillan, Salsich and Williams in 1889, McKenna grew to a population of 200. When the timber played out six years later the company moved the mill to Star Lake, and McKenna became a ghost town. (Courtesy of the State Historical Society of Wisconsin, neg. number WHi[V24]1733)

of logs and stacks of lumber occupied the greater part of the site of any company town (Figure 11.6). In most period photos, the mill complex dominates the entire view, often appearing to lord over the town. Generally it included the sawmill, planing mill, shingle mill, dry kiln, refuse burner, mill pond, water tank(s), machine shop, lumberyard, horse barn, and perhaps a roundhouse. The mill or mills were usually located directly on a riverbank or lakeshore, while the rest of the town occupied a higher area nearby. The lumberyard was usually located between the mill and the riverbank or lakeshore.[31]

The company towns had quite limited business interests. The residents of some towns, in fact, had many of their needs met by traveling salesmen, who came regularly to sell everything from suits, photographs, watches, and jewelry to sewing machines. Virtually every town, however, had its company store. The company store typically was located near the town's center and served as a gathering place for exchanging and disseminating news and gossip and just passing the time of day. Often it contained the post office and company office and sometimes an express and telegraph office. Occasionally the business district included a boot and shoe shop, a meat market, drugstore,

Figure 11.7. When the Goodman Lumber Company established Goodman in 1908, it literally had to hack a townsite out of the virgin forest. The houses that line the rough street are typical of those erected in many company towns. (Author's collection)

land office, blacksmith shop, barbershop, or bank. A few even had an opera house, billiard hall, movie house, community hall or clubhouse, or hospital.[32]

Company towns invariably contained a church and school. Many lumber companies saw their establishment as yet another means to attract families and thereby obtain a stable and permanent work force. While a company hoped to attract as many married men as possible, in reality a good percentage of its work force was single. So, besides individual residences, the lumber companies usually built a hotel, boardinghouse, or cabins for single men. The boardinghouse was usually a quite substantial structure—two to three stories high, often with one or two wings, and a porch along its front and sometimes around three sides. It typically dominated a town's business district. The residential area often consisted of row after row of identical houses, sometimes all even painted the same color (Figure 11.7). While architectural style often varied from town to town, the dwellings of a given town generally conformed to a single style.[33]

Typically, the most imposing dwellings of a company town were the homes of the mill owner or superintendent, which characteristically occupied a prominent location in the town. Company towns typically contained no sections that were markedly older, all parts displaying an obviously contemporaneous character. The nature and function of the lumber company town likewise accounted for the lack of certain structures. Most lumber companies allowed no saloons or liquor within the limits of their towns, which resulted in fewer disciplinary problems and a more productive labor force.[34]

Lumbering's Urban Legacy

With the end of lumbering, towns dependent on it faced economic disaster. The company towns felt it the hardest. The closing of the mill(s) meant a precipitous decline in population and the decline, if not the entire disappearance, of the town. Witness the example of Spider Lake. In 1898, Philadelphia and Tonawanda lumbermen formed the Spider Lake Saw Mill & Lumber Company and erected a sawmill about 19 miles west of Ashland. Within a year, travelers on the Northern Pacific Railroad between Ashland and Duluth noticed quite a little village along the shore of Spider Lake, "including homes of the employees, boarding house, and two large and comfortable residences in which reside the members of the company and their families." The town soon had a population of 250. In less than four years, however, the company had cut its timber, and in October 1902 a clambake followed by a formal celebration marked the passing of the town. The town was torn down and most of the people moved to Lovejoy. Spider Lake was not unique. In Wisconsin, lumbering left in its wake upwards of 75 ghost towns. Even some of the larger, more prosperous and diversified mill towns struggled with varying degrees of success to withstand the decline and then the end of lumbering. Some, like Marinette, which lacked the necessary transportation facilities and agricultural land, continue to struggle. Marinette's population totaled 11,523 in 1890. A hundred years later it stood at 11,843. Even Oshkosh, although still a thriving manufacturing city, never regained its rank as Wisconsin's second city. Regardless of whether a lumber town survived or not, it left an imprint. In the case of some lumber company towns, it might be only a place-name or two. In other instances, towns and cities still show the influences of the lumber era through their location, street patterns, population characteristics, architectural styles, and extant lumber company structures.[35] In so many ways then, lumbering influenced the pattern and characteristics of urban places in northern Wisconsin.

Notes

1. C. F. Watson, "The Lumbering and Wood-Working Industries," *Journal of Geography* 12 (1914): 235.

2. R. Rohe, "Settlement Patterns of Logging Camps in the Great Lakes Region," *Journal of Cultural Geography* 6 (Fall-Winter 1985): 79, 93.

3. *Green Bay Advocate*, March 9, 1871.

4. *Jefferson Banner*, July 10, 1879.

5. *Northwestern Lumberman*, January 13, 1877.

6. The following discussion of changes in the spatial pattern of sawmills is based on data drawn from the pages of *The Wisconsin Lumberman, The American Lumberman*, and the *Wisconsin State Gazetter and Business Directory*. See also the maps in R. Rohe, "The Landscape and the Era of Lumbering in Northeastern Wisconsin," *Geographical Bulletin* 4 (April 1972): 17–18; and M. Williams, *The Americans and Their Forest: A Historical Geography* (New York, 1989), 215.

7. R. F. Fries, *Empire in Pine* (Sister Bay, Wis., 1989), 41.

8. R. H. Whitbeck, "The Geography of the Fox-Winnebago Valley," *Wisconsin Geological and Natural History Survey*, Bulletin 42 (1915): 46–48; R. Rohe, "Oshkosh: The Rise and Decline of a Lumber Center," *Proceedings of the 13th Annual Meeting of the Forest History Association of Wisconsin* (Wisconsin Rapids, 1988), 31–41.

9. *Marinette and Peshtigo Eagle*, June 17, 1882; C. Krog, "Marinette: A Lumber Camp Becomes a City, 1880–1910," *The Old Northwest* 6 (Spring 1980): 31; *The Eagle*, May 27, 1880, June 18, 1881, August 22, 1881, June 17, 1882; *Door County Advocate*, May 26, 1881, *Milwaukee Sentinel*, May 9, 1881; *Oconto County Reporter*, March 16, 1878.

10. *The Eagle*, August 1, 1885; U.S. Congress, *Reports of the Examinations of the Saint Croix and Chippewa Rivers*, 43d Cong., 2d sess., 1875, H. Doc. 75, part 6, serial 1645, 6.

11. *Green Bay Advocate*, August 10, 1865.

12. *Northwestern Lumberman*, August 28, 1880; *Marinette and Peshtigo Eagle*, May 22, 1880.

13. *Green Bay Advocate*, January 11, 1877.

14. R. Rohe, "The Upper Great Lakes Lumber Era," *Inland Seas, Journal of the Great Lakes Historical Society* 40 (Spring 1984): 23.

15. R. Rohe, "Lumbering's Impact on the Landscape of the Wolf River Area of Northeastern Wisconsin," M.A. thesis, University of Colorado, 1971, 175; *Oshkosh City Times*, October 13, 1868, August 31, 1869.

16. *Oshkosh City Times*, October 13, 1868.

17. D. A. Peterson, "Lumbering on the Chippewa: the Eau Claire Area 1845–1885," Ph.D. diss., University of Minnesota, 1970, 94; *La Crosse Independent Republican*, February 25, 1857; L. Barland, *Sawdust City: A History of Eau Claire, Wisconsin* (Stevens Point, Wis., 1960), 16, 21; R. N. Current, *The History of Wisconsin*, vol. 2: *The Civil War Era, 1848–1873* (Madison, 1976), 77, 427.

18. I. Voleger, *Wisconsin, a Geography* (Boulder, 1986), 65; Current, *The His-*

tory of Wisconsin, 445–446, 420, 422–423; C. Krog; "Marinette: The Origin and Growth of a Community, 1850–1870," *The Old Northwest* 3 (December 1977): 396–397, 402–403; *United States Manuscript Census of 1870, Winnebago County;* J. R. Hollingsworth and E. J. Hollingsworth, *Dimensions in Urban History* (Madison, 1979), 67; *Tabular Statement of the Census Enumeration and the Agricultural, Mineral, and Manufacturing Interests of the State of Wisconsin* (Madison, 1895), 72, 87, 107–108, 110–111.

19. P. A. Lautenschlager, "The West Siders: The Development and Disintegration of the Volga German Community in Oshkosh, Wisconsin," senior thesis, Lake Forest College, 1977, 4, 52–53, 55–58, 62.

20. M. Taylor, *Final Report: Intensive Historic/Architectural Survey of the City of Eau Claire, Wisconsin* (1983), 41.

21. R. Rohe, "Myths & Realities: Life in Wisconsin's Boom & Bust Lumber Towns," in *Wisconsin and Its Region, Proceedings of the Annual Institute of Wisconsin Studies Conference,* part 3 (n.p., 1988), 289; N. Olson, *Time in Many Places* (Saint Cloud, Minn., 1980), 81; *Tabular Statements of the Census Enumeration . . . of the State of Wisconsin* (1895), 73.

22. Krog, "Marinette: The Origin and Growth of a Community," 396; *United States Manuscript Census of 1870, Winnebago County; The Legislative Manual of the State of Wisconsin* (Madison, 1871), 266, 270, 274; *The Blue Book of the State of Wisconsin* (Milwaukee, 1887), 341, 345; R. W. Finley, *Geography of Wisconsin* (Madison, 1965), 91; R. C. Nesbit, *The History of Wisconsin*, vol. 3: *Urbanization & Industrialization, 1873–1893* (Madison, 1985), 333.

23. Rohe, "Lumbering's Impact on the Landscape of the Wolf River Area," 178.

24. J. Brady Foust, A. R. de Souza, and I. Vogeler, "The Region," *Wisconsin Natural Resources,* Special Northwoods Edition 7 (July-August 1983): 5.

25. Krog, "Marinette: The Origin and Growth of a Community," 393–395, 397–399; Taylor, *Final Report Intensive Historic/Architectural Survey of the City of Eau Claire,* 6–7, 26, 33, 42; quotation is from *Eau Claire Free Press,* February 2, 1860, see also February 26, 1873, issue; *Green Bay Advocate,* November 1, 1877, June 26, 1879; *Mississippi Valley Lumberman,* March 2, 1877.

26. Taylor, *Final Report: Intensive Historic/Architectural Survey,* 27, 34, 36, 38, 41; *City of Merrill, Wisconsin Intensive Survey Report Architectural and Historical Survey Report* (La Crosse, Wis., 1992), 68–69, 71, 74; Dale A. Peterson, "Lumbering on the Chippewa, the Eau Claire Area, 1845–1885," Ph.D. diss., University of Minnesota, 1970, 460–461; Krog, "Marinette: A Lumber Camp Becomes a City," 23.

27. *Oshkosh Journal,* November 9, 1872; Barland, *Sawdust City: A History of Eau Claire, Wisconsin,* 60, 73–74; *Oshkosh City Times,* April 29, 1876; Rohe, "Oshkosh: The Rise and Decline of a Lumber Center," 35.

28. Quoted in the *Chippewa Herald,* July 12, 1873.

29. Quoted in the *Northern Wisconsin News,* February 9, 1884; see also its August 2, 1884, issue and the *Mississippi Valley Lumberman,* August 8, 1884.

30. Otto Christianson, *Plat Book of Sawyer County, Wisconsin* (n.p., 1897), no pagination.

31. Rohe, "Myths & Realities: Life in Wisconsin's Boom & Bust Lumber Towns," 288–289.

32. Rohe, "Myths & Realities: Life in Wisconsin's Boom & Bust Lumber Towns," 290, 295–296; R. Rohe, "Lumber Company Towns in Wisconsin," *The Old Northwest* 10 (Winter 1984–1985): 417.

33. Rohe, "Myths & Realities: Life in Wisconsin's Boom & Bust Lumber Towns," 291–292, 294, 296–297.

34. Rohe, "Myths & Realities: Life in Wisconsin's Boom & Bust Lumber Towns," 292–295.

35. Quotation is from *American Lumberman*, November 4, 1899; *Mississippi Valley Lumberman*, October 24, 1902; C. Krog, "Marinette: Biography of a Nineteenth Century Lumbering Town, 1850–1910," Ph.D. diss., University of Wisconsin, 1971, 307–308; *The Blue Book of the State of Wisconsin* (Milwaukee, 1891), 412; *State of Wisconsin, 1991–1992 Blue Book* (Madison, 1991), 755.

12

Homes on the Range
Settling the Penokee-Gogebic Iron Ore District of Northern Wisconsin and Michigan

Arnold R. Alanen

The Penokee-Gogebic Iron Range spans the East Branch of the Montreal River, which flows between northern Wisconsin and Michigan. The Wisconsin portion of this linear iron ore formation began to be termed the Penokee by the mid-nineteenth century, whereas the Michigan section was referred to as the Gogebic. For some years thereafter the entire range was called the Penokee-Gogebic, but it has been identified as the Gogebic since the late 1800s.[1]

Iron ore made the Penokee-Gogebic a distinctive natural resource region within Wisconsin and Michigan, but the social and cultural characteristics of the people who settled the range and of the communities that accommodated them also left an indelible imprint upon its landscape. While the ensuing discussion will provide an overview of the Gogebic's natural characteristics, early exploration, and production history, the population groups and settlements of the district will receive the greatest emphasis. The Wisconsin portion of the range will serve as the primary focus for the discussion, but the physical and cultural unity displayed by the entire Gogebic demands that consideration also be given to the Michigan section. The discussion will be limited to the time period of the early 1800s to the late 1920s—the years that mark the establishment and flourishing of mining and settlement activities on the Gogebic.

241

Iron Ore Mining in Wisconsin and the Lake Superior Region

Iron ore was known by European Americans to exist in northern Wisconsin as early as the 1840s, but their actual mining activities did not commence on the Penokee until the early 1880s; this was well after iron ore had been extracted from two small pockets in south-central Wisconsin. The state's first iron ore mine, situated in the Mayville district of eastern Dodge County, supplied a local market from 1849 to 1928, and another mining endeavor took place at Ironton in Sauk County between 1850 and 1880. Iron ore mining of a very minimal nature was also initiated in the Baraboo Hills of Sauk County (Freedom Township) from 1889 to 1899, followed by the extraction of some higher-grade ores during the 1904–1925 interim. (Much later, from 1969 to 1982, a low-grade iron ore called taconite, which requires considerable beneficiation before it can be used in steel production, was extracted from a small mine in the vicinity of Black River Falls.) It was the Penokee-Gogebic, however, which proved to be the state's most productive iron ore district, and which made Wisconsin the fourth highest iron ore–producing state in the country during the early twentieth century.[2]

The Gogebic Iron Range is one of seven mineral-production districts that compose the Lake Superior mining region of northern Michigan, Wisconsin, and Minnesota (Figure 12.1). When ore shipments from the Gogebic commenced in 1884, some 40 years had already passed since northern Michigan first entered the region's mineral-production ranks. During the early 1840s, the first ore in the copper range, or "Copper Country," of the Keweenaw Peninsula was mined by European-American settlers, while some 75 miles to the southeast iron ore mining activities were also occurring on Michigan's Marquette Range. The Menominee Iron Range shipped its first ore from Michigan in 1877, followed three years later by the opening of a few mines at Florence and Commonwealth on the small portion of the Menominee that extends into Wisconsin.[3]

When the first trainload of ore departed the Gogebic Range for a Great Lakes port in 1884, the primary mining frontier for the Lake Superior region already had moved farther west to Minnesota. The initial shipment of ore from the Vermilion Iron Range of Minnesota also occurred in 1884, but it was the development of the Mesabi Iron Range after 1892 that quickly changed the mining complexion of the entire region. By 1903, Minnesota's Mesabi exceeded the combined output of all the other Lake Superior iron ore ranges, and its domination increased steadily thereafter. The last of the region's ore deposits to be

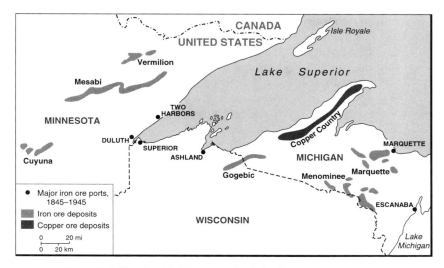

Figure 12.1. The Gogebic Range and the Lake Superior mining region

exploited was the small Cuyuna Iron Range of east-central Minnesota, which shipped its first ore in 1911.

Natural Features and Iron Ore Production

The iron-bearing strata of the Gogebic and Lake Superior region were deposited 1.9 billion to 1.8 billion years ago during the middle Precambrian age—the same era that contributed to the formation of the world's other major iron deposits in Australia, Canada, India, Russia, Venezuela, and elsewhere. During the middle Precambrian, the rocks of the Lake Superior region experienced massive folding, tilting, and faulting, actions which resulted in the emergence of highlands and even mountainous terrain. The Penokee Range, in its present form, has been described by geographer Lawrence Martin as similar to the Blue Ridge of Appalachia, only "not so high."[4]

During the Keweenawan period (about 1.1 billion years ago), the strata of the Gogebic started to tilt, with the angle sometimes exceeding 60 degrees. From a geological standpoint, all the tilted strata and surface features of the Gogebic are considered part of the "range." Most observers, however, define the Gogebic as the thin band of hills and ridges that extends from eastern Gogebic County in Michigan and then proceeds across Iron County, Wisconsin, to west-central Ashland County; thereafter, the rock formations are less visible, but they actually continue even farther to the southwest until reaching Namekagon Lake in Bayfield County. Overall, the Gogebic is some 80 miles in length, with 53 miles, or two-thirds, being in Wisconsin; nonetheless, less than one-fourth of all the ore (70 million of 318 million tons) produced by the entire range was derived from the Badger State.[5]

A few open pit mines were developed along the Gogebic, but the majority of ore came from underground operations such as the Montreal Mine in Wisconsin, which was recognized as one of the world's deepest shafts (4,335 vertical feet) when it closed in 1962. Production figures for the Gogebic peaked during both world wars (1914–1918 and 1941–1945) and the consumer-oriented 1920s, whereas marked downturns occurred during the 1930s depression era. The depletion of the Gogebic's higher-grade deposits, which was hastened by the huge iron ore demands of World War II, culminated in the closing of Wisconsin's Cary Mine in 1965, followed one year later by the shutdown of the Peterson Mine near Bessemer, Michigan. Iron ore reserves still remain throughout much of the Gogebic, but no mining has taken place since the mid-1960s because of high extraction costs and the poor quality of the remaining deposits.

Early Explorations and Speculation

In 1842, the Mississippi and Lake Superior Ojibwa ceded their lands in northern Wisconsin and Michigan, including the mineral wealth, to the United States government. Almost immediately, intensive explorations were undertaken by surveyors and geologists. The first mineral claim made in this area of Wisconsin was established in 1846 by the Montreal River Mining Company at a point where a copper deposit was exposed several miles north of the Penokee Range; nevertheless, little more than minor explorations occurred at the site, and all the operations were abandoned in favor of the much richer copper deposits of Michigan's Upper Peninsula. In 1847, geologist J. G. Norwood noted an outcrop of iron ore in the vicinity of present-day Upson, and two years later Charles W. Whittlesey made a thorough examination of the district's iron-bearing rocks. During the 1850s, the well-known scientist Increase Latham made visits to the Penokee; and Whittlesey returned in 1860, this time under the auspices of the Wisconsin Geological Survey. When making his initial 1849 reconnaissance, Whittlesey coined the term *Penokee* to identify northern Wisconsin's linear iron ore range. (Whittlesey determined that the Ojibwa word for iron was *pewabik*, but the term was misspelled as *penokee* on the maps that were subsequently prepared.)[6]

In 1856, the first serious attempt to exploit the iron ore deposits of the western Penokee was undertaken by the Wisconsin and Lake Superior Mining and Smelting Company of Milwaukee. The company platted a townsite (Ironton) along the Lake Superior shoreline at present-day Saxon Harbor, Wisconsin, and also planned settlements at Penokee Gap and along the gorge of the Tyler Forks River. Warehouses and docks were constructed at Ironton, and a route was surveyed for an envisioned railroad that would extend from the Penokee Range to Lake Superior. The enthusiasm of potential investors was quelled by the financial panic of 1857, however, and both the company and Ironton faded into oblivion within a few years.[7]

Another firm, the La Pointe Iron Company, was organized by four midwestern investors in 1859, and soon this firm was acquiring properties along the western Penokee. When seeking to raise funds in 1871, the company's promotional materials claimed that its holdings included "ore enough to span the world with railroads." Small amounts of mineral were extracted in 1873 and 1874, but transportation limitations, coupled with the poor quality of the deposits, resulted in the production of no merchantable ore.[8]

During the 1870s, geological surveys by Raphael Pumpelly confirmed that Michigan's Gogebic Range was not a separate iron formation, but a continuation of the Penokee. Following discoveries of higher-quality ores in both states, a Wisconsin group of investors organized the Northern Chief Iron Company in 1882 and sought to develop a mine site; almost simultaneously, another company was organized, which included New York financier Charles Colby among its backers. The firm achieved success in 1883 at the Colby Mine, but no ore was exported until October 1884, at which time the Milwaukee, Lake Shore, and Western Railway reached the site. The first shipment of ore (1,022 tons) left on flatcars for Milwaukee, where it was then transferred to a lake steamer that departed for Erie, Pennsylvania. Once the rail line was extended to Ashland in 1885, the ores of the Gogebic could be shipped to eastern iron and steel mills via a Lake Superior port. This rail link was constructed in a northeast-southwest direction along the lowland that parallels the hills of the Gogebic Range. After reaching Mellen the line joined with the Wisconsin Central Railroad, which had pierced the hills of the range in 1877 at a point where the Bad River flows through the Penokee Gap; the Wisconsin Central continued northward for about 40 miles until terminating at Ashland's recently constructed ore dock. In 1891, the Duluth, South Shore, and Atlantic Railroad completed another line that connected Hurley and Ashland over a more northerly route.[9]

The year 1884, Walter Havighurst has stated, "began a land rush unparalleled in the history of the north country." Several additional mines emerged in Michigan, as did the diminutive Germania Mine by Hurley. Many miners working in Wisconsin continued to apply their picks and shovels to the relatively exposed but low-grade iron ore formation at the far western end of the Penokee—the same area that had proved unproductive in the 1870s; soon, however, it was recognized that the richest ores were located farther to the northeast between Iron Belt and Hurley. In fact, more than 80 percent of all ore produced in Wisconsin came from just two mines, the Cary and the Montreal. Meanwhile in Michigan, some mining was attempted at the far eastern terminus of the Gogebic, but it was revealed that the state's most productive deposits were situated along the section of the range that stretches from Wakefield to Ironwood.[10]

By 1886, a report noted that the Gogebic Range was serving as the site for many "embryo mining schemes," and another account stated that the area was "literally knee deep in stocks." Claims were made at such a brisk pace that, during a portion of the year 1886, one or two new mining stock companies emerged each day. In Milwaukee, where the number of companies reached 50, three-fourths of the city's professional men and seven-eighths of its businessmen were reported to be involved in the Gogebic fever. One year later, investors could choose from close to 200 stock offerings. Observing this frenzy, the *Chicago Tribune* admonished prospective investors to "visit the range and pick out a mine, and see that there is really one as represented by the stock before purchasing"; other newspapers pointed out that, since only a few Wisconsin mines had been successful, people should avoid "wildcat schemes" and "fraudulent companies."[11]

As could be expected, the speculative bubble was short lived; by late 1887, scores of stock companies had declared bankruptcy. While most ventures failed to result in the extraction of even one shovelful of Gogebic ore, other companies were forced into insolvency because they lacked sufficient funds to hire skilled laborers or to purchase the kinds of equipment that deep underground mining required. By 1889, only 15 companies remained to pursue mining throughout the entire span of the Gogebic Range.[12]

In assessing the Gogebic's problems, economist Harold R. Mussey contrasted its frantic development to the orderly progression of mining activities on Minnesota's Vermilion Range. The latter district, he wrote in 1905, represented "an industrial undertaking, the Gogebic a financial; the Vermilion mines were worked to make profit by mining, the Gogebic by selling stocks; the Vermilion mines belonged to one corporation, the Gogebic to threescore."[13]

The Settlements

Because of the isolation and remoteness that defined the Gogebic Range during its infancy, the mining companies found it necessary to provide their employees with virtually everything needed for exis- tence, including housing and day-to-day provisions. While mine tim- bers and building construction materials were derived from the adja- cent coniferous forests, other supplies had to be hauled great distances over treacherous trails by horses, mules, oxen, and men. Many of the first miners camped in tent communities erected next to individual mine sites.[14]

Locations

Once the productivity of the ore deposits was assured and transporta- tion links had been established to the district, a number of more sub- stantial and permanent communities were established for mining em- ployees and their families. Unlike many other mining districts of the United States, the spectrum of settlements in the Lake Superior region included very few full-fledged company towns. Instead, the companies sponsored numerous residential enclaves, termed *locations*, through- out the region. The word itself was used initially in the United States, during the early nineteenth century, to define the areal limits of spe- cific mining claims. In northern Michigan, Wisconsin, and Minnesota, however, a location soon came to mean the entire mining complex, including its residences, streets, shafts, and related structures. The ac- tivities that accompanied early location development on the Gogebic have been described vividly by Havighurst: "Each location had a noisy camp," he wrote, "axes thundering, timber crashing down, powder blasting in the rock, men and mules scraping away the overburden. On the rivers sawmills were swirling through pine and hardwood, turn- ing out lumber for new railroad beds, new towns, new mine works."[15]

At least 65 locations punctuated the landscape of the Penokee- Gogebic (20 in Wisconsin and 45 in Michigan), and hundreds of others emerged throughout the Lake Superior mining region. Since the pri- mary purpose of these enclaves was to provide residential accommo- dations, separate locations generally were developed within walking distance of each productive mine site. Most locations—with names such as Annie, Germania, Hennepin, Kakagon, Minnewawa, Section 33, Shores, Trimble, Tyler Fork, and Windsor—existed on the place-name map of Wisconsin for a few years only. Other locations, like Montreal in Wisconsin and Anvil, Asteroid, and Puritan in Michigan, continue to reflect the mining legacy of the Gogebic Range.

Although the locations served strictly as residential communities, they did display some differences in spatial form and social organiza-tion. At one end of the spectrum were the *unplatted* or *squatters' loca-tions,* which sprang up when miners constructed hastily built shanties and houses on unsurveyed, company-owned land. According to one Gogebic Range observer, these locations displayed a "loosely joined series of very irregular housing areas" and streets that appeared "thor-oughly haphazard" as well as "rudimentary and unfinished." The most common settlement type, the *company location,* was laid out by mining engineers in a gridlike pattern. With their similar housing forms and uniform lot sizes and street setbacks, claims were made that the engi-neers had followed the same mathematical formula in laying out each unit. The third settlement type, the *model location,* was represented by only three examples on the Gogebic: Montreal in Wisconsin, and Aster-oid and Plymouth in Michigan. Designed by professional town plan-ners, these locations were intended to house mining supervisors and highly skilled workers; they included well-built housing, community centers, playgrounds, parks, schools, and "streets proportionate to use, varying setbacks, architectural massing . . . and sidewalk plantings."[16]

Location residents generally acquired their housing in one of two ways: either through a low-cost rental or by building a home on land leased from the company. Corporate executives anticipated that, by providing housing and some services in the locations, employ-ees would not search for work elsewhere and would be less likely to question corporate practices, join labor organizations, or participate in strikes. Occasionally a schoolhouse, clubhouse, or temperance hall was constructed in one of the locations, but labor and socialist halls, con-sumers' cooperatives, and similar communitarian organizations were not allowed. Nothing, however, was regulated more stringently than vice: saloons and prostitution were rigidly restricted to areas outside the location boundaries.[17]

Montreal, situated four miles west of Hurley, serves as the premier version of location planning in Wisconsin and as one of the foremost examples in the entire Lake Superior region. Prior to the early 1920s, Montreal was a typical company location that included about 70 frame houses built between 1907 and 1917, and 50 prefabricated bungalows dating to 1918. Company supervisors and key employees resided in the

larger and more elaborate dwellings, while miners and laborers inhabited the smaller and less ornate structures. The cows that grazed along
the streets and roadways of the location were kept out of the residents' yards and gardens by wire fences, while small barns, saunas (for
the Finns), sheds, and privies were arrayed at the back of each lot. In
1921, when Oglebay, Norton & Company made a decision to expand its
Montreal mining operations, corporate executives voiced concern that
the location would not attract the skilled, dependable laborers they
wished to hire. Therefore, Albert Taylor, a nationally known landscape
architect from Cleveland, Ohio, was commissioned by the company to
prepare an expansion and improvement plan for Montreal.[18]

Taylor began his assignment by undertaking an intensive reconnaissance of the range that focused upon the aesthetic shortcomings
of Montreal; he quickly noted the absence of any "softening effect"
typically provided by trees and shrub growth. To address "the general problem of the cow," Taylor called for the development of communal pastures and barns, and recommended that a cowherder and
fence commissioner be appointed. Because the increasing presence of
the automobile during the 1920s soon made it unnecessary to provide additional housing within walking distance of the mine site, Taylor's expansion plans for Montreal were shelved; however, his planting plan had a lasting impact. Proposing that community gardens and
"trees, vines, shrubs, and hardy flowering plants" be introduced, Taylor correctly reasoned that these features would "provide the least expensive and most effective way to modernize and beautify the entire
range." Over the ensuing years the company carried out many of Taylor's suggestions: a community landscape supervisor was appointed,
fences removed, sidewalks built, nurseries developed to provide trees
and shrubs for residents, and 100,000 seedlings planted throughout the
area. Soon, Montreal was known as one of the most attractive mining
communities in the entire region.[19]

In addition to Montreal's relatively attractive physical features, residents were provided with a clinic as well as a clubhouse (the Hamilton
Club) that included bowling lanes, a library, an auditorium for viewing movies, a barbershop, and recreation rooms. The maintenance of
yards and grounds and the painting and repair of houses (including
the yearly changing of screens and storm windows) were also conducted by the company. Despite the benefits they received, Montreal's
inhabitants resided in a highly paternalistic and regulated social environment: criticism of company policies could lead not only to the
termination of employment but even to the loss of one's home. An
early-twentieth-century Finnish-American labor organizer expressed

frustration because the mining companies operating in and around Montreal were able to "dictate whom the boardinghouse owner can keep as a boarder, [and] when meetings can be held. . . . The mines belong to the steel trust, the men are its, too—the whole town is in its care."[20]

Townsites

To accommodate the many residents who could not or chose not to settle in a location, and to provide the business services that the region demanded, numerous *townsites* were promoted and developed by various individuals (including mining company executives) who sought to capitalize upon the booming real estate market afforded by the Gogebic. Despite the speculative origins of the townsites, it was only in such communities that commercial enterprise occurred and some level of political independence was achieved.

At least 22 townsites were platted and marketed throughout the entire Penokee-Gogebic Range (12 in Wisconsin, and 10 in Michigan). With very few exceptions, all were platted during the 1884–1891 interim. A few ventures led to the evolution of relatively large townsites such as Hurley, Ironwood, Bessemer, and Wakefield; some, like Iron Belt and Pence, Wisconsin, never grew larger than village or hamlet size; several faded away only a few years after they were platted; and others resulted in little more than lines on a surveyor's map. In Michigan, the platted townsites were, from east to west, Tobin, Fink, Wakefield, Hillville, Ramsay, Bessemer, Jessieville, Monticello, North Park, and Ironwood. Thereafter the townsite array continued westward into Wisconsin via Hurley, Gile, Pence, Hoyt, Iron Belt, Upson, Iron Center, Finney/Plummer, Benjamin, Hoppenyan, Mellen, and Pratt. Except for the railroad center of Mellen, all the Wisconsin settlements west of Upson were abandoned or vacated when nearby ore deposits proved inadequate to support mining operations. In Michigan, Tobin, Fink, and Hillville experienced a similar fate, whereas Monticello and North Park were incorporated into the city limits of Ironwood.[21]

Platted in 1884, Hurley was the first townsite to emerge on the range, and it served as the initial economic and social center for the district. The federal population census of 1890, however, revealed that Hurley's permanent population of 2,655 was already far exceeded by the 7,750 people who lived in Ironwood. Nevertheless, "early Hurley" achieved both fame and infamy as a sin center for the Gogebic Range and much of northern Wisconsin and Michigan. Large numbers of miners regularly descended upon Hurley during their free-time hours, and lumberjacks arrived from the Pineries in the spring with "a winter's pay in their pockets and pent-up hungers in their blood." This transient population swelled Hurley's numbers well beyond those recorded by census enumerators. Hurley became the Gogebic's "good-time town, with gambling, girls, and piano music, with rows of saloons, variety shows, and amusement halls." No wonder a popular northern Wisconsin anecdote made the following claim: "The four toughest places in the world are Cumberland, Hayward, Hurley and Hell"—with Hurley being the toughest of the lot.[22]

No single, definitive factor explains why Hurley's reputation as the "hell-hole of the range" surpassed all other mining centers of the Lake Superior region, but a major reason undoubtedly was related to its location along a state boundary. Since laws governing the sale of alcoholic beverages in Wisconsin were not as strict as those in Michigan, Hurley's taverns attracted a larger clientele than most mining towns of similar size. Mining and local governmental officials in Michigan also sought to concentrate as much vice as possible, especially prostitution and gambling, on the western side of the Montreal River. "While other nearby towns traded in the necessities of life," one writer commented several years ago, "Hurley's chief commodity was sin."[23]

By 1893, Hurley's spectrum of commercial enterprises included 60 saloons, or 1 establishment per every 38 residents. Ironwood supported 65 outlets, but because of its significantly larger population size, the ratio was 1 per 120 people. By way of comparison, the highest figure displayed on the Mesabi Range of Minnesota occurred at Bovey, where 1 saloon per 68 inhabitants existed during the early twentieth century.[24]

The vast majority of Hurley's saloons were distributed along Silver Street, which extended westward from the Wisconsin-Michigan border for five blocks; a streetcar line ran down the middle of the street and provided regular service between Hurley and the Michigan mining towns (Figure 12.2). The saloons and variety theaters concentrated proximate to the state border were especially infamous for their bawdiness and ribaldry. A number of Silver Street's establishments accommodated three vertically separated spaces within their confines: a saloon occupied the ground floor, gambling was sequestered in the basement, and prostitution occurred on the second level. No other community throughout the entire Lake Superior mining region even came close to displaying such a significant concentration of watering holes along a single street. When the Prohibition Amendment took effect in 1920, many saloons were converted into so-called soft-drink

Figure 12.2. Silver Street, Hurley, Wisconsin, ca. the early 1900s (Courtesy of the Michigan Historical Collections, Bentley Historical Library, University of Michigan)

parlors; however, since they often served as "blind pigs," where illegal alcoholic beverages were sold, the establishments experienced regular visits by federal and state agents.[25]

Whatever may have been responsible for Hurley's reputation, outside observers were more than willing to criticize the community for its perceived excesses. As early as 1886, the *Ironwood Times* claimed that Hurley was the headquarters for thugs and swindlers (Ironwood's newspapers sometimes displayed their disdain for "hurley" by refusing to capitalize the first letter of its name). The countless, sensationalized accounts of Hurley made by Chicago and various Wisconsin newspapers were termed "outrageous slanders" by the town's defenders. Nevertheless, Hurley's newspapers also reported regularly on the presence of vice in their community, as in 1889 when 18 of Hurley's "fallen angels" were arrested and presented with the option of a $60 fine or 90 days in jail. In 1890, a police crackdown netted 26 "soiled doves," and a number of Silver Street saloons were raided later in the year on the assumption that they maintained houses of ill fame. Hurley may have been the foremost example of an entertainment center in the Lake Superior region, but it must be emphasized that many mining centers throughout the district and elsewhere displayed similar characteristics during their formative years of existence. In Hurley, however, these activities persisted for many decades, rather than several years, and were concentrated along one infamous street.[26]

Immigrant Imprints

As employment opportunities became increasingly available during the late nineteenth and early twentieth centuries, significant numbers of foreign-born immigrants made their way to the Penokee-Gogebic. From the early 1800s to the mid-1890s, "old immigrants" (northern and western Europeans, as well as Canadians) predominated among the foreign-born population; thereafter and continuing to 1914, when large-scale immigration to the United States ended, it was "new immigrants" (eastern and southern Europeans) who became increasingly evident throughout the range.

The 1900 federal population census, which provides the first detailed data for both the Wisconsin and Michigan sections of the Gogebic, reveals that roughly equivalent numbers of people had been born in the United States (11,450) and in foreign countries (10,500). Finns (2,900) constituted the largest immigrant group, followed by the Swedes (1,585), British (1,155), Canadians (875), Italians (730), "Austrians," or South Slavs (primarily Slovenes and Croatians, 615), Germans (555), Poles (550), Norwegians (405), Irish (310), and four other groups represented by 85–200 people. Just under half of the foreign-born were new immigrants in 1900.[27]

The number of immigrants on the Gogebic reached its census-year zenith in 1910 (14,780 people); this figure represented 47 percent of the entire population. Almost three-fourths of the entire foreign-born group displayed new-immigrant origins (Figure 12.3). The sizes of the three largest groups—Finns (4,440), Italians (2,650), and Poles (1,460) —demonstrated the especially noticeable gains made by new immigrants during the 1900–1910 interim, while the number of people in the fourth and fifth largest groups—the Swedes (1,370) and British (930)— reflected the declines experienced by all the Gogebic's old-immigrant groups.

Three of four immigrants resided in Michigan. As the largest city on the range, Ironwood served as home to the most significant concentrations of individual immigrant groups. The only exception was the Italians, who were primarily concentrated in Hurley and the small mining enclaves situated on the Wisconsin side of the range; the Italians played an especially important role in the commercial life of these communities.[28]

Throughout the early period of mining activity, the Gogebic clearly was the domain of male immigrants (Figure 12.4a). In Knight Township, Wisconsin, which included both mines and lumber camps in 1900, 7 of 10 immigrants were males; nearly 40 percent of the entire foreign-born

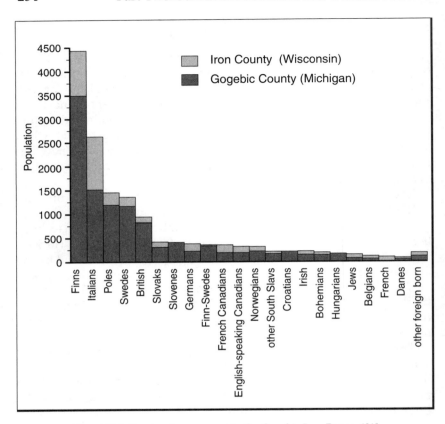

Figure 12.3. Foreign-born groups in the Gogebic Iron Range, 1910

population were young men who ranged from 20 to 34 years of age. To accommodate these men, numerous boardinghouses were operated throughout the range, and much of the saloon life that characterized early Hurley and other mining communities developed in response to these skewed gender and age compositions. It was not until the second and third decades of the century that greater balance in the social structure of the Gogebic was achieved.

To a great extent, the occupational hierarchy of the range was also organized along ethnic lines (Table 12.1). In Knight Township, half of all adult, immigrant males worked as miners in 1900, but only 5 percent of the American-born men did likewise. Furthermore, new immigrants were especially concentrated in these jobs: 81 percent of the Italians, 72 percent of the South Slavs, and 57 percent of the Finns

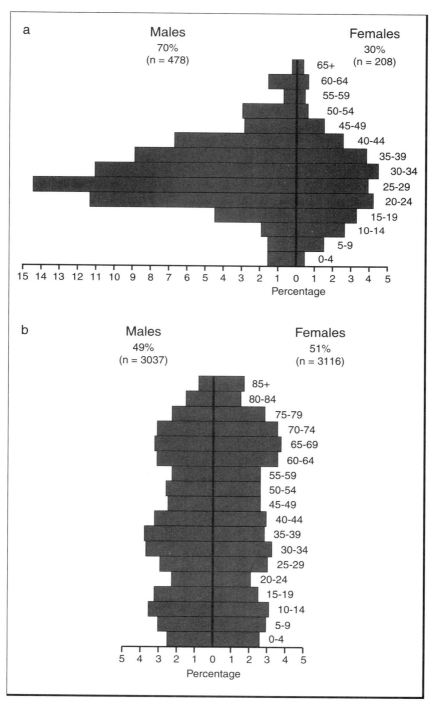

Figure 12.4 (*a*) Age and gender groups for the foreign-born populations at Knight Township, Iron County, 1900 (*b*) Age and gender groups in Iron County, 1990

Table 12.1. Occupations of adult males, by country of birth, Knight Township, Iron County, Wisconsin, 1900

	Occupation														Totals	
	Miners and trammers		Lumberjacks and teamsters		Unskilled laborers		Skilled workers		Proprietors (other than saloon keepers)		Saloon keepers		Supervisors and professionals			
Country of birth	n	%	n	%	n	%	n	%	n	%	n	%	n	%	n	%
"Austria" (Slovenia and Croatia)	31	72.1	–	–	6	13.9	3	7.0	–	–	1	2.3	2	4.7	43	100
Canada	1	2.4	13	30.9	15	35.7	10	23.8	–	–	1	2.4	2	4.8	42	100
Finland	83	56.9	28	19.2	23	15.8	6	4.1	1	0.6	5	3.4	–	–	146	100
Great Britain and Ireland	15	48.4	–	–	2	6.4	7	22.6	2	6.5	–	–	5	16.1	31	100
Italy	61	81.4	1	1.3	4	5.3	2	2.7	1	1.3	5	6.7	1	1.3	75	100
Norway	6	27.3	2	9.1	7	31.8	4	18.2	–	–	2	9.1	1	4.5	22	100
Sweden	12	30.8	2	5.1	9	23.1	7	17.9	1	2.6	3	7.7	5	12.8	39	100
All other foreign countries	16	36.3	1	2.3	12	27.3	5	11.4	4	9.1	4	9.1	2	4.5	44	100
United States	8	5.5	28	19.4	41	28.5	40	27.8	5	3.5	3	2.1	19	13.2	144	100
Knight Township totals	233	39.8	75	12.8	119	20.3	84	14.3	14	2.4	24	4.1	37	6.3	586	100

Source: Manuscript schedules for the 1900 federal census of population

worked as miners. The British, who had close to half of their men employed in mining, served as the only old-immigrant group highly represented in such work, but most of these individuals held highly skilled positions. (Of the British, it was men from Cornwall who played an especially essential role in underground mining.) Participation in the skilled and unskilled labor force was dominated by old immigrants and Americans; supervisory and professional positions were clearly the domain of British, American, and Swedish males. Immigrants owned or worked in stores and shops, and every major, non-English-speaking group was represented by its quota of saloon keepers.

Some women worked as domestics and clerks throughout the range, and a very small number were employed as teachers and professionals, but overall, few females pursued employment outside the home, especially if they were immigrants. Many women, however, provided room and board for the large number of single males who resided throughout the Gogebic. In Knight Township, 40 percent of the 190 households included boarders in 1900; the highest rates occurred among Italian (56 percent) and Finnish households (50 percent). An average household with boarders included 4.4 men, but several accommodated many more. Mary Kyro, a 27-year-old Finnish woman married to an Iron Belt saloon keeper, cared for her two young children as well as 16 Finnish boarders (assisting her were two female cooks, one a Finn, the other a Pole). Elsewhere in the township, Mary Bergman, a 64-year-old Swedish widow, supported herself by housing a total of seven boarders from Sweden, Norway, and Finland.

The work associated with the operation of a boardinghouse was acknowledged as especially demanding and time consuming. A typical day might begin at 4:00 A.M., when the women arose to prepare breakfasts and lunches for the men and to bake bread. The remainder of the day was spent in further food preparation and in pursuing the endless task of cleaning, not only the boardinghouse itself, but especially the bed linens and the miners' iron ore–caked clothing. Some women worked more than 100 hours per week in the boardinghouses; as late as 1917, the average wage for these women seldom exceeded 10 cents per hour.[29]

In mining regions, laborers constantly placed their lives in danger, even as they slept. In fact, the earliest disaster on the Gogebic Range did not occur in a mine, but took place in March 1887 when a Bessemer boardinghouse fire killed nine men. This was followed, in July, by a fire that consumed a large part of Hurley's Silver Street and killed 11 performers in the Alacazar Variety Theater.[30] Mining, of course, has long been noted as an extremely hazardous occupation. Unlike Michigan and Minnesota, where thousands of men died in mining accidents, no official agency was formed in Wisconsin to maintain an annual listing of injuries and deaths; nonetheless, a review of newspaper accounts offers poignant insights into the calamities that occurred.

On the Penokee, where at least 55 men were killed in mining accidents from 1888 through 1892 alone, and unknown numbers of others suffered serious injuries and illnesses, four grisly but typical accounts serve to summarize the situation. An Italian laborer, who was riding up the shaft of the Montreal Mine in an ore bucket, plunged to his death when the lift experienced clutch failure; another miner, survived by his wife and two children in Finland, died after making a misstep and falling 60 feet into a shaft; a Swede was killed after being buried alive by several tons of earth in the Cary Mine; and the body of an unidentified miner was described as being "literally blown to atoms" by a premature dynamite blast. Juries were rapidly impaneled after each accident, ostensibly to determine responsibility for the catastrophes, but in virtually every case the verdicts ruled that such deaths were accidental, with "no blame to attach to the mine management for carelessness or otherwise."[31]

Throughout the early decades of Gogebic Range history, the dangers and unpredictable nature of mining led many residents to seek other employment opportunities, including efforts to farm the cutover land of the Lake Superior region. "Let's flee this place!" stated the wife of a Finnish miner in Wisconsin, "[and go] where there are no mining companies and bosses, where no whistle's shriek awakens us, where we can rise when it pleases us. We are free people." Despite the difficulties they encountered in converting the refractory land into farms, thousands of people were more than willing to depart the Gogebic and other mining ranges for the agricultural districts of the region: life on a cutover farm might have been economically precarious, but it was safer and less regulated.[32]

Postscript

As the last underground mines of the Gogebic Range were abandoned during the 1960s and allowed to fill with water, local communities and people entered the throes of an often traumatic social and economic transition. Population numbers, which began to drop during the 1940s, continued to fall; by 1990, the combined population figure for Iron and Gogebic counties (24,205 people) approached the number displayed at the beginning of the century. The 1990 totals for the four largest cities of the Gogebic Range—Ironwood (6,850), Wakefield (2,320), Bessemer (2,270), and Hurley (1,780)—reveal especially significant decreases, although smaller central places such as Iron Belt, Montreal, Pence, and Upson have experienced even greater relative declines. Also, the overall age of Iron County's population has increased significantly over the century. By 1990, the median age of Iron County's population was 42.5 years—the second highest figure for any of Wisconsin's 72 counties, and a figure that exceeded the state median by 10 years (Figure 13.4b)[33]

Today, it is the defunct buildings, waste rock piles, small houses, and place-names of the Gogebic Range that reflect its mining legacy. The Plummer Headframe by Pence now serves as the only remaining example of a once common structural form that marked the site of virtually every underground mine on the Gogebic. Ethnic identities and traditions are still expressed in certain churches and organizations, in some of the foods prepared in local homes and restaurants, and in the surnames of many residents. Another example of a local event with an ethnic tie is the Paavo Nurmi Marathon—named for Finland's famed long-distance runner of the 1920s—which has been run annually from Hurley to Upson since 1968.

No longer on Hurley's Silver Street, however, may one find saloons that cater to individual nationality groups, even though 80 saloons could still be counted among its 115 business establishments as late as 1938. The night life of Hurley continued to thrive into the 1940s and 1950s, so much so that a group of Finnish journalists, when encountering Hurley for the first time in 1949, exclaimed that the little town "has something to open up the eyes of a visitor from the Old World— a long street lined with nothing but taverns." Nevertheless, by the late 1960s, a local publication noted that historic Silver Street was "in its waning days."[34]

In 1995, 15 taverns and bars still operated on Silver Street. The building facades that line both sides of the street exhibit only a few dim reminders of the bright lights and exotic activities that once distinguished Hurley, but some of the tavern interiors still retain their original decor and ambiance. After disavowing the more notorious aspects of Hurley's history for several decades, a growing number of residents now appear ready to recognize and capitalize upon its unique and colorful past, albeit in a relatively sanitized manner.

In nearby Montreal, the paternalistic hand of the mining company no longer extends over the community. When its mine closed in 1962, the Oglebay Norton Company began to dispose of Montreal's housing stock. Within relatively few years, a majority of properties had been purchased—most by nonresidents who converted them into seasonal homes, rental units, and bed-and-breakfast operations—primarily for use during the winter skiing months. This transition, from extractive industry to recreation and tourism, defines the evolving pattern of Gogebic Range activities that has occurred since the 1960s. In all likelihood it is such scenic and recreational amenities, when combined with the recognition now being given to the economic benefits of historic preservation and interpretation, which will become increasingly important on the Gogebic Range as it enters the twenty-first century.[35]

Acknowledgment

I would like to thank Dean Proctor for sharing his insights into the Gogebic Range with me through numerous discussions and experiences in the field.

Notes

1. H. R. Aldrich, *The Geology of the Gogebic Iron Range of Wisconsin,* Wisconsin Geological and Natural History Survey, Bulletin 71, (Madison, 1929), 20–22. During the nineteenth century, it was believed that iron ore might also be found along another ridge of hills located a few miles south of the Penokee. This ridge was termed the South, or Messembria, Range.

2. G. Schultz, *Wisconsin's Foundations: A Review of the State's Geology and Its Influence on Geography and Human Activity* (Dubuque, Iowa, 1986), 67 and 135; E. C. Holden, "Mineral Industry of Wisconsin," *Wisconsin Engineer* 17 (1913): 162–166.

3. All production figures and comparative data used in this chapter have been derived from the following sources: Lake Superior Iron Ore Association, *Lake Superior Iron Ores* (Cleveland, 1952); M. R. Alm, *Mining Directory Issue: Minnesota, 1965* (Minneapolis, 1965); unpublished data from the Wisconsin Geological and Natural History Survey; and A. R. Alanen, "Years of Change On the Iron Range," in Clifford E. Clark, Jr., ed., *Minnesota in a Century of Change: The State and Its People since 1900* (St. Paul, 1989), 157–159.

4. Schultz, *Wisconsin's Foundations,* 51–52; L. Martin, *The Physical Geography of Wisconsin* (Madison, 1965), 377. Martin's monograph was originally published in 1916 as Bulletin 36 of the Wisconsin Geological and Natural History Survey.

5. Schultz, *Wisconsin's Foundations,* 51–52.

6. Aldrich, "Geology of the Gogebic Iron Range," 20–22; J. R. St. John, *A True Description of the Lake Superior Country* (New York, 1846), 61; D. J. Mladenoff, "Vegetation Change in Relation to Land Use and Ownership on the Gogebic Iron Range, Wisconsin," unpublished M.S. thesis, University of Wisconsin–Madison, 1979, 76–77; R. N. Satz, *Chippewa Treaty Rights: The Reserved Rights of Wisconsin's Chippewa Indians in Historical Perspective* (Madison, 1991), 33–49; D. Connors, *Going for the Iron* (Friendship, Wis., 1994), 15–18.

7. Mladenoff, "Vegetation Change," 77–78; *History of Northern Wisconsin* (Chicago, 1881), 66; Connors, *Going for the Iron*, 19.

8. *La Pointe Iron Company of Ashland County, Lake Superior, Wisconsin* (Chicago, 1871), 7; *Ashland Press*, November 1, 1873, November 22, 1873, February 7, 1874; Mladenoff, "Vegetation Change," 78–79.

9. W. Havighurst, *Vein of Iron: The Pickands Mather Story* (Cleveland, 1958), 40; C. Techtmann, *Rooted in Resources* (Park Falls, Wis., 1993), 22; Aldrich, "Geology of the Gogebic Iron Range," 5; Schultz, *Wisconsin's Foundations*, 53; Mladenoff, "Vegetation Change," 84.

10. Havighurst, *Vein of Iron*, 40, 55; Writers' Program of the Work Projects Administration in the State of Michigan, *Michigan: A Guide to the Wolverine State* (New York, 1941), 548.

11. Havighurst, *Vein of Iron*, 56; J. A. Lake, *Law and Mineral Wealth: The Legal Profile of the Wisconsin Mining Industry* (Madison, 1962), 91–95; *Chicago Tribune*, September 15, 1886; *Montreal River Miner* (Hurley), February 3, 1887; *Portage Lake Mining Gazette* (Houghton, Mich.), March 10, 1887; *Miner and Manufacturer* (Milwaukee), June 4, 1887; *Bayfield County Press* (Bayfield, Wis.), November 26, 1887.

12. Havighurst, *Vein of Iron*, 56.

13. H. R. Mussey, *Combination in the Mining Industry: A Study of Concentration in Lake Superior Iron Ore Production* (New York, 1905), 93.

14. Havighurst, *Vein of Iron*, 39–40; V. Lemmer, "Reminiscences," in the Victor Lemmer Papers, Bentley Historical Library, University of Michigan, n.d.

15. A. R. Alanen, "The 'Locations': Company Communities on Minnesota's Iron Ranges," *Minnesota History* 48 (1982): 95–96; Havighurst, *Vein of Iron*, 55.

16. Alanen, "The 'Locations,'" 96–98; A. D. Taylor, *Improvements Report on the Montreal Mining Company Properties in the Ironwood District, Michigan* (Cleveland, 1921), 6–9, 16.

17. Alanen, "The 'Locations,'" 103–104.

18. Oglebay, Norton & Company, *Report on Housing, Water & Sewer Requirements for the Montreal Mining Company's Employees in the Village of Hamilton, Wisconsin* (Ironwood, Mich., 1923), i; A. R. Alanen, "The Planning of Company Communities in the Lake Superior Mining Region," *Journal of the American Planning Association* 45 (1979), 267–269; Taylor, *Improvements Report*, 10.

19. Oglebay, Norton & Company, *Report on Employee Housing at Montreal* (Ironwood, Mich., 1939), 1–2; Alanen, "The Planning of Company Communities," 269; Taylor, *Improvements Report*, 35, 41–50.

20. J. I. Kolehmainen and G. W. Hill, *Haven in the Woods: The Story of the Finns in Wisconsin* (Madison, 1951), 119–120.

21. U.S. Census Bureau, Department of Commerce, *Federal Population Census of 1890* (Washington, D.C., 1893). The information about townsite origins has been derived from plats filed in the Ashland, Iron, and Gogebic county courthouses.

22. Havighurst, *Vein of Iron*, 56–57 (quotation); Writers' Program of the Work Projects Administration in the State of Wisconsin, *Wisconsin: A Guide to the Badger State* (New York, 1941), 375 (quotation).

23. Iron County WPA Historical Project, "History of Iron County," unpublished manuscript in possession of the Iron County Historical Society (1937 and 1938); Writers' Program in the State of Michigan, *Michigan*, 550; L. C. Reimann, *Hurley—Still No Angel* (Ann Arbor, Mich., 1954), 3.

24. Saloon numbers for the Gogebic Range have been derived from A. G. Wright, *Wright's Directory of Ironwood Together with Hurley and Bessemer for 1893–94* (Milwaukee, 1893), whereas the population figures are from the U.S. Census Bureau, *Federal Population Census for 1890*. Other saloon counts for Silver Street come from Sanborn Insurance Maps (1888, 1891, 1898, 1905), on file in the Archives and Manuscripts Division, State Historical Society of Wisconsin, Madison; the information for Minnesota is found in Alanen, "Years of Change," 168.

25. Iron County Board of Supervisors, *Iron County 75th Jubilee Edition* (Hurley, 1968), no pagination; Techtmann, *Rooted in Resources*, 148–149; Writers' Program in the State of Wisconsin, *Wisconsin*, 375; interview with Larry Peterson, May 10, 1995, Hurley, Wisconsin (Peterson is the local historian of the Gogebic Range).

26. *Montreal River Miner*, December 2, 1886, January 6, 1887, November 21, 1889, May 22, 1890, October 9, 1890; Alanen, "Years of Change," 168–169.

27. All the information on immigrant numbers has been derived from the manuscript schedules for the federal population censuses of 1900 and 1910.

28. P. A. Sturgul, "Italians on the Gogebic Range," in R. J. Vecoli, ed., *Italian Immigrants in Rural and Small Town America* (Staten Island, N.Y., 1987), 172–173.

29. Alanen, "Years of Change," 176–177; P. Stofer, "Shared Beds, Shared Bread: Boardinghouse Life on the Michigan Iron Frontier," *Michigan History* 78 (1994), 43.

30. Stofer, "Shared Beds, Shared Bread," 45; *Ashland Press*, July 16, 1887.

31. *Montreal River Miner*, September 27, 1888, June 6, 1889, July 11, 1889, January 30, 1890.

32. Kolehmainen and Hill, *Haven in the Woods*, 36–37.

33. Bureau of the Census, U.S. Department of Commerce, *1990 Census of Population and Housing: Population and Housing Unit Counts* (Washington, D.C., 1993).

34. A. Tamminen et al., *Kamara kiertää Amerikan suomalaisten parissa: A Camera Tour among the Finns of America* (Helsinki, 1949), no pagination; Iron County Board of Supervisors, *Iron County 75th Jubilee Edition;* Workers of the Writers' Program in the State of Wisconsin, *Wisconsin*, 375.

35. D. Proctor, F. Sancar, and A. Alanen, *Preservation Planning on the Gogebic Iron Range of Wisconsin and Michigan: A Report on the Historic Resources Workshops*, September–November 1988, Hurley, Wisconsin (Madison, 1989).

13

Polish Routes to Americanization
House Form and Landscape on Milwaukee's Polish South Side

Judith T. Kenny

On "Raising the Flag Day" in the spring of 1918, the residents of Milwaukee demonstrated support for the American war effort with parades in neighborhoods around the city. Among those marching in front of their homes on old south Tenth Avenue[1] were members of the Mazur, Neuman, and Nisiewiecz families (Figure 13.1). The residents of this primarily Polish neighborhood turned out in numbers to express solidarity and community. Although it was a difficult time for families of German heritage even in a place once described as the most German city in America, Milwaukee's Poles could demonstrate both patriotic commitment to the United States and hope for the rebirth of Poland in their support for the war. All the symbols of a Fourth of July celebration were on display, but the flag given primary importance was the blue one that carried a star for each of the neighborhood's American "fighting boys."

This *American* image provides a starting point for an examination of the area traditionally called Milwaukee's Polish South Side. Looking past the marchers and their banners in Figure 13.1, you see a streetscape of wood frame cottages and store fronts that appears similar to other Milwaukee working-class neighborhoods of the period. This might be the prototypical landscape for a place that was once called the workingman's city and was widely noted for its large foreign-born

Figure 13.1. "Raising the Flag Day" on old South Tenth Avenue, spring 1918 (Photograph courtesy of Bernice Mazur)

population. On the basis of such an inspection, if we were to judge whether there are features of ethnic, immigrant culture surviving in the landscape of Milwaukee's major Polish neighborhood, the answer would be no. The Poles embraced new building patterns and outwardly conformed to American styles appropriate to their economic status.

Yet, when this picture was taken, there were—and to a large extent continue to be—assumptions of *difference* about the qualities of this neighborhood when compared with the city's other working-class areas. Only within the last 20 years has the stereotype of the South Side as a separate, culturally homogeneous world of the Polish blue-collar worker begun to break down.[2] Popular images associate the ethnically distinct population with distinctive features of the built environment. Milwaukeeans identify the Polish flat as a house form unique to Milwaukee's old Polish neighborhoods, suggesting that it is a remnant of traditional Polish material culture. Again, despite its name, the house shown in Figure 13.2 actually has little to do with vernacular build-

Figure 13.2. Polish flat, 1994 (Photograph by the author)

ing traditions brought by the Poles. What then is Polish about the Polish flat?

As suggested in its title, "Polish Routes to Americanization," this chapter examines the mechanisms by which strongly held ethnic values helped the Poles negotiate their transition to American life and build their community in Milwaukee. At the same time, however, the representation of the Polish South Side from the perspective of Milwaukee's dominant society has influenced local interpretations of the landscape and, as a consequence, suggests the need for study as well. The issues surrounding the representation of the Polish South Side are embodied in the term *Americanization*. Although today the term may be regarded as a neutral one, during the late nineteenth century and early years of the twentieth century, it had a specific, normative meaning related to what it meant to be American. Contemporary reformers used the term to express a range of opinions about the prospects for transforming the *new immigrants* from eastern and southern Europe into citizens. In this context, *Americanization* refers to the debates on methods for the assimilation of these immigrants into an American middle-class culture defined as having its roots in northern and western Europe. In the industrial cities of the Northeast and Midwest, links between the immigrants, the qualities of their neighborhoods, and American values were analyzed in a manner suggestive of environmental determinism. As stated in a 1906 Milwaukee housing study: "Healthy home life is necessary to make good men, and also good citizens."[3] Such views are an important part of the puzzle when trying to resolve what was Polish about the Polish flat or the landscape of the Polish South Side.

Milwaukee's Poles

One of the earliest descriptions of Poles in Milwaukee was provided by a *Milwaukee Sentinel* reporter in 1874. Sounding much like an explorer among an exotic people, the reporter related that he was "detailed to sojourn among them, to ascertain their number, study their modes of life, and avoid pronouncing their names."[4] His sketch of an estimated 7,000 Poles combined a background on their origins and an observation of group traits that is noteworthy in its consistency with the set of characteristics associated with the community over many years.

The majority of his study population had arrived within the last decade from the area known as Prussian Poland and were described as "active, industrious, but very poor." Commenting on other aspects of life for these new residents, he referred to them as "clannish." They clung to their compatriots, language, and (Roman) Catholic faith with a "remarkable tenacity." Voting was cited as the only custom of this country readily adopted by the Poles. Yet the reporter noted that strong nationalist feelings still were expressed in their hope for the reconstitution of Poland.

The men found employment by laboring on the "streets, sewers and various other public works" and, during the winter, by sawing wood. The reporter summarized the constraints on their employment by saying, "The men are perhaps the most unskillful laborers who reach these shores, especially for work in cities." Having a primarily peasant background, their training was not appropriate to the more skilled positions in an industrial city. Yet, despite this less than positive analysis, the reporter also observed that, in spite of their low incomes, they "have a strong prejudice against paying rent . . . usually the first money they can call their own is put into the purchase of a lot or part on which they mean to erect a house as soon as possible. . . ."

Both the peasant background and the desire for land typified the Polish immigrant to the United States. During the nineteenth century, a veritable revolution had transformed the social and economic systems in a Poland divided among Prussian, Austrian, and Russian rulers.[5] The consequence was a high level of rural unemployment and dislocation. This process was experienced first in the Prussian-controlled area early in the century. When Bismarck launched his *Kulturkampf* to Germanize the Polish population in the 1870s, the incentives to migrate were further accelerated, and the Prussian government did little to discourage the growing number of emigrants. The earliest members and the largest portion of Milwaukee's Polish community came from the countryside of East and West Prussia and the Prussian-dominated Poznan District. Poles from Galicia in Austria and from Russia joined this well-established "German"-Polish community during successive waves of migration, beginning circa 1895 and 1900, respectively.

Landownership was the goal of many of the Polish migrants, whether they were part of a family hoping to purchase land abroad or single men who viewed their stay in America as a means of obtaining sufficient funds so that they might return to their farming villages. Thomas' and Znaniecki's major study of the assimilation of immigrants, *The Polish Peasant in Europe and America*, devotes considerable attention to what is described as the *land hunger* of the Poles and its influence on their social and economic motivations.[6] By the 1870s, when their migration began on a large scale, desirable farmland was no longer plentiful in Wisconsin or many other parts of the Midwest, and the city offered the best possibility for income. The traditional desire for farmland was transferred to the goal of homeownership in the city. A financial leader in Chicago's Polish community, Albert Wachowski, explained in 1913 that "the Polish arrivals quickly comprehended the necessity of land ownership, knowing from past experience that . . . the man without his own land and home had always been a serf, a slave."[7]

Just as changing conditions in Europe had set their journey in motion, the new arrivals entered Milwaukee during a period of notable change and growth. The Polish immigration into Milwaukee paralleled the city's post–Civil War transformation from a commercial trade center to an urban economy based on processing and heavy industries. The city became a metropolis, increasing fivefold between 1870 and 1910, growing from 71,000 to 374,000.[8] Over a similar period, Milwaukee's Polish community grew almost 10-fold, reaching approximately 70,000 in 1910. In that year, Milwaukee contained more persons of foreign birth than at any other time in the city's history. The Poles constituted 20 percent of that foreign-born population and apparently accounted for a much higher percentage of the recent arrivals.

The timing of their arrival, as well as their rural background, had significant implications for their establishing themselves in industrial Milwaukee. Compared with the Scandinavians and Germans, with whom they shared the South Side, the Poles could not rely on a significant group of predecessors to help them adjust to their new life. Gerd Korman's work on the city's industrial and immigration history argues that early employment practices shaped occupational patterns along ethnic lines, especially for the Poles.[9] Although the German-speaking Poles received some initial benefit from the managerial positions of Germans and German Americans in Milwaukee's labor hierarchy, advancement above the rank of foreman was difficult to achieve. Poles became "typed" as common laborers and unskilled and semiskilled industrial workers. The *Milwaukee Sentinel* reporter's description of Polish men as being "strong and healthy and possess[ing] great powers of endurance" is remarkably similar to the characterization almost 40 years later, in a study of living conditions in American towns, that stated that Poles were best fitted for "dirty tasks" and jobs that required strength.[10]

Despite such limitations on economic mobility, the Poles' ability to achieve within one generation the common goals of homeownership and the establishment of traditional community organization in the form of the Catholic parish was acknowledged as remarkable. In part because of the homogeneity of their community as German Poles and the leadership of their church, the Poles gradually became prominent socially and politically as well as numerically. As a measure of this, Milwaukee historian Bayrd Still noted that by 1890 no party dared ignore the political strength of the Poles by leaving a Polish-American candidate off its ticket.[11]

The Polish "Colony"

Characterizing immigrant life in Milwaukee as a process of "slow and orderly, yet natural and certain" assimilation, William George Bruce, one of Milwaukee's prominent citizens, gave particular attention in his 1937 memoirs to the Polish-American community. In observing their advance, he commented on his vivid boyhood memory associated with the arrival of a growing number of Poles during the 1870s:

We boys deemed it one of the sights to behold these anxious immigrant men and women and abashed children nestled among the boxes, bundles and bedding of an old-world household, awaiting transfer from the railway station to a more permanent abode. Usually a rickety wagon, drawn by one horse, took them to the southern limits of the city, which, up to this time, had only been sparsely populated. They were travel-stained, poor, and uninformed, but they were hopeful, courageous, and ambitious as they looked toward the wooded lands south of Greenfield Avenue, which they later transformed into a vast area of cottages with high basements, each accommodating two families, with gardens in the rear and some shrubbery and a rest bench in the front.[12]

The progress recounted by Bruce makes particular note of the transformation of a sparsely settled area on the southern outskirts of the city into a landscape of homes—the "vast area of cottages with high basements." Wards 11, 12, and 14 became the core of the Polish South Side (Figure 13.3). In 1905, Ward 14 contained over 53 percent of all the German Poles in Milwaukee, with 75 percent of the heads of household born in Poland and an additional 10 percent born to Polish-born parents.[13]

In part because of Milwaukee's river-divided topography, industrial growth resulted in urban expansion and decentralization of the city's manufacturing jobs. Prominent new development occurred along the Milwaukee River, in the Menomonee River valley, and at the mouth of the Kinnickinnic River. On the South Side, Wards 5, 12, and 17 became centers of heavy industry and thus major sources of employment for unskilled laborers.[14] Judged as a less desirable section of the city, the area south of the Menomonee River valley had lower land values. Ward 14 proved to be attractive real estate for the prospective working-class homeowner. There the worker could benefit from lower land costs while still being close enough to walk north to the industrial jobs in the Menomonee River valley or east to the steel mills near the lake.[15]

To a large extent, the history of the Polish neighborhoods' growth is reflected in the dates of construction of their churches. St. Stanislaus was the first Polish parish. Located initially in 1866 in a former Lutheran church on Mineral and Grove (later renamed as First Avenue), the St. Stanislaus congregation grew sufficiently large by 1871 to build an imposing new church on Mitchell and Grove. By the early 1880s, the population west of First Avenue supported the establishment of St. Hyacinth. The financial sacrifice entailed in church construction was a reflection of both a strong faith and the important role of the church as a transmitter of culture, values, and nationalism. In partitioned Poland, the church had exerted similar influence in its support of Polish culture. Each parish church became a social as well

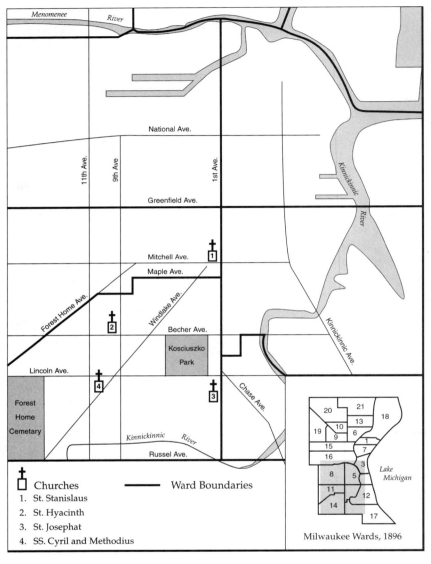

Figure 13.3. Milwaukee's Old Polish South Side—location map, 1896 (with street names changed to the present system)

as a religious center for surrounding residents, as evidenced by the Polish names attached to various neighborhoods. In the case of the St. Stanislaus Parish, the area was referred to as Stanislawowo, much as a village in Poland would be named.[16] With a degree of irony, Father Waclaw Kruszka, a leading religious figure in Milwaukee and a historian of Poles in America, praised America as a place where Poles could be Polish.[17]

Immigrants from Prussian Poland brought with them practices and institutions that provided means for achieving their goals in America while allowing a certain independence from the larger community. Michael Kruszka—Waclaw's half-brother, editor of the *Kruyer Polski* (a Polish-language newspaper) and later a state senator—commented on the political organization of the Poles, citing their training in Prussia, where solidarity was required to protect their rights.[18] To accomplish the important goal of homeownership, Polish building and loan associations were established as a critical form of mutual aid and cooperation brought with them from Prussian Poland. There the operation of building and loan associations had been in place since at least the early nineteenth century.[19] One observer of the community noted in 1911 that five cooperative building and loan societies had been established among the Poles, and that "nearly all save the poorest are said to be members."[20]

A Landscape of Workers' Homes

Such mechanisms of mutual aid helped the Poles achieve their goals of homeownership, yet the degree to which they created an ethnically distinctive or Americanized landscape remains to be answered. Land division and architectural practices represented a distinct separation from Old World practices, being radically different from the experience of all but the wealthiest immigrant Pole. Milwaukee's residential landscape, whether in the Polish South Side or in other working-class districts, reflected the prevalence of the speculator's land division system and American modern construction technologies, including balloon-frame construction and a national distribution system of machine-made components such as windows and doors. Thus, housing patterns became more uniform throughout America.

During the late 1870s and the 1880s, the South Side began to assume its present physical form.[21] In Ward 14, most land was subdivided during the real estate boom of the 1880s. The process was completed during the 1890s. Since 1857, when the state of Wisconsin virtually wrote the grid plan into law, the subdivision plat became standard through-

Figure 13.4. Mazur cottage on old south Tenth Avenue, 1899 (Photograph courtesy of Bernice Mazur)

out most of Milwaukee. Governed by the cost of footage along the street, variation was generally limited to the width of the lot. Based on similar practices in other larger cities, including New York and Chicago, the standard lot in working-class neighborhoods was long with a narrow frontage, measuring 20–30 feet wide by 100–150 feet deep.

On these lots, speculators constructed units of housing appropriate to the local neighborhood market. The single-family dwellings that accommodated working-class families in the mid- to late nineteenth century were generally small, three- or four-room cottages much like the Mazur's home shown in Figure 13.4. The small worker's cottage became one of Milwaukee's earliest speculatively built house forms, providing a housing solution that was flexible, quickly achieved, and inexpensive.[22] Although otherwise modest, the cottages could be attractively trimmed with machine-tooled window frames and architectural ornamentation which have come to be labeled Victorian.

Anton and Katarzyna Mazur bought their cottage and lot on old Tenth Avenue from Joseph and Emily Fehrer in 1894 for $800.[23] Less than a decade old, the cottage served as their first home. A marriage arranged by her maternal uncle brought the 26-year-old Austrian steel worker and 21-year-old Katarzyna together soon after she arrived from the Poznan section of Poland. The newlyweds moved into their home with Katarzyna's father and younger sister and brother. By 1899, when the photograph (Figure 13.4) was taken, the Mazurs had two children. Five more would follow by 1914. To accommodate the needs of the family in the three-room cottage, an additional sleeping area for the boys and their grandfather was accessed through a trap door to the cellar. Windows in the cedar-post foundation gave some light to the basement.

Just as their cottage was typical of housing in Ward 14, the Mazurs were representative of the area's new residents. Ninety-five percent of Ward 14 was identified as German in 1880, but by the 1905 census the area was almost exclusively Polish.[24] Heads of household in the ward were primarily employed as unskilled or semiskilled workers. The age of the majority of immigrants placed them in the child-bearing stage. As a consequence, in addition to the difficulties of establishing themselves in a new country, there were children to care for, not to mention other members of their multigenerational families. These larger families placed demands on living space. Children, however, contributed to the family by leaving school at 13 or 14 to begin work. Frequently when the children's incomes began to supplement the family budget, improvements to the cottage could be considered.

This was the case with the Mazur cottage. Although a somewhat uncommon form of incremental building, the improvements to the Mazur house reflect the frugal, additive approach to building that was used to obtain more living space. In June 1917, the Mazurs bought "Uncle Joe" Nisiewiecz's cottage, which was lifted off its foundation and rolled down the street from its site one block away. A larger home

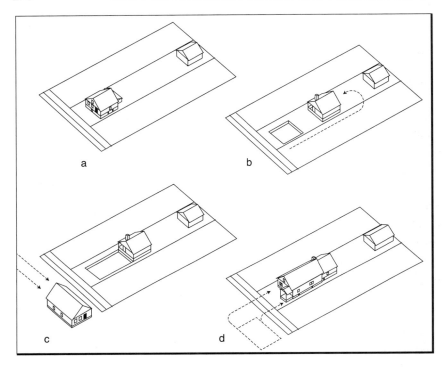

Figure 13.5. Expansion of Mazur cottage, 1917

was constructed by joining the two cottages back to back (Figure 13.5). At the same time, the cedar-post foundation was replaced with cement blocks. By 1895, a large number of Polish contractors in the city had facilitated such incremental building. Nineteen Poles were listed as contractors at that time, and it was noted that there was a "blooming business" for house movers as well within the Polish community.[25]

Late in the nineteenth century, in other Milwaukee neighborhoods, the two-family flat began to replace the cottage as the most common form of working-class housing in Milwaukee. The duplex, one flat upstairs and one downstairs, was described as the "rule" throughout Milwaukee in a 1911 housing study, but it never became that common in the Polish neighborhoods.[26] The advantage of this house form was the rental income derived from the second unit, which was useful for paying off the mortgage. Located in the old 800 block of Grove Street on the South Side, the structure shown in Figure 13.6 housed members of five German families and their boarders.[27] Recently arrived immigrants

Figure 13.6. "Duplex" on old 800 block of Grove Street, 1914 (Photograph courtesy of Henry J. Wojcik)

created a market for rental housing that often strained the supply. The desire for homeownership combined with the need for rental apartments in a growing industrial city to create a densely populated landscape.

The benefits of a duplex could be obtained through incremental ar-

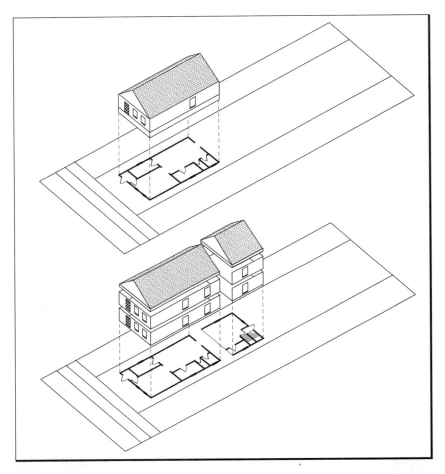

Figure 13.7. Expansion of Potrykus cottage on old Garden Street to create a Polish flat, 1914

chitecture, as described in the 1911 study mentioned above: "In the south side district, where a large class of the poorer section of the Poles live, the custom is to erect first a four-roomed frame dwelling. When this has been paid for, it is raised on posts to allow a semi-basement dwelling to be constructed underneath. . . . This basement or the upstairs flat is then let by the owner, who, as soon as his funds permit, substitutes brick walls for the timber of the basement. . . ."[28] The raised cottage form, the poor man's version of the duplex, was known in the local vernacular as the Polish flat. Instances of one family eventually occupying the entire structure indicate that this house-building strategy accommodated various demands without relinquishing the dream of a single-family home.

The expansion of the Potrykus house offers a dramatic example of the physical adaptation of cottage space to the changing needs and values of a family (Figure 13.7).[29] Joseph and Josephine Potrykus and their seven children lived on old Garden Street in the St. Josaphat Parish. With the marriage of their eldest child in 1914, they began a great reconstruction of the house to provide a first-floor flat where their son might begin his married life. The cottage was "exploded" in all directions. The structure was raised to build a new brick first floor, the roof was raised about four feet to add a bedroom in the front of the attic, and additions were made to the rear of the house. The cottage floor plan changed considerably, reflecting not only the necessity of new living space but also changes in life style. The lower flat was a self-contained unit reminiscent of the early cottage. The "new" second story—the original cottage structure plus addition—contained a dining area and a kitchen to be used on special occasions. Josephine Potrykus did her cooking and laundry downstairs in a large kitchen at the back of the house. This "summer kitchen," as her grandson called it, separated the messier household tasks from the living area upstairs and provided direct access to the family garden in the backyard. The entire family, including the occupants of the downstairs flat, would gather in the large kitchen for shared meals and evening get-togethers. The house was a hybrid creation reflecting the American middle-class dream of separate rooms for separate uses and the Polish emphasis on providing for family.

The ultimate sign of neighborhood prosperity was represented in improvements to the houses and yards. As one observer noted: "The ambition of a Polish house-owner is not crowned until he is able to have cement walks and iron railings in front of his house."[30] This public "front" did not preclude an intensive use of the backyard for vegetable gardens and animals, such as chickens, rabbits, and the occasional cow. The persistence of agrarian practices have led some to conclude that the Polish neighborhoods' greatest cultural influence on the landscape could be seen in the way they arranged neighborhood space rather than in the architecture.[31] These uses did not conform to the middle-class aesthetic. Descriptions of the *foreign* landscape frequently cited villagelike scenes of animals in the roadway.[32] Balanced against the "picturesque," other commentaries cited the evidence of the residents' thrift and industry.

Assimilation and "Americanization"

The Polish *colony* on Milwaukee's South Side shared many American dreams, including the goal of homeownership and the commitments to civic participation and to the neighborhood centering itself on church. Urban reformers of the late nineteenth and early twentieth centuries would have expected such traits of citizenship to be linked to the domestic rewards of homeownership. What was not generally anticipated was the Poles' rate of success. Even in a city where it became an undisputed belief that "no other city of its size in the world contained as many workingmen who owned their own homes," the Poles' desire for property ranked them highest among Milwaukee's ethnic groups in terms of residential ownership.[33] A foreign researcher noted this commitment, saying: "Amongst the Germans and Poles this habit is more marked than is the case with any other nationality, the determination to possess their homes leading them to practice great self-denial, and to accept considerable risks in the confidence that the rapid and continuous development of the city will greatly increase site values."[34] This same researcher drew the rather *un*-American conclusion that, given their low wages and the efforts required of the entire family, perhaps the Poles made sacrifices for homeownership that should have been discouraged.[35] The house was a means of attaining upward mobility for the family and a sign of optimism in assessing the city's potential.

The extent to which the Polish South Side existed as a separate settlement, however, was a sensitive issue during a period when assimilation and Americanization were major concerns on the part of the dominant society. The early-twentieth-century "scientific" field of eugenics, which justified selective breeding to protect the "social organism," lent credibility to racist stereotypes as support grew for the restrictive immigration laws that would be adopted in the 1920s. A national urban reform journal expressed such views of other "races": "The Italian, the Hebrew and the Slav, according to popular belief, are poisoning the pure air of our otherwise well regulated cities, and if it were not for them there would be no filth, and no poverty. . . ."[36] It is within this context that we might interpret a Milwaukee housing report's attack on the phenomenon of Polish flats: ". . . on each side of every street or avenue is an almost continuous line of basements, miles and miles of gloomy, poorly lighted, damp, unventilated, overcrowded rooms, thousands upon thousands of homes fatal to infants, debilitating to children. . . . The 12th and 14th wards are more than any others the regions of the modern cave dwellers. The basements are occupied from choice and long fixed habit, as well as, in some cases, to reduce the cost of living. . . ."[37] The racist content of this health worker's comments is obvious but even more so when one recalls William George Bruce's image of the Polish South Side.

Although often chosen by the Poles, the Polish flat is not uniquely their own. The attachment of *Polish* to the Milwaukee name for a raised cottage construction, however, is unique—and apparently reflected a judgment on this house type and the residents of the area by Milwaukee society. Certainly, the desire to provide common housing for several generations of family was not uniquely Polish. The Neuman family, German neighbors to the Mazur family on old Tenth Avenue, also raised their cottage and created a first-floor flat when the oldest child married. Similar actions by Italian, German, and Slavic families of various kinds residing in Polish flats were not uncommon. The land hunger of the Poles and their incremental approach to financing and building their homes, however, made this relatively inexpensive house form particularly prevalent in the neighborhoods of Polish unskilled workers. Many of the sacrifices, including the acceptance of crowding, were considered necessary to achieve stability and a certain status in the Polish community. As early social worker Edith Abbott argued, investment in a house was the first step toward becoming an American.[38]

Today's residents celebrate Milwaukee's diverse European ethnicity along with homeownership and cohesive neighborhoods as valued traditions, while attaching a certain disdain and assumptions of "foreignness" to other urban immigrants. These new groups fit uneasily into the city and once again challenge our notions of what it means to be an American citizen, to share an active role in American society. Whether they are Hispanic, Hmong, new Polish residents of the South Side, or the African Americans of the near North Side, the modern history of Milwaukee's inner-city neighborhoods holds many parallels with, as well as striking differences from, Milwaukee's Polish South Side at the turn of the century. Sorting out the features of ethnic, immigrant culture from the social and economic status of the particular groups that reside in the city is an ongoing task.

Notes

Research for this chapter was conducted with support from an Urban Research Initiative Grant from the University of Wisconsin–Milwaukee. I wish to acknowledge the contribution of Tom Hubka, the project codirector, in the design of the project's research program and for comments on drafts of this work. Thanks are also given to Herb Childress, Victor Greene, John Gurda, Jason Nyberg, and Paula Oeler for their helpful suggestions on the chapter's contents. Jason Nyberg and Paula Oeler assisted with fieldwork and figure preparation. Bernice Mazur and Henry Wojcik were particularly generous in providing personal photographs to illustrate this chapter.

1. All addresses referred to in this chapter are the old pre-1930 street names.

2. J. Gurda, *A Separate Settlement: A Study of One Section of Milwaukee's Old South Side* (Milwaukee, 1974), 5.

3. Wisconsin Bureau of Labor and Industrial Statistics, "The Housing Problem in Wisconsin," *Twelfth Biennial Report* (Madison, 1906), 275.

4. "The Polacks," *Milwaukee Sentinel*, November 30, 1874.

5. J. Bukowczyk, *And My Children Did Not Know Me: A History of the Polish Americans* (Bloomington, Ind., 1987), 1–15.

6. W. Thomas and F. Znaniecki, *The Polish Peasant in Europe and America* (Chicago, 1927), 190–196.

7. Cited in V. Greene, *For God and Country: The Rise of Polish and Lithuanian Ethnic Consciousness in America, 1860–1910* (Madison, 1975), 57.

8. B. Still, *Milwaukee: The History of a City* (Madison, 1948), 570–571.

9. G. Korman, *Industrialization, Immigrants and Americanizers: The View from Milwaukee, 1866–1921* (Madison, 1967).

10. United Kingdom, Board of Trade, *Cost of Living in American Towns* (London, 1911), 258.

11. Still, *Milwaukee: The History of a City*, 271.

12. W. G. Bruce, "Contacts with the Polish-American Element," in his *I Was Born in America* (Milwaukee, 1937), 322.

13. R. Simon, "The City-building Process: Housing and Services in New Milwaukee Neighborhoods 1880–1910," *Transactions of the American Philosophical Society*, vol. 68, part 5 (Philadelphia, 1978), 37.

14. Simon, "The City-building Process," 13.

15. Such a residence pattern appears to be the case in many Great Lake and midwestern industrial cities where the residential demands of urban growth gave Slavic immigrants access to newer housing on the outskirts of urban development. See O. Zunz's study *The Changing Face of Inequality: Urbanization, Industrial Development, and Immigrants in Detroit, 1880–1920* (Chicago, 1982).

16. Bukowczyk, *And My Children Did Not Know Me*, 40.

17. W. Kruszka, *A History of the Poles in America to 1908* (Washington, D.C., 1993), 185.

18. M. Kruszka, "Forty Thousand Polanders," *Milwaukee Sentinel*, October 16, 1895.

19. J. Grzemski, "Thrift among the Poles," in his *Poles of Chicago* (Chicago, 1937), 183.

20. United Kingdom, Board of Trade, *Cost of Living in American Towns*, 258. See Greene, *For God and Country*, 54, for a description of a building and loan association's operation.

21. Simon, "The City-building Process," 258.

22.   Landscape Research, *Built in Milwaukee: An Architectural View of the City* (Milwaukee: 1981), 60.

23.   Personal interview with Bernice Mazur, Milwaukee, Wis., April 18, 1994.

24.   Simon, "The City-building Process," 36.

25.   Kruszka, *A History of the Poles to 1908*, 189; Still, *Milwaukee: The History of a City*, 273.

26.   United Kingdom, Board of Trade, *Cost of Living in American Towns*, 266.

27.   Personal interview with Henry Wojcik, Milwaukee, Wis., September 2, 1994.

28.   United Kingdom, Board of Trade, *Cost of Living in American Towns*, 266.

29.   Personal interview with Leonard Zurkowski, Milwaukee, Wis., June 16, 1994.

30.   United Kingdom, Board of Trade, *Cost of Living in American Towns*, 266.

31.   C. Reisser, "Immigrants and House Form in Northeast Milwaukee," M.A. thesis, University of Wisconsin–Milwaukee, 1977.

32.   W. Bishop, *The Golden Justice* (Boston, 1887), 135.

33.   Simon, "The City-building Process," 18.

34.   United Kingdom, Board of Trade, *Cost of Living in American Towns*, 266.

35.   United Kingdom, Board of Trade, *Cost of Living in American Towns*, 266.

36.   E. A. Goldenweiser, "Immigrants in Cities," *The Survey* 25 (1911): 596–602.

37.   Wisconsin Public Documents, "Basement Tenements in Milwaukee," *Fifteenth Biennial Report*, vol. 5 (Madison, 1912), 174.

38.   E. Abbott, *The Tenements of Chicago* (Chicago, 1937), 381.

14

Religious Identity as Ethnic Identity
The Welsh in Waukesha County

Anne Kelly Knowles

Every immigrant group passes down archetypal stories to its descendants that capture some essence of the group's ethnic character. From Welsh Americans, the story one hears time and again is that the first thing a group of Welsh immigrants did upon reaching their destination was to build a chapel. Although in truth it usually took a few years for Welsh pioneers to find the money and time to build a chapel, the stereotype signifies how profoundly important religion was to rural Welsh immigrants across the Midwest during the mid-nineteenth century. Nothing was more natural to the Welsh Nonconformists who settled in Wisconsin than to offer thanks to God for their safe arrival on the frontier and to organize religious meetings as soon as possible, even if the only building available for worship services were a drafty barn or a crowded log cabin.

Fundamentalist Protestant religion had a special place in the lives of Welsh immigrants. Many of them had undergone the powerful emotional experience of religious conversion in revivals. The immigrants brought with them strongly developed institutions of chapel organization, self-governance, and discipline that were remarkably well suited to structuring community life on the American frontier. At the same time that their Protestant convictions made Welsh immigrants "good citizens" in the eyes of American society, the uniquely Welsh qualities in their particular brand of Calvinism became the focal point around which rural Welsh ethnic identity crystallized during the first and second generation of settlement.[1]

The Welsh farm families who settled in Waukesha County, Wisconsin, in the 1840s were typical of their people and their time. They came from a limited number of regions in Wales, notably from the counties of Cardiganshire and Anglesey. Almost all of them belonged to one or another of two distinctly Welsh denominations—the Calvinistic Methodists or the Congregationalists—both of which conducted worship in the Welsh language, the mother tongue of virtually every member of the Waukesha community. Although this chapter focuses on the immigrants in Waukesha, their experiences represent the ways in which religion commonly structured Welsh immigrant communities and served as a refuge for Welsh language and culture. I will also consider how cultural change registered in the immigrants' religious lives and ultimately how religion came to epitomize key aspects of Welsh-American ethnic identity. For *Cymry* (Welsh people) from Utica, New York, to Tacoma, Washington, religious identity became ethnic identity through a process in which a single aspect of Welsh culture assumed heightened importance in a new context.

Religion in Welsh Culture

Welsh society underwent two great changes between 1750 and 1850. One of these was the rapid development of heavy industry in a few geographically concentrated regions, notably the iron industry in Glamorgan and Monmouthshire, slate mining in Merioneth and Carnarvonshire, and coal mining in Denbighshire and Flint. The other great change was a flowering of Welsh religious denominations, called Nonconformist because they did not adhere to the tenets of the established Anglican church or recognize its authority. By 1850 as much as 80 percent of the Welsh population claimed membership in a Nonconformist chapel, a figure which the Welsh themselves embraced as proof of their exceptional religiosity.[2]

Welsh Congregationalists, Baptists, and Calvinistic Methodists disagreed about the sacrament of baptism and matters of church governance, but they shared important beliefs, including the primacy of the Bible and of one's personal relationship with God, as well as the Calvinistic conviction that God's will, not an individual's actions, determines whether one will be saved or damned. The Calvinistic streak in Welsh religion had a profound effect upon popular culture. In their eagerness to appear respectable, believers embraced a strict new

Figure 14.1. "Nain" (Grandma) Breese was a pillar of Tabernacle, the one Congregational chapel in the Waukesha Welsh community. In this photo, taken circa 1860–1865, she wears her Sunday best and holds a small Welsh Bible. (Courtesy of James Southcott)

morality that forbade folk traditions such as dancing, singing, and outdoor games. They found pleasure instead in more respectable, bourgeois pastimes, including the innovation of Welsh-language competitions in poetry and music called *eisteddfodau*. The most serious among them devoted their free hours to studying the Welsh Bible, the Welsh translation of John Bunyan's *Pilgrim's Progress,* and biblical commentaries by Welsh and English authors. No wonder that revered members of the Nonconformist community in Wales were praised as "old Israelites" and "Puritans."

By no means was every Welsh household orthodox in its beliefs or behavior in the mid-nineteenth century. Yet "Israelites" and "Puritans" figured prominently as lay leaders and people of influence in new Welsh-American communities, and people with strong religious convictions may have formed a larger proportion of the immigrants than of the Welsh population at large. Welsh immigrant obituaries from the middle of the nineteenth century—the period of heaviest Welsh migration to the rural United States—suggest that the great majority of those who settled in rural areas originally came from the strongholds of Nonconformity in North and West Wales. The men and women who founded Welsh settlements were often former chapel elders and experienced believers.

Elinor Breese was one such member of the Waukesha community (Figure 14.1). She emigrated in 1847 from the hamlet of Llanbryn-Mair, a place famous in Wales for producing Congregational ministers and ideas about preserving Welsh culture by transplanting it overseas. Elinor experienced conversion under the preaching of John Roberts, one of the great Welsh Congregational preachers of the early nineteenth century. While her kindly manner made her a favorite among local children, she also set a firm example as a devoutly religious woman. Catherine Williams, a staunch Calvinistic Methodist from north-central Cardiganshire, was greatly admired for her ability to recite "the whole of the New Testament with the exception of certain portions of the Book of Acts. And the Old Testament from the beginning of the Book of Job to the end of Malachi with the exception of some portions of Jeremiah and Ezekiel." Women such as these inspired the girls' class at Jerusalem Chapel to memorize the entire book of Matthew and to recite it, in four installments, at Sunday services.[3]

As was true for immigrants from other parts of northern Europe, the Welsh who settled in Waukesha County during the 1840s predominantly came from the middling class of farmers, shopkeepers, and craftspeople. They were the folk who had sufficient means to emigrate and to buy at least a little land in the States. As the lay leaders of rural

Welsh society, they also tended to be the people most committed to religious values and most experienced in organizing chapel life. Welsh immigrants' religiosity was often reinforced and perhaps heightened by the trials of emigration, beginning with the experience of traversing the ocean in close quarters with people of other nationalities. Letters about the transatlantic journey make a special point of contrasting the piety and propriety of Welsh passengers with the drunken, ribald behavior of the Irish. "During the whole journey when the sea was quiet there was nothing to be heard except their singing, dancing, shouting, and noise," a Welshman wrote of his Irish shipmates. The Welsh on board "showed that they knew what seemliness was. We found the Welsh singing the hymns of old Wales with spirit and taste." As the Welsh saying goes, *Meddu pwyll, meddu'r cyfan* (who possesses discretion possesses everything).[4] Such encounters confirmed the Welsh in their religious convictions while also planting the seeds of a public, ethnic persona based on the notion that the Welsh were an exceptionally godly people.

Religion as the Keystone of the Immigrant Community

Newly arrived Welsh families commonly held weekly fellowship meetings in one another's homes virtually from the moment they reached their destination. Soon these rotating domestic gatherings developed into the more formal *seiat* (a Welsh corruption of the English "fellowship society"), where one or more men elected as elders would lead the meeting in worship and assess the faith of those in attendance. When an itinerant preacher passed through the settlement, one or more *seiadau* would gather to hear a sermon. If they liked the preacher's style they would try to persuade him to stay and perform more of his repertoire. A *seiat* often eventually became a congregation with its own chapel once its membership gained sufficient numbers and resources to build a sanctuary. This process of building religious institutions from within the home and the local neighborhood reiterated the steps that Welsh Nonconformists had followed for generations in Wales. The use of lay leaders to establish cells of new believers in sparsely populated areas neatly coped with the problems of scarce, itinerant clergy and even scarcer family resources. It was a practice that helped make Methodism the leading denomination of the American frontier and Calvinistic Methodism the most popular denomination in Wales by 1840.

Among the Welsh in Waukesha County, this practice of using lay leaders nicely illustrates the process of building institutional religion from a local base. Between 1842 and 1860, increasing numbers of Welsh immigrants from North and Mid Wales bought land and established themselves as farmers on the hills and vales of the Kettle Moraine. The greatest number arrived between 1843 and 1847. Many of them bought land from Richard Jones, known as "King" Jones because of his many services to the community, including his acting as a land broker and host for new arrivals, who found temporary accommodation in his barn. Jones also hosted the settlement's first religious services from 1842 to 1845 and donated the land for the first chapel, Jerusalem.[5] By 1860 the settlement had grown to a total of 944 Welsh-born and their children and had eight chapels, all but one of them Calvinistic Methodist. The ninth and last Welsh chapel was built in 1868 on the northeast edge of the settlement. All but one had begun as a *seiat* anchored in one of the neighborhood clusters of Welsh farms that together formed a contiguous ethnic territory across northern Genesee and southern Delafield townships and a more scattered Welsh presence in Ottawa Township (Figure 14.2).

The King Jones farm, Bronberllan (named after his home farm in Cardiganshire), also figured prominently in the early history of Calvinistic Methodism in the state. The first three statewide *cymanfaoedd pregethu*, or preaching assemblies, were held at Bronberllan in 1844–1845. Not long afterward, Welsh ministers and lay leaders organized four Calvinistic Methodist districts (the functional equivalent of presbyteries) under the umbrella of the Welsh Synod of Wisconsin. The districts were centered on the main Welsh settlements in the state—Waukesha, Welsh Prairie (around Cambria), Dodgeville, and the Bangor–La Crosse area (Figure 14.3). By the 1870s, the Wisconsin synod was the largest Calvinistic Methodist body in the Welsh General Assembly in America.[6]

Religion gave rhythm to each week, just as farming determined the daily work patterns and seasonal life in the Waukesha community. The biblical commandment to rest on the Sabbath was strictly enforced in many Welsh homes as late as the middle of the twentieth century. Owen Williams, the fourth generation on his family's farm in Delafield Township, told me that his father was never allowed to do field work on Sundays. There scarcely would have been time, as the family walked a mile and a half to and from chapel three times on Sundays, once to attend the worship service, once again for Sunday school, and a third time for prayer meeting in the evening.[7] A former member of Tabernacle Chapel told an interviewer in 1978, "You were not expected to shine your shoes [on Sunday]. That was done the night before. My mother always peeled her potatoes the night before. If she ever for-

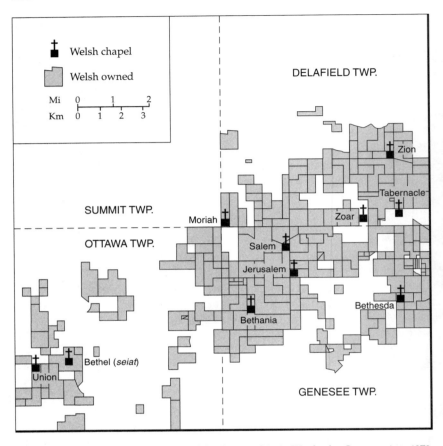

Figure 14.2. Map of Welsh chapels and landownership in Waukesha County, circa 1873 (Sources: U.S. Manuscript Census, 1870; *Waukesha County Atlas* [(Madison?), 1873]; Williams, 1926, 26–108)

got, she would boil them in their jackets on Sunday." Her parents forbade card playing and dancing and frowned on smoking. Nor could she think of any chapel families who drank: "It just wasn't done," on Sunday or any day. Prayer meetings were still sometimes fervent after the turn of the century, although old believers' outpourings in emotional, biblical Welsh baffled their grandchildren, many of whom did not understand the language. "One old woman used to pray and cry, tears running down her cheeks," Mildred Hughes Southcott told me. "We had no idea what she was saying. She had the *hwyl*."[8]

Welsh immigrants' religious turn of mind also imbued the physical landscape with symbolic meaning. The mother chapel built on King Jones's land was christened Jerusalem. An old man told Daniel Jenkins Williams, historian of the Waukesha Welsh community, that leaders of Jerusalem Chapel wanted to be sure the sanctuary was properly located when it was enlarged in 1859. "Dr. Perry relates that when he was a small boy, a delegation from the Jerusalem neighborhood came to his home in Jericho in the Township of Mukwonago to ascertain the distance from Jerusalem to Jericho, having a desire to build their new church on a site which would be the same distance from Jericho as the distance from Jerusalem to Jericho was in the Holy Land."[9] Every chapel in the community was named after a place in the Bible, as was customary in Wales. In larger settlements that were dominated by immigrants from a particular region in Wales, such as "Little Cardiganshire" in southern Ohio, new chapels bore the same names as the settlers' home sanctuaries in the old country. The Waukesha Welsh did not follow this pattern, perhaps because their community contained a mix of immigrants from North and South Wales, with no one locality of origin overshadowing the others in importance. Many settlers did, however, name their American farms after their homes in Wales. These naming traditions had sentimental value but also may

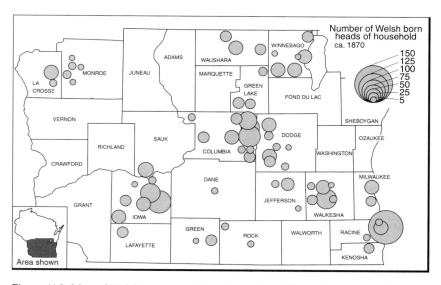

Figure 14.3. Map of Welsh-born population in southern Wisconsin, ca. 1870 (Sources: U.S. Manuscript Census, 1870; Wisconsin county atlases, ca. 1861–1880)

have helped soften the strangeness of the immigrants' new surroundings, since a familiar name carried some of the rich connotations of its original place.

Religion and Language

Welsh remained a living tongue in rural Wisconsin communities for two or three generations. It was the language of the home for many families until after the turn of the century. But it lasted longest as the language of ritual in the weekly recital of psalms, Bible verses, and hymns during religious services and Sunday school. Churches were sanctuaries of language and cultural traditions for every immigrant group in Wisconsin. For the Welsh, the use of their native tongue in chapel was particularly important because they made few efforts to preserve it in other spheres. Unlike language-proud Germans, Welsh immigrants educated their children in English-language schools. Although Welsh-American children in late-nineteenth-century Waukesha "played in Welsh, quarrelled in Welsh, and talked Welsh all the while except when making recitations [in class]—and that was partly in Welsh," their teachers were Yankees and their primers were in English.[10] This was consistent with long-standing attitudes toward education in Wales, where virtually all formal education was conducted in English. Children who used their native language in Welsh schools were punished and humiliated by those teachers who either despised it as an inferior tongue or were themselves convinced that only by learning English could a Welsh child have any hopes of succeeding in the world. The latter notion naturally carried over to the idea that immigrants' children should learn English to succeed in America. Given the lack of status Welsh enjoyed in education and the relatively small size of most Welsh-American settlements, the surprising thing is that Welsh remained in use as long as it did.

It was the special relationship between language and religion that made the latter the most significant and enduring institutional focus for rural Welsh-American ethnicity, an identity that always included a strong element of exclusivity. Welsh Calvinistic Methodists did not merge with the very similar Presbyterian Church of the U.S.A. until 1920. The Welsh churches in Wisconsin kept a distance even then, maintaining their own synod until 1934 and their own presbytery until 1954, longer than any other Welsh-American branch of the denomination. At least one Bible class, in Randolph, Wisconsin, still conducted its discussion in Welsh as late as 1954.[11]

From the pioneer period onward, the public face of Welsh identity in Wisconsin was of a people who tended to keep to themselves but whose religious values, manifest in an outward modesty and hard work, made them welcome additions to American society. Like other good Americans, though perhaps with special zeal, the Welsh in Waukesha County enthusiastically supported the temperance movement. In 1873, all the members of Bethania Chapel took the pledge to be "total abstainers," as did 26 of Zoar's 28 members and 44 of Bethesda's 48. The one exception was Salem Chapel, where only 13 of 37 members signed the pledge.[12] (I will return to this harbor of Welsh liberalism.) The community also generally kept clear of politics. No Welshman held a township or county office in Waukesha before 1860 and only nine were elected up to 1880, a tiny fraction of all officeholders.[13] A local newspaperman wrote in 1897, "The Welsh voters of this county . . . are cautious, conservative. . . . They are largely outside the machinery of politics and are not easily 'worked,' either for candidates or political ends." He went on to say that the Welsh "are on the most friendly terms with their neighbors of other nationalities and are first-rate American citizens. But they cling close to those of their own blood, they like to live near together, and are apt to marry each others [*sic*] sons and daughters."[14]

Abstinence from both alcohol and politics indicates the extent to which the Welsh were a self-regulating community. According to the Calvinistic Methodists' *Rules of Discipline* no member of the denomination was to take another to court, "but that they [first] place the matter in dispute before the Church, and that they be submissive to the judgment of the Church in every such affair." What was said of the Welsh in Jackson County, Ohio, was equally true for the Waukesha community: "The Welsh of the county were at first almost a state by themselves. They established their own churches, in which only the Welsh language was used. They cared for their own poor, and they settled their disputes in their churches."[15] The clannishness of the Welsh was perfectly acceptable because they kept their noses clean.

Old Forms Change in New Circumstances

Another reason for considering religion as the key to Welsh ethnic identity is the extent to which religion became hyper-developed in Welsh-American settlements. Rural Welsh communities supported many more churches than was typical of rural European settlements in the Midwest. For example, Lutheran or Catholic churches established by Scandinavian, German, or Dutch immigrants in Wisconsin normally had several hundred members, as did parish churches in each nationality's homeland. The location of a large sanctuary in the physical center of an ethnic settlement, often on the brow of a hill, symbolized the central role of religion and the need to make the church as accessible as possible to a dispersed agricultural community. The resident priest or pastor was the key figure of authority in these communities. Welsh chapels, in contrast, were small buildings tucked inconspicuously into the landscape, and they housed equally small congregations. When the Waukesha district of the Welsh Synod conducted a survey of Calvinistic Methodist chapels in 1873, a period when chapel membership was probably at its peak, the highest recorded attendance at any one chapel was just 150, and membership ranged from 18 to 66, with an average of 41 members per chapel (Table 14.1). Welsh preachers and ministers were respected and admired as men of exceptional merit, but only the most charismatic exercised special authority over community affairs, in part because so much of chapel governance lay in the hands of the elected elders and the votes of the congregation.

Like other rural Welsh-American communities, the Welsh in Waukesha produced a plethora of chapels because of the ways in which three forms of enhanced institutional support altered the Welsh tradition of decentralized, fairly autonomous congregations. First, the community had a surfeit of men capable of leading a *seiat*, as well as more than enough preachers and ministers to lead Sunday services. Itinerant Welsh preachers from as far away as Ohio and Pennsylvania stopped in Waukesha en route to and from settlements farther west. Thus leadership was no bar to the formation of Welsh congregations as it was for immigrant groups who were obliged to wait for their home church to send them a minister. Second, the Welsh could afford to build chapels with contributions from a small number of members, thanks to the standard of living they achieved in the United States as owners of productive farmland and the austere Nonconformist aesthetic that called for unadorned, modest sanctuaries. Third, an element of pride and neighborhood identity may explain why one *seiat* after another became a chapel. Groups of kin and friends tended to buy adjoining parcels of land. They helped one another through the difficult phase of pioneering homesteads and together created a new life in a foreign land. The strong sense of neighborhood identity that developed between them, combined with ample means and adequate leadership, made it a natural step for clusters of Welsh families to found their own small chapels.

The chapel also took on new importance in Waukesha as the one institution and the one physical structure where virtually all Welsh social events took place. Chapel was naturally the focus of family ac-

Table 14.1. Welsh chapel membership and attendance, Waukesha County, 1873

Chapel	Year established[a]	Denomination	Members	Hearers[b]	Sunday school
Jerusalem	1842	C. Methodist	66	150	85
Zoar	1844	C. Methodist	28	70	35
Tabernacle[c]	1845	Congregational	38	—	—
Bethel	1848	C. Methodist	18	90[d]	60
Salem	1851	C. Methodist	37	55	35
Bethania	1852	C. Methodist	31	75	40
Bethesda	1857	C. Methodist	48	115	95
Moriah	1860	C. Methodist	60	115	75
Zion	1868	C. Methodist	45	75	75
Total			371	745[e]	500[e]

Sources: "Mynegiad o'r Gofyniadau, a'r Atebion gafwyd iddynt" (Statement of the questions, and the answers received to them), Welsh [Calvinistic Methodist] Synod Records, 1849–1877 Waukesha, Wisconsin; Delafield, Wisconsin, First Congregational Church Records, 1866–1890.

[a] The year when members of a *seiat* decided formally to organize a chapel. The chapel building was usually constructed some years later.

[b] Given the seating capacity of the chapels, it seems most likely that "hearers" counted everyone present for the main service, including members.

[c] Data for 1872.

[d] The handwriting is unclear; figure may be less than 90.

[e] Adding probable attendance at Tabernacle and Union, the Congregational chapel in Bark River, the total number of hearers would have been at least 820–850 and the number of Sunday school attendants at least 550–575.

tivities on Sunday, but it also was the location for midweek religious gatherings, choir rehearsal, and meetings of the ruling elders. By the 1870s, the chapels in Waukesha were taking turns hosting quarterly Sunday school competitions, where adult and children's classes from across the community vied for the Sunday school banner in group recitations of the Bible and the Welsh catechism and written essays on theological themes. Early in the twentieth century, Jerusalem Chapel became host to the major Welsh gathering of the year, the annual oyster supper at Christmastime.[16] Meetings at chapel, whether sacred or secular, reminded the Welsh of their ethnic identity, because it was in chapel where they heard Welsh, saw Welsh friends, followed the Nonconformist liturgy, and remembered the passing generations whose Welshness no one could question. Throughout the nineteenth century, only people of Welsh origin became chapel members in Waukesha. Even though the plain clapboard buildings did not physically resemble the stone-walled, slate-roofed chapels of Wales, the society they housed had direct, meaningful connections to the old country.

Despite these elements of continuity, changes also occurred, eventually weakening what was distinctly Welsh in Welsh-American religious life. The Sunday school competitions, for example, probably developed from the Mid Wales tradition of *y Gymanfa Bwnc* (literally, "the topic assembly"), an annual service in which classes recited portions of the Welsh Bible and catechism in haunting rhythms, sometimes in harmony. In Waukesha, the *pwnc* became more openly competitive and hence more secular and American in nature. The genuine zeal of religious conversion faded, in some cases becoming a stale righteousness. One local character could not abide the piousness of his fellow Calvinistic Methodists. This Richard Jones was an abrasive farmer-poet who signed his poems Cymro Cloff (the lame Welshman). In protest against what he considered his kinspeople's hypocrisy, he declared himself an atheist in old age and insisted that his body be buried in a cemetery outside the Welsh settlement.[17]

Three events in the community's religious history further mark significant shifts away from Welsh cultural traditions. The first was the sole case of schism in the Waukesha settlement, which occurred in 1851. Local people recall two versions of the story. In the innocuous version, a fracas started between two members of Jerusalem Chapel when someone opened his umbrella in the chapel's horse stall and frightened a neighbor's horse. Bad feelings festered between the two, people took up sides, and eventually the minority group hived off to build their own chapel, Salem, less than one mile from Jerusalem. The more credible story is that the schism arose when Jerusalem's conservative congregation refused to allow an illegitimate baby to be buried in the chapel cemetery. The mother's parents and other sympathetic members left to form their own congregation. The infant was the first person buried in Salem Cemetery, which lies on the other side of the road within sight of Jerusalem Cemetery. Salem became known for its literary society (at whose meetings members reportedly drank sherry!) and for its members' progressive outlook.[18] Salem closed in 1878, but its creation indicated that at least some in the Welsh community had relaxed their grip on Calvinistic morals and begun to adopt more liberal standards.

The second event suggests the influence of prosperity upon Welsh religious feeling. In 1857, a young Cardiganshire immigrant named Humphrey Rowland Jones was converted at the Fulton Street revival in New York City. He became a revivalist himself and spent the next two years preaching, mainly in Wisconsin Welsh settlements. Jones claimed in his biography that, during his visit to Waukesha County, "the dawn broke and there happened the greatest revival of my ministry in America." When he returned to Wales in 1858, Jones and his friend David Morgan initiated the great midcentury revival that swept across the Welsh countryside, swelling chapel membership so much that hundreds of congregations built grand new chapels like the Calvinistic Methodists' in Pontrhydfendigaid (Figure 14.4). Extant records for the Waukesha chapels, however, indicate that Jones exaggerated his impact on the people here. Only Bethania Chapel's membership registered a jump in 1857, when Jones's revival moved about 40 people to join. Membership declined soon thereafter.[19]

Historians often interpret outbursts of revivalism as expressions of anxiety over unstable social conditions.[20] The Waukesha community's muted response to Humphrey Jones suggests that they were a fairly contented people, materially and spiritually. The Welsh enjoyed a steadily improving standard of living in the 1850s. Their control over territory was continuing to expand as parents helped their children buy land and establish farms in the area. Chapel membership was already near the saturation point. If anything, God was confirming their faith with the blessings of prosperity.

The third event was a symbolic geographic change that again focused on Jerusalem Chapel. Early in the twentieth century, the congregation struggled with the question of whether to relocate its sanctuary to the center of the village of Wales, which had sprung up in 1882 when the Chicago, Milwaukee and Northwestern Railroad came through the county. Traditionalists and many farmers wanted to keep the plain wooden chapel where it was on the outskirts of the village. Families engaged in business in the village argued that the congregation should build a new, up-to-date sanctuary that reflected the community's prosperity. According to Mildred Southcott, the village contingent eventually lost patience and decided to force the issue. Late one night, several men sneaked into the old chapel and stole the communion cup and plate. They kept the sacred objects hidden until a majority of the congregation capitulated to the new plan and voted to begin construction. Needless to say, committed traditionalists were outraged by this tactic. Only years of gentle diplomacy by Daniel Jenkins Williams healed the rift.[21]

Religion as the Vessel of Ethnic Culture

I have argued here that rural Welsh-American ethnicity crystalized around the single institution of religion. I should pause, in conclusion,

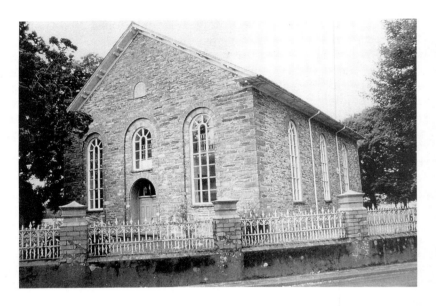

Figure 14.4. Calvinistic Methodists in Pontrhydfendigaid, Cardiganshire, responded to the 1859 revival, led by Humphrey Jones and David Morgan, by building a handsome new sanctuary nearly three times the size of their former chapel. The new chapel seats several hundred people in polished wooden pews. The plain wooden benches in Waukesha's Bethania Chapel, now a private residence, would have been crowded with its 30 members and 40 new converts in 1857 (Photographs by the author)

to explain what I mean by ethnicity. It is not the transplantation of a foreign culture to American soil but a distinctive process of change in that culture, from its being the medium in which all aspects of daily life take place to its being restricted to or reserved for particular social functions. Another way to put it is that a culture becomes ethnic when it is confined to certain activities, relationships, and places. In the case of the Welsh in Wisconsin and many other rural regions, the chapel became the repository for most of the social functions conducted in Welsh outside the home. As a social center it also generated new activities, such as the Sunday school competitions, that the Welsh themselves recognized as something that distinguished them from other immigrants and old-stock Americans.

Scholars have argued that immigrant communities in rural America were simplified versions of the complex societies from which the immigrants had come, precisely because they tended to have fewer institutions and less complex social relationships.[22] One could say this was true of the Welsh in Waukesha County. As immigrants, most of them achieved a comfortable, middle-class standard of living that far exceeded what they had known in Wales. The very complexity of society in Wales worked against the prosperity and self-advancement of the Welsh-speaking, indigenous population, for a deeply entrenched class structure of English-speaking, or at least culturally anglicized, landlords over an impoverished Welsh tenancy severely restricted the latter's social mobility and access to land. The difficulty in obtaining land to rent, let alone to buy, was one of the chief motivations for emigration. Yet religion became a more complex institution in the United States than it had been in Wales. Rather than seeing immigration and settlement as processes that stripped away social complexity, we may better understand Welsh immigrants' focus on chapel life as an elaboration and enrichment of a single institution, making it the vessel of ethnic culture and indeed a crucible in which that culture was transformed without losing its original identity.

The chapel tradition has remained important to Welsh Americans in Wisconsin and has undergone something of a revival in recent years. Every summer, *cymanfaoedd canu* (hymn-singing assemblies) in little Welsh chapels across the state attract full congregations of singers, many of whom are learning to pronounce Welsh for the first time in order to sing the moving old hymns in their original language. Members of the Wisconsin Gymanfa Ganu Association are concerned over the relatively few young people who attend these singing services, a worry shared by Welsh-American organizations across the country. Yet so long as the twin interests of genealogy and ethnic heritage compel

Americans to examine their pasts, Welsh-American ethnicity of some kind is in no danger of disappearing or of losing connection to its religious roots.

Notes

1. Urban Welsh settlements, although shaped by a strong religious element, had other foci for ethnic identity, particularly the dominance of the Welsh in certain industries, notably coal mining and iron making. On the industrial Welsh, see W. D. Jones, *Wales in America: Scranton and the Welsh circa 1860–1920* (Cardiff, 1993).

2. A. H. Dodd, *Life in Wales* (London, 1972), 135. As Henry Richard memorably wrote, "The Welsh, amid poverty, isolation and discouragement, have provided themselves with more ample means of religious worship and instruction than can be found, perhaps, among any people under the face of heaven." From *Letters on the Social and Political Condition of Wales* (1866), 23, quoted in C. Turner, "The Nonconformist Response," *People and Protest: Wales 1815–1880*, ed. T. Herbert and G. E. Jones (Cardiff, 1988), 85.

3. D. J. Williams, *The Welsh Community of Waukesha County* (Columbus, Ohio, 1926), 133 (quoting from Catherine Williams' obituary in *Y Cyfaill* [The Friend], July 1867, p. 134).

4. A. Conway, ed., *The Welsh in America: Letters from the Immigrants* (Minneapolis, 1961), 48–49. The Welsh saying is from a list of proverbs in H. M. Evans and W. O. Thomas, eds., *Y Geiriadur Newydd, the New Welsh Dictionary* (Llandybïe, 1953), 226.

5. Williams, *The Welsh Community*, 146; Register of Deeds, Waukesha County, 1840–1850.

6. D. J. Williams, *One Hundred Years of Welsh Calvinistic Methodism in America* (Philadelphia, 1937), 171–172.

7. Interview with Owen Williams at his home, Waukesha, Wisconsin, July 30, 1988. The order of the service and Sunday school would vary according to the minister's schedule of sermons in various chapels.

8. Interview with Gwen Davies (b. 1899), conducted by Edward Wicklein, December 26, 1978, tape on file at the Presbyterian Historical Association, Philadelphia; interview with Mildred Hughes Southcott at her home, June 24, 1988. *Hwyl*, related to the Welsh word *hwylio* (to sail), describes fervent prayer as well as the classical, emotive style of delivery to which Welsh preachers aspired.

9. Williams, *The Welsh Community*, 148.

10. Williams, *The Welsh Community*, 227–228, 238. The names of Waukesha teachers from the earliest years of the school were checked in the 1850 U.S. Manuscript Census, which includes individuals' place of birth, and were found to be American-born, not of known Welsh descent.

11. *Milwaukee Journal*, June 11, 1954, p. 1; E. Edwin Jones Papers, manuscript minutes from a meeting of the Welsh Board of Missions of the Presby-

terian Church of the U.S.A., September 29, 1936. My thanks to D. Clair Davis, Ft. Washington, Pennsylvania, for sharing the Jones Papers with me.

12.   Welsh Synod Records, Llyfr Perthynol i Gyfarfod Dosbarth . . . y Tref-nyddion Calfinaidd, Eu Dosbarth, Waukesha, Wisconsin, 1849–1877, vol. 1.

13.   *The History of Waukesha County* (Chicago, 1880), 366–371, 926–929.

14.   "Early Welsh Settlers," *The Waukesha Freeman* (April 29, 1897), p. 1.

15.   *The History, Constitution, Rules of Discipline, and Confession of Faith, of the Calvinistic Methodists in Wales,* 3/e (Mold, Wales, 1840), rule no. 20, p. 37; E. B. Wilard, ed., *A Standard History of the Hanging Rock Iron Region of Ohio,* vol. 1 (n.p., 1916), 448.

16.   Williams, *The Welsh Community,* 155; interview with Mildred Hughes Southcott, June 24, 1988.

17.   Interview with Mildred Hughes Southcott, June 24, 1988; interview with Enid Hoffman at her home, Madison, Wisconsin, July 18, 1988.

18.   Interviews with Mildred Hughes Southcott, June 24, 1988, and Owen Williams, July 30, 1988; Williams, *The Welsh Community,* 154–155. Interestingly, Williams says nothing to explain the location of Salem—"the largest and most convenient church in the community when it was built." *The Welsh Community,* 155.

19.   J. J. Morgan, *Dafydd Morgan a Diwygiad 1859* (David Morgan and the revival of 1859) (privately published by the author, 1906), 20–21; Williams, *The Welsh Community,* 157.

20.   See, for example, M. P. Ryan, *Cradle of the Middle Class: The Family in Oneida County, New York, 1790–1865* (Cambridge, England, 1981).

21.   Interview with Mildred Hughes Southcott, June 24, 1988; Williams, *The Welsh Community,* 149–150.

22.   R. C. Harris puts the classic case in "The Simplification of Europe Overseas," *Annals of the Association of American Geographers* 67 (December 1977): 469–483.

15

Community Building, Conflict, and Change
Geographic Perspectives on the Norwegian-American Experience in Frontier Wisconsin

Ann Marie Legreid

Norway's emigrants were drawn to America by the promise of a better material life, yet they also expressed in their behavior a desire for stability, community, and continuity with their past. The strongest expression of this desire was, quite inarguably, the transplantation of religious faith and practice. "The very process of adjusting immigrant ideas to the conditions of the U.S.," wrote Oscar Handlin, "made religion paramount as a way of life."[1] Religious observance was, from the beginning, a fundamental aspect of life in Norwegian immigrant communities, providing social as well as spiritual solace, and it was often the source of the first emergence of a sense of community. These church-centered communities often reflected strong provincial, parish, and neighborhood allegiances transplanted from the homeland as well. The intensely streamlike chain migrations of Norwegians from specific valleys and fjords in Norway to specific settlements in America created a kind of cultural homogeneity and sense of belonging in Norwegian immigrant settlements that was conducive to the implantation of a fervently supported parish church as a community focus.

But religious conflict was also a recurring theme in Norwegian social life in America. Immigrant families, neighborhoods, and communities often splintered over social, procedural, and doctrinal issues that came to light in local congregations. The clean outlines of the Norwegian communities were thus disrupted by internal strife and the splintering of congregational memberships. The reorganization of local congregations in the face of these conflicts suggests that the Church could also become a destabilizing force in many immigrant communities.

Norwegians and Norwegian Americans are well known for their intense theological disputes over the formalistic approach to religion espoused by the state church, the Church of Norway, over popularly led revivalist challenges to the establishment, and over the difficult issue of predestination. We know less about the local particulars of those confrontations and their long-term impact on evolving ethnic settlements. Was community, in some sense, redefined? What role did community discord play in the broader context of community development? Drawing from the formative experiences of several Norwegian settlements in western Wisconsin, this chapter aims to broaden our understanding of both conflict and cooperative interaction in developing ethnic communities.

The spatial foci of this chapter are the three Norwegian immigrant communities of Trempealeau Valley, North Beaver Creek, and Blair, which straddle the Trempealeau-Jackson County line, and a fourth community, Coon Valley, which is located to the south in Vernon County. Settled by chain migrations of kinship groups, beginning in the 1840s and 1850s, each community displayed a pattern of spatial exclusivity based on specific origins in Norway, which extended down to the provincial, *bygd* (culture region), and parish level. Trempealeau Valley drew its residents heavily from parishes in Gudbrandsdal and Østlandet. North Beaver Creek displayed a bias for the parishes of Inner Hardanger. Blair, more diverse in character, pulled its people from Gudbrandsdal, Telemark, and assorted parishes in central and southern Norway. Coon Valley was dominated by people from Toten, Gudbrandsdal, and Telemark. Unlike some neighboring Yankee villages, Blair was unmistakably *en norsk by* (a Norwegian town).[2] Focused on Lutheran churches of the Norwegian Synod, each community exhibited a remarkable degree of cultural uniformity and cohesion. From the beginning, these church-centered communities were bound together as sister congregations in multimember pastorates. Each became embroiled in secular as well as theological controversies, splintered, and then reorganized. They represent, in microcosm, the strife that plagued Norwegian-American Lutheranism and Norwegian-American communities throughout the latter part of the nineteenth century.

The Norwegian Ethnic Church

The nineteenth century was a time of intense religious activity in Norway. One of the strongest movements to grip the Norwegian peasantry was the revival, led by country lay preacher Hans Nielsen Hauge. For the state Church of Norway, Hauge's efforts ushered in decades of internal discord between the traditionalists, or religious elite, and the strongly evangelical pietists led by Hauge. The picture was complicated in the last quarter of the century by the Johnsonian Revival, which combined the doctrinal orthodoxy of the state church with a pietistic spirit in a form that was copied and became dominant among Norwegian Lutherans in America. In contrast with the Haugean movement, the Johnsonian Revival was directed by clergy within the state church, and subsequently drew the Haugean and state church factions closer together. The last major revival, the Western, was highly critical of the established church and sometimes extreme in its low church views. In this atmosphere of ecclesiastic struggle, Norwegians left in large numbers for America.[3]

This same religious enthusiasm that animated Norwegian society characterized and influenced the founding of congregations on the American frontier. State churchism and revivalist Haugeanism clashed in Norwegian settlements of the Middle West along the same lines as they had in Norway. Circumstances in America, furthermore, triggered new issues and debates, which, together with the conflicts inherited from Norway's state church, helped to shape the patterns of Norwegian immigrant Lutheranism.

The Norwegian Synod, representing the high church point of view; the Conference, representing the broad church; and the Haugean Synod, representing the revivalist, or low church, point of view, were the three dominant synods in Norwegian-American Lutheranism by the 1880s. Only the broad church extended across ethnic lines to any significant degree, primarily to the Danes in the Conference and Swedes in the Augustana Synod. The Norwegian Synod, a true "ethnic synod," was composed almost exclusively of Norwegians, enjoyed preeminence among the synods, and was largely unrivaled until internal dissension over the predestination issue fragmented its ranks in the late 1880s.[4]

The Norwegian Synod had no long-term interests in proselytizing among other national groups. Rather, they confined their mission efforts to group conversion and to the improvement of their clerical forces within existing congregations. The low church Haugean Synod, on the other hand, made a strong effort to promote evangelistic work, although only within its own districts. The Norwegian Synod's insistence on a resident ministry and the congregation as the only tool for evangelism, therefore, hindered the expansion of the synod into areas void of Norwegian immigrants. This conservative strategy was quite unlike the Methodists, for example, who expanded rapidly across the frontier through the efforts of the circuit riders.

A pressing issue among Norwegian Lutherans, throughout the nineteenth century, was the place of English in their churches, and the major question was how to preserve Norwegian from eventual extinction. The Norwegian Synod championed the use of Norwegian and only Norwegian in their congregations, a major deterrent to membership from other language groups. Their strict insistence on the use of the mother tongue was inextricably bound to the high church theology. While the leadership of the synod recognized the need for English in the immigrants' larger world, they believed it equally important to "see that the faith of the fathers was preserved in its purity."[5] To the members of the Norwegian Synod this meant the "language of the heart," and they feared that American doctrinal discrepancies would be absorbed into the Norwegian churches along with the adoption of English. In this vein, however, the synod differed markedly from the revivalist and broad churches, the latter more anxious to make the language transition. As early as 1841, Elling Eielsen, revivalist lay preacher, had *Luther's Small Catechism* translated and published in English. Language, slavery, and parochial education were but a few of the myriad of social concerns that tore at immigrant congregations.[6]

As immigrants flowed onto the frontier, church leadership became painfully aware of the need to staff these developing communities with a proper clerical force. The Norwegians, by and large, did not find homes in other national or ethnic churches, depending instead on their fellow countrymen for religious guidance. The state Church of Norway, moreover, was reluctant to respond to the educational and missionary needs of its countrymen in America. As Norwegian-American historian, Theodore Blegen, has observed, ". . . no clergymen of the State Church went with them until nearly two decades after the emigration was launched by the Restauration and its sloop folk" in 1825.[7] Circuit ministries, in which a clergyman served from several to a dozen or more congregations, usually answered the needs of settlers in the earliest years, or, in the case of the Haugeans, lay persons were called on to provide a less formal ministry. The official stance of the Norwegian Synod was uncompromising in that "no congregation should call any but an ecclesiastically approved clergyman."[8]

Thus, the task of staffing the frontier with a proper ministry rested

with the religious leadership in America. With regard to rites and ceremonies, furthermore, synod congregations were obligated to follow
the orthodox ritual of the Churches of Denmark and Norway dating
from 1685, with few allowances made for the special circumstances of
the frontier.

In the mid-nineteenth century, members of the German Missouri
Synod formed the largest, most powerful body of American Lutheranism, and the Norwegian Synod naturally called on them for assistance in the preparation of its clergymen. As early as 1859, prospective
pastors of the Norwegian Synod were being trained at the German
Lutheran Seminary in St. Louis. Since some Missouri Synod views
were controversial, this action displeased a large segment of the immigrant population and caused a cooling of relations between the Norwegian Synod and the state Church of Norway.[9]

The most damaging controversy generated by the association of the
Norwegian Synod with the German Lutherans was that of election or
predestination. The debate on predestination has erupted at various
times and with varying intensity throughout the centuries. It flared
up among Lutherans in America in the 1870s and 1880s and lingered
for more than a half-century before finally becoming a dead issue. The
predestination doctrine holds that God has chosen or elected only a
few individuals to eternal blessedness and that the remainder, regardless of their faith and merits, are predestined for eternal damnation.

The issue caused a great conflagration among Norwegian church-
men, prompting them to polarize into Missourian (belief in predestination) and anti-Missourian factions. The years of bitter debate
culminated in the withdrawal, in 1887, of Friedrich Schmidt and his
associate, P. A. Rasmussen, of the Norwegian Synod, and about a
third of the membership of the synod to form the Anti-Missourian
Brotherhood. The other Norwegian Lutheran synods were unmistakably anti-Missourian in their sentiments, but did not engage actively
in the controversy.

The Anti-Missourian Brotherhood immediately created a movement toward union between the Conference, the Augustanans, and
themselves. Their goal was realized with the formation of the United
Norwegian Lutheran Church in America (UNLCA) in 1890.[10] With
more than half of all Norwegian Lutherans in America in its registers,
it easily surpassed the mother Norwegian Synod in numerical size
and influence. As tensions escalated, religious realignments were triggered among the Norwegian Synod congregations. The most common
form of realignment was reaffiliation, which meant that a dissenting
congregation simply broke its relationship with the Norwegian Synod

and reaffiliated with the Anti-Missourian Brotherhood. Schism was the more radical and common form of realignment in which a group of dissenters broke away to form their own independent congregation.

Norwegians in Western Wisconsin

Vernon, Trempealeau, and Jackson counties lie within the embrace of Wisconsin's Driftless Area, a vast unglaciated region of rugged relief that extends from the northwestern corner of Illinois to the Chippewa River valley in west-central Wisconsin. They also form part of a swath of Norwegian settlement that encompasses much of the western part of the state. Norwegian immigrants flowed into Vernon and Crawford counties in the 1840s, settling first in the vicinity of Coon Prairie, and then moving on into Coon Valley in 1848 under the leadership of Even O. Gullord, who was one of the first emigrants from Biri Parish. Among the principal towns in the area was nearby Viroqua, a thriving village and county seat, founded by the Yankees. The towns of De Soto, Victory, and Geneva on the Mississippi were the primary markets to which farmers shipped their grain. In addition, three small villages developed along the length of the Coon Valley. Stoddard, which lies below the mouth of the valley on the Mississippi, had almost no Norwegian population. Chaseburg, lying about eight miles up the valley, had a mixed population of mostly Yankees, Germans, and Norwegians. And Coon Valley Village, another eight miles up the valley, was almost exclusively Norwegian and the true heart and core of the Coon Valley settlement.

The fact that Norwegians dominated the early phase of settlement, of course, was favorable to the emergence of a pattern of strong consolidated landholding. The accompanying map of landownership (Figure 15.1) shows the pronounced concentration of Norwegians along the length of Coon Valley. The parish groupings depicted on the map show that clustering also occurred on a very intimate level, where small groups from common parishes took up lands in a contiguous pattern.[11] Migrants from Sannikedal Parish, Skien, Telemark, as one example, congregated in Haagadalen, a small coulee at the westernmost edge of the valley near Stoddard. The strongest evidence of parish clustering can be seen in the northeastern portion of the valley, Skogdalen; this area was occupied almost exclusively by people from Nordre Fron Parish, North Gudbrandsdal. Various degrees of group mixing are also evident, most notably in the areas immediately adjacent to Chaseburg and Coon Valley villages. The mixing of groups suggests that in some areas the spread of settlement was so rapid and

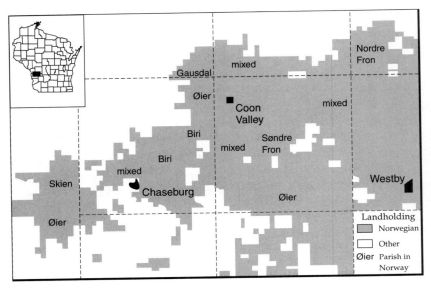

Figure 15.1. Map of major parish groupings by landownership, Coon Valley, late nineteenth century (Compiled from the U.S. Census of Population, 1900; plat maps of Vernon, La Crosse, and Monroe counties; and local congregational registers)

the demand for land so intense that a contiguous pattern of landholding was impossible to achieve. Clearly, immigrants acted on their desires to take up lands adjacent to fellow Norwegians, whose customs and language were familiar. Growth of the settlement was more intensive than extensive. In the words of sociologist Peter Munch, the "Norwegians have actually managed to withdraw the whole area from the economic and social control of the Yankee-dominated . . . Viroqua."[12]

In Trempealeau and Jackson counties the patterns of ethnic and parish concentration were repeated. A few Norwegian farmers, chiefly from Muskego and other early Norwegian settlements in southern Wisconsin, arrived between 1854 and 1857 and settled among the Yankee population in the valley of the Trempealeau River. In 1857 the first settlers arrived in Beaver Creek by ox cart from the Koshkonong settlement in Dane and Jefferson counties. In the succeeding years Norwegians poured into the townships, most of them direct from Norway, overflowing the main valleys into surrounding coulees and spreading onto the bluffs and slopes, mingling with the Germans, Swedes, Poles, English, Scots, and Irish.

Local Religious Realignments

The growth of congregations in western Wisconsin paralleled the development of the communities themselves. The first efforts to organize a congregation in Coon Valley date from 1854, although actual incorporation came later and unity was short-lived. Dissension over the construction of the church building led to the secession of residents in lower Coon Valley, who protested the travel distance. The residents of upper Coon Valley, also irritated by the decision, wrote a new constitution and seceded in protest to form another new congregation. The weakened parent congregation, middle Coon Valley, thus lay between its dissenting progeny. The same doctrinal issues that rocked other Norwegian Lutheran congregations, however, played a relatively minor role in the religious history of Coon Valley. Fearing that tranquility in the valley was threatened by the predestination debates, the congregations withdrew from the Norwegian Synod and reaffiliated with the Anti-Missourian Brotherhood. Schism was avoided. The valley, however, remained free of religious strife for only a short time. On the basis of a vote concerning the retention or dismissal of a pastor in 1898, some members of the upper valley congregation broke away and formed the Norwegian Evangelical Congregation for the Skogdalen District. Thus, the simple pattern of 1858 gave rise to a more intricate pattern by 1898 (Figure 15.2).[13]

A similar pattern developed among the Trempealeau Valley, Blair, and Beaver Creek settlements. In 1857, only three years after the first group of Norwegian immigrants entered the Trempealeau Valley, Pastor H. A. Stub was called to bring them together formally into a Norwegian Synod congregation. Through a series of major divisions, this mother church parented no fewer than six area congregations, including some of the Haugean and Conference persuasion. The Blair and North Beaver Creek mother congregations were less prolific; these communities were fractured only by the predestination issue in the 1880s. The spatial outlines of these congregations in 1880 were quite distinct, with each mother congregation exhibiting a spatially well-defined membership field. These congregational memberships were ethnically pure, for example, 96 Norwegian souls and 1 German; likewise, there were no nearby Lutheran congregations competing for members. In the wake of schism, however, these spatially well-defined patterns were transformed into a complex pattern of overlapping memberships, counter to the expectation that membership patterns would become more compact and exclusive over time, which would

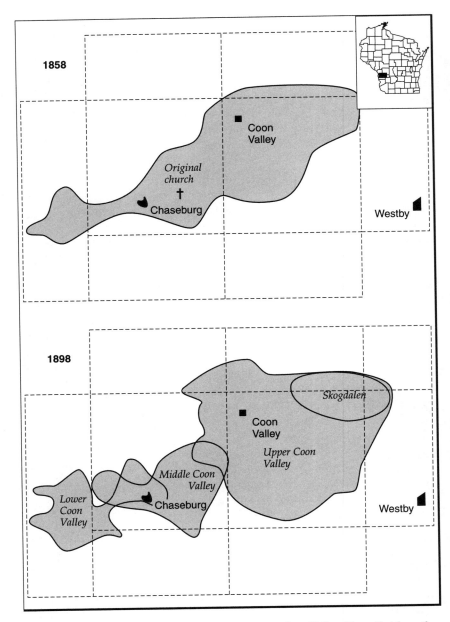

Figure 15.2. Maps of religious affiliation, 1858 and 1898, Coon Valley (Compiled from the U.S. Censuses of Population, 1860 and 1900; plat maps for Vernon, La Crosse, and Monroe counties; and local congregational registers)

have implied a more tightly bound community. The realignment of memberships, however, suggests that the immigrant church assumed a divisive role in these frontier communities (Figure 15.3).

Prelude to Schism

While admittedly partisan in interpretation, the more than 400 pages of special congregational and council minutes, covering the period 1881–1892, enlighten us to the special conditions and consequences of the predestination controversy. The secretarial entries provide specifics on the doctrine itself, arguments, charges, and countercharges, as well as the sad spectacle of attacks against some individuals.[14]

During the predestination debates, the Blair, Beaver Creek, and Trempealeau Valley congregations were served together in a circuit of churches by Pastor Brynjulf Hovde of the Norwegian Synod. Congregational minutes give this picture of the prelude to schism:

North Beaver Creek (NBC), October 26, 1887. A newly-drafted Constitution is presented to the NBC congregation for their examination. It is met with strong opposition by Pastor Hovde [Missourian in his views] ". . . it will have to go as it will," he said, "I will have no part in redoing or drafting a new congregational Constitution." A member of the Anti-Missourian group assails the pastor's integrity, accusing him of building upon false principles and of "poisoning the lifeblood of the congregation." Hovde is angered by the charges and asserts that the congregation would have no sound basis for dismissing his services, whereby K. K. Hallanger, spokesman for the Anti-Missourians, replies: "We think that dissatisfaction with you, unrest in the congregation, and in turn, so little of your pastoral service are grounds enough!" The main reason for parishioner discontent is revealed publically for the first time by layman Torkelson [January 1888]: "At a pastors' conference in Decorah, 1884, Hovde knowingly and willingly defected to the Missourian party, without consulting the congregation and without regard for the wishes of the majority . . . and that he kept his actions secret for as long as possible . . . and it is this action that has brought about such a stir of discontent." Hovde is asked to relinquish his ministry in the community.

North Beaver Creek, February 16, 1889. Pastor Hovde's presence at the congregational meeting is viewed as "malicious interference." He is dismissed of his duties; the Synod is notified that the basis for termination is Hovde's stance on predestination and conversion [Missourian]. A parishioner refers to the congregational discord as "wrought by the Pastor and the devil himself." In a letter to Chairman Preus, Hovde described the actions of the Anti-Missourians in Beaver Creek as "with total disregard for the (Synod and) Constitution," and

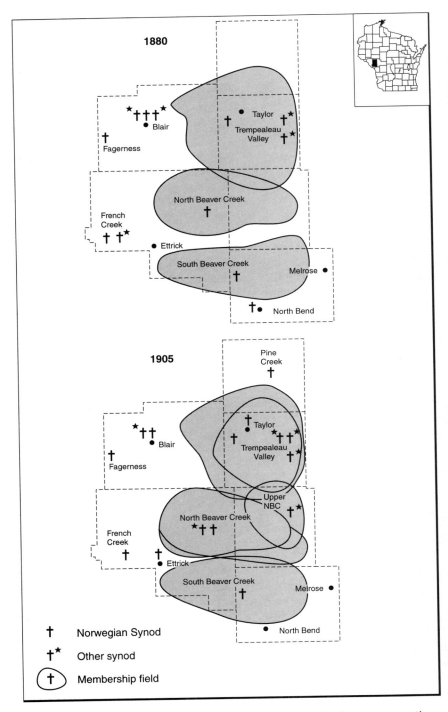

Figure 15.3. Maps of patterns of affiliation with Norwegian Lutheran congregations, townships in Trempealeau and Jackson counties, 1880 and 1905 (Compiled from the U.S. Census of Population, 1880; Wisconsin State Census of Population, 1905; plat maps for Trempealeau and Jackson counties; and local congregational registers)

"in Trempealeau Valley the situation is even worse." At one point he describes them as "the most ungodly people in the world. . . ." While absent from congregational volumes, references to religious fervor do occasionally appear in area newspapers. At times the religious derangement was so intense that local doctors diagnosed it as "insane" and had individuals committed to the county asylum.

The mechanics of schism could be difficult. In Beaver Creek, Hovde led a minority of Missourians out of the mother church and built a new church of the Norwegian Synod farther up the valley. In Trempealeau Valley, Hovde and his Missourians were for a time denied access to the church building, meeting instead in member homes and outbuildings. When the Anti-Missourians left the valley to join their counterparts in Blair, Hovde and his followers were able to resume use of the original church building.

In the Aftermath of Schism

Were the patterns of community life seriously disrupted by the conflicts surrounding schism? Were social networks reordered and communication eclipsed between the differing congregations? The evidence, albeit fragmentary, indicates no. Take this example from the North Beaver Creek community (1892):

A motion was made to give cemetery rights to the Synod congregation/those who had seceded. A suggestion is made that the majority also aid the Synod congregation with a gift of money. A gift of two hundred dollars was given to the Synod church in an act of "true Christian brotherhood." With this gesture, the recorded history of rivalry between the two congregations comes to a close, and the congregational minutes are then dominated by plans for building and program expansion.[15]

The congregations, former combatants, worked in concert in multiple ways: parochial school, cemetery maintenance, church anniversary celebrations, and through several war efforts. Neighbors continued to aid neighbors in barn raising, cornhusking, and threshing, and parishioners integrated socially in birthday clubs, sewing circles, literary and debate societies, and even the *bygdelag* in Blair. They fooled together as *julebukkere,* and cooperated with the larger community of Norwegians in the Fraternal Insurance Company of Ettrick and in several farm cooperatives.[16] A small chapter of the Ku Klux Klan, reacting to the presence of the Catholic Irish, existed for a time between the world wars. No formal membership lists are known to exist, but

312 Part Two. Settlement Processes and Cultural Patterns

oral interviews confirm that KKK members traversed the congregational lines. Children were integrated in both public and parochial instruction. Even during the worst of the doctrinal conflict, parishioners continued to hold several months of common classes in the *religionskole*, or parochial school. Strong marriage, boarding, and labor linkages were maintained between the congregations. In cases of intermarriage, however, it was generally widely known and remembered that the spouse had hailed from "the other congregation." Many parishioners may have disapproved of the theological wrangling, but few members' households (only one, in North Beaver Creek)—actually defected from the congregations in the true sense of the word.

Apart from community churches, there were other focal points for social exchange. The old country store, supplement to the larger towns, was a place where parishioners mingled in the exchange of business, gossip, and personal opinions. There were two such hamlets in Beaver Creek, each with a post office–store, owned and operated by local Norwegians. Since most purchases were made in an interlocking system of credit, the records of these general stores, the day ledgers, help one to flesh out the business and social networks of the community. For example: O. N. Herreid credited his account in 1891 by bringing in eight bushels of grain, while K. K. Hallanger bought a portion of the grain along with a scythe and bridle. These commercial patterns do not appear to have changed in the face of the religious restructuring. People still patronized the local general store. Certainly kinship, need, and proximity played greater roles in the patterns of patronage than religious affiliation did.

Letters from America, few in number, deal foremost with family matters, wages, and the purely practical details of pioneering. While largely devoid of theological rhetoric, several letters do refer to the "disorder" created by the Norwegian Synod pastors.[17] Depositing the blame on the clerical leadership, passing the buck, made it possible for the congregants to dismiss the matter and go on with their lives. But how was it, then, that so many allowed themselves to be misled by a Norwegian Synod pastor?

If one must give a final verdict, the weight of evidence does seem to lie on the side of the parishioners. Rev. Hovde, rather than dwell on the need for congregational harmony, fed the fires of debate. By insisting that false teachings had poisoned the congregation, he lent a certain urgency to thoughts of secession. At times, personality conflicts nearly obscured the doctrinal issue, and Hovde, like many of his parishioners, was not endowed with the gift to differ gracefully. Moreover, the heat of the debates made it virtually impossible to be either passive or neutral.

Was the Reverend Hovde a fomenter of schism or simply a preacher who brought to light the tensions inherent in his congregation? By virtue of his training at the Missourian seminary, Concordia in St. Louis, Hovde's loyalties to the Norwegian-German alliance were to be expected. Congregations whose pastors were trained in the homeland, where the Missourian views were neither taught nor tolerated, tended to join the Anti-Missourian alliance. On the other hand, congregations with Concordia (Missourian)-trained pastors were most likely to remain loyal to the mother synod, whose Missourian views they shared.

While some ministers did not stir great emotion, others burned with religious conviction and took up the issue (predestination) at every opportunity. Since most pastors served multiple pastorates, it was likely that the issues were argued with similar intensity among sister congregations and often with very similar results. Sister congregations, those served by the same pastor, responded in nearly identical fashion, in fact, both locally and throughout the Midwest.[18]

As tensions deescalated, the issue slowly dissolved from community consciousness. "Tensions were great and not always gentle," writes Belgum, "but tensions within the church may make for life and progress, whereas easy argument often leads to somnolence, complacency, and loss of vigor."[18] In the end, the controversy had not caused a current of defection or excommunication. Indeed, parishioners may have found themselves established more solidly in their doctrinal beliefs, more involved in church activities, and driven by a new zealousness for the mission of the church at large. Many Missourian and Anti-Missourian congregations have stood side by side for years, engaging occasionally in minor squabbles, their parishioners bound by a common sense of heritage and purpose as Norwegian Americans and as Lutherans. Although the doctrinal strife did cause missionary efforts to languish and debts to grow, the synods were quick to restore internal harmony and to resume their Lutheran mission.

The Demographics of Schism

Schism in these Wisconsin communities was not completely religious in motivation, but was also compounded by personal conflicts, socioeconomic differences, and cultural differences and latent tensions imported from the homeland. While it is impossible to isolate all the factors that weigh upon a person's decision-making, certain factors lend themselves to simple quantification, such as intergenerational differences, kinship ties, recency of arrival, and cultural (*bygd*) or parish background.

The conflict between the generations is basically a conflict between Old World values, as integrated into the parental value system, and American values, which became a part of the value system of their children through American language and institutions. The evidence suggests that the first-generation Norwegians were quite pietistic and vitally concerned with the affairs of their congregations. The narrow, ascetic views of some older church members may have alienated younger parishioners, who were dedicated to different goals and cared little about doctrinal specifics. One might expect the heads of household within the dissenting congregations, which emerged later in the century, to have had a lower average age than the parishioners of the mother congregations. Comparisons of mean age between mother and daughter congregations, however, did not reveal any striking contrasts.

More discernable was the effect of recency of arrival on religious affiliation. The churches in this region were established very early and by persons who were imbued with the deep spiritual feeling of nineteenth-century Haugeanism. Immigrants who arrived later in the century were likely to have a different religious outlook or perhaps be more secularized and less zealous about spiritual matters. The gap between these value systems might well have found expression in the local doctrinal altercations. The mother congregations in these Wisconsin settlements retained a high percentage of first-generation worshipers; the greater share of their members immigrated prior to 1870—they were the community pioneers. The daughter congregations, on the other hand, had a much greater proportion of recent arrivals, that is, those who had immigrated after 1870. In North Beaver Creek, for example, 77 percent of all households were headed by recent arrivals. Similar patterns were evident among congregations in Coon Valley.

Was schism selective with respect to *bygd* or parish background? In North Beaver Creek, the mother congregation was dominated by people from the parishes of Ulvik and Ullensvang, Hardanger. The daughter congregation, however, included a disproportionate number of householders from parishes other than Ulvik and Ullensvang, that is, places along Sogne Fjord, Telemark, and Gudbrandsdal. The original Coon Valley congregation was composed largely of people from Biri Parish in Toten. The lower Coon Valley group, which broke away in 1859, had a strong representation from parishes in Gudbrandsdal (41 percent). The upper valley church had by far the most diverse population, more than half the settlers coming from parishes throughout Gudbrandsdal, while the Skogdalen group originated overwhelmingly (76 percent) from Nordre Fron in Gudbrandsdal (Figure 15.4). In the first years, a desire for cooperation and companionship may have overshadowed very subtle differences of religious opinion, but as the community grew, doctrinal and cultural differences were probably expressed more overtly. In each of the cases studied here, daughter congregations emerged in response to the growth of peripheral settlements, while mother congregations retained their hold on the long-established "community cores." The secession of dissenting elements left behind mother congregations that were smaller in absolute numbers, presumably diminished in power, yet more culturally homogeneous and spatially well-defined.

Community Building, Conflict, and Change among Norwegian Immigrants in Western Wisconsin

As more and more immigrants streamed onto the frontier, the religious patterns of one decade proved inadequate for the next. Churches proliferated in response to the needs of a burgeoning frontier population. Building space and travel distance, as in Coon Valley, were very real and practical concerns. The frontier itself gave free rein to secessionists; the state churches had no power here, "for a minority was always free to secede and drift off in its own direction."[19]

The Norwegians who settled in western Wisconsin carried forth a collective purpose and consciousness that reinforced a strong sense of community, even in the face of the worst of these controversies. Moreover, the resolution of the predestination issue among these Norwegians might be viewed as a form of intragroup tension management, the release of tension and emotion among people that were otherwise quite reserved in manner and lifestyle. Schism cut deeply across kinship lines, a condition that was unavoidable given that kinship was the primary basis for the formation of these communities. In communities, like families, relationships of conflict can be as normal and binding as relationships of cooperation.[20]

The decision to secede and reorganize probably gave rise to renewed trust and communication, whereby the larger goals of the church and the ethnic community could be achieved. Seen in these terms, schism can be viewed as a part of the process of community restructuring or redefinition, not as the basis for community disintegration or collapse.

The descendants of these debates are mostly ignorant of the issues that inflamed their forefathers. Some of the reasons they offer for

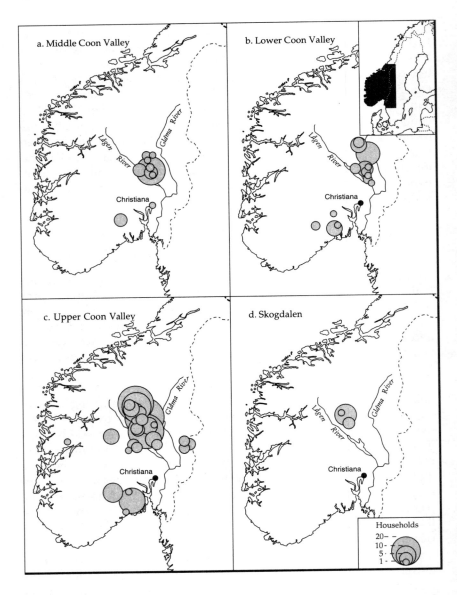

Figure 15.4. Maps of parish origins, the Coon Valley congregations (a) Middle Coon Valley (b) Lower Coon Valley (c) Upper Coon Valley (d) Skogdalen (Compiled from Holand, 1928; and local congregational registers)

schism: "The pastor was a troublemaker." "The parishioners needed more space." "Norwegians are strong-willed and tend to argue." Perhaps there was a certain satisfaction in argument, in asserting one's identity and opinion. Perhaps it was important in the process of building confidence to meet the challenges of the New World. While the ethnic church served as the primary arena for intragroup warfare, at least for many Norwegian immigrants it also provided a measure of order and stability that superseded those conflicts. This schismatic nature was true of the immigrants themselves; that is, they were inwardly Norwegian while outwardly in tune with their American world.

Notes

1. O. Handlin, *The Uprooted* (Boston, 1951), 117.

2. Detail on the formative years of the Blair, Beaver Creek, and Trempealeau Valley communities is provided in the author's dissertation, "The Exodus, Transplanting, and Religious Reorganization of a Group of Norwegian Lutheran Immigrants in Western Wisconsin, c. 1836–1900," University of Wisconsin–Madison, 1985. The Coon Valley community was scrutinized in the author's unpublished paper, "Coon Valley: The Formative Experience of a Norwegian-American Immigrant Community," 1990. Migration and community building have occupied other historical geographers; see, for example, R. C. Ostergren, *A Community Transplanted: The Trans-Atlantic Experience of a Swedish Immigrant Settlement in the Upper Middle West, 1835–1915* (Madison, 1988).

3. Developments in Norwegian Lutheranism are detailed in M. Nodtvedt, *Rebirth of Norway's Peasantry* (Tacoma, Wash., 1965); J. M. Rohne, *Norwegian American Lutheranism up to 1872* (New York, 1928); and E. C. Nelson and E. L. Fevold, *The Lutheran Church among Norwegian-Americans*, vol. 1 and 2 (Minneapolis, 1960).

4. For elaboration see Nelson and Fevold, *Lutheran Church among Norwegian-Americans*, vol. 1, 126–133 and 149–150.

5. Nelson and Fevold, *Lutheran Church among Norwegian-Americans*, vol. 1, 299; see also, 271–273 and 277–279.

6. Nelson and Fevold, *Lutheran Church among Norwegian-Americans*, vol. 1, 300. Congregational minutes show that all these issues were debated at one time or another in the communities under discussion.

7. T. C. Blegen, *Norwegian Migration to America: The American Transition* (Northfield, Minn., 1940), 101.

8. Nelson and Fevold, *Lutheran Church among Norwegian-Americans*, vol. 1, 115.

9. Nelson and Fevold, *Lutheran Church among Norwegian-Americans*, vol. 1, 161–163.

10. Nelson and Fevold, *Lutheran Church among Norwegian-Americans*, vol. 1, 123–124, 161–163, and 253–270. History will say that June 9, 1917, was the biggest day in the saga of Norwegian-American Lutheranism. Three major synods (Norwegian, Haugean, and the United-UNLCA), comprising more than 90 percent of the Norwegian Lutherans in the country, were united in the Norwegian Lutheran Church in America.

11. The classic history is H. Holand's *Coon Valley* (Minneapolis, 1928). Holand's rich biographical material, in combination with local congregation and census records, has made it possible to map landholding by parish of origin. See also E. M. Roger, ed., *Memoirs of Vernon County* (Madison, Wis., 1907); and *History of Vernon County, Wisconsin* (Springfield, Ill., 1884).

12. P. A. Munch, "Segregation and Assimilation of Norwegian Settlements in Wisconsin," *Norwegian-American Studies and Records* 18 (1954): 114. See also, P. A. Munch's "Social Adjustment among Wisconsin Norwegians," *American Sociological Review* 14 (1949): 780–787; and P. A. Munch, "Authority and Freedom: Controversy in Norwegian-American Congregations," *Norwegian-American Studies and Records* 28 (1980): 3–34. Coon Valley is a prominent extension of a larger block of Norwegian settlement that occupies much of the north-central portion of Vernon County. The county's second distinct block of Norwegian settlement is contained in the south-central townships and has its center of gravity at Folsom and West Prairie. Munch's studies show that the two communities experienced marked differences in growth. The south-central block underwent extensive growth as Norwegians succeeded the Yankee elements, while the Coon Prairie and Coon Valley settlements had almost no territorial growth after 1878, their development largely in the form of consolidation. Westby, a town founded and dominated by Norwegians, functioned as the pivotal point of the Norwegian population in this region.

13. For a fuller explication of these congregational histories, see Holand's *Coon Valley*, 60–141.

14. In 1875 the Norwegian settlement in the parent Trempealeau Valley covered an area about 30–40 miles east to west and 10 miles north to south (including Blair). The Beaver Creek settlement bordered it to the south, stretching east-west about a dozen miles. Congregational histories include *Our Heritage* (Black River Falls, Wis., 1959), and *This, Our Beloved Valley* (Blair Wis., 1925).

15. Congregational minutes, North Beaver Creek, December 15, 1892, located in the congregational parsonage in rural Ettrick, Wisconsin.

16. *Julebukkere* is a term for celebrants in costume at Christmastime. *Bygdelag* is an association of persons either born in or descended from a common region in Norway, such as the Hardanger Bygdelag.

17. Luther College Archive, North Beaver Creek Congregation manuscripts, Trempealeau County, Wisconsin. Data in support of these statements can be found in O. M. Norlie's *Norsk Lutherske Menigheter i Amerika*, 2 vols. (Minneapolis, 1918). See also chapter 7 of the author's dissertation, "Exodus, Transplanting, and Religious Reorganization."

18. G. L. Belgum, "The Old Norwegian Synod in America, 1853–1890," Ph.D. diss., Yale University, 1957, 142.

19. Handlin, *The Uprooted*, 114–115.

20. In North Beaver Creek, for example, 20 of the 26 charter households were linked in kinship, most of those linkages attributable to two large, extended families from Hardanger, western Norway.

16

Americans by Choice and Circumstance
Dutch Protestant and Dutch Catholic Immigrants
in Wisconsin, 1850–1905

Yda Schreuder

An interesting aspect of nineteenth-century frontier settlement in the Midwest is the idiosyncractic behavior of different immigrant groups. Even among groups of the same nationality background, differential behavior occurred. Dutch Protestant and Dutch Catholic immigrants, for example, lived totally separate lives, lived in different settlements, and followed different paths of adaptation and assimilation on the Wisconsin frontier. In fact, while many Wisconsinites today might be able to name several Dutch Protestant communities, most would be hard pressed to identify even one or two Dutch Catholic settlements. From a historical-geographic perspective this is an interesting point, given the fact that the two groups first arrived at about the same time in the mid-nineteenth century and were of similar socio-economic background in the Netherlands.[1] Both groups settled on what was then the western frontier, established rural communities, and engaged primarily in farming. However, after two or three generations of settlement in Wisconsin, their patterns of behavior in terms of rural/urban residence, occupation, marriage, and adherence to Dutch culture and traditions were very different. The marked contrast that is so noticeable in Wisconsin between the strong and persistent ethnic communities maintained by Dutch Protestant immigrants and the weaker proclivity toward ethnic association and identity among Dutch Catholic immigrants is found among Dutch immigrants in other parts of the country as well.[2] Dutch immigration historians have often made the observation that Dutch Protestant immigrants—in particular the orthodox Dutch Calvinist, or Dutch Reformed, immigrants—tended to isolate themselves within their exclusive religious-ethnic communities. Henry Lucas, in *Netherlanders in America*, noted that "Reformed principles" gave the Calvinists their conception of life and helped them to organize their social and economic activities.[3] Dutch Catholic immigrants, on the other hand, assimilated easily and rapidly, principally because there was no separate Dutch Catholic church.

Scholarly research on Catholic immigrants in the United States has shown that this pattern was common, though not universal, among Catholic immigrant groups.[4] Where Catholic immigrants of a particular nationality were present in large numbers, the Catholic church hierarchy sometimes decided to establish so-called nationality parishes. In general, however, Catholic church policy dictated the integration of different Catholic immigrant groups. That is not to say that everyone integrated happily with Catholics of other nationality backgrounds. The well-known conflicts between Irish and German Catholics in Wisconsin in the mid-nineteenth century is a classic example of ethnic discord in the Catholic church. However, most of the nationality parishes were short-lived and served only to accommodate the needs of the newly arriving immigrants rather than to promote long-term ethnic separation.

This chapter focuses on the divergent ethnic experiences of Dutch Protestants and Dutch Catholics in Wisconsin. It delimits the areas in which immigrants of both groups settled, and traces the often different ways in which the two groups formed communities, married, worked and urbanized, in short, the way in which they assimilated or resisted assimilation into the American mainstream. In particular, it assesses the role of religion in these processes.

From Immigration to Urbanization

Around 1850, large numbers of Dutch Catholic immigrants began to settle in east-central Wisconsin along the Fox River in Brown and Outagamie counties. Most of the immigrants came from the northeastern part of the province of Noord-Brabant and the northern part of the province of Limburg in the Netherlands. Simultaneously, Dutch Protestant immigrants began to settle to the south and along the shores of

Lake Michigan, particularly in Fond du Lac and Sheboygan counties. Both groups, at the time, chose to live in rural areas, and most of them became farmers.[5]

By 1870, however, east-central Wisconsin was rapidly making the transition from a farming frontier to a more specialized and urbanized society. Integration within the national railroad network had changed the competitive position of Wisconsin within the national economy, which in turn led to shifts in agricultural and industrial specialization. In agriculture, dairy farming replaced wheat production; in industry, paper manufacturing became one of the main sources of employment in the Green Bay area.[6] The transition from wheat to dairy farming called for a substantial adjustment and capital investment, and many farmers failed to complete the transition. Simultaneously, opportunities and higher wages made urban employment more attractive. The combined effect was a cityward movement of rural people to fast-growing centers like Green Bay, Appleton, Oshkosh, Fond du Lac, and Sheboygan.[7]

These developments had a differential effect on the two elements of Wisconsin's Dutch population (Figure 16.1). By 1905, after about one generation of settlement, there was a striking contrast in the rural-urban distribution pattern between Dutch Catholics and Dutch Protestants in their respective settlement areas. The Dutch Catholics, who settled in Outagamie and Brown counties, were spread over quite a large area along the Fox River and in and around Green Bay and Appleton. On the other hand, the Dutch Protestant immigrants, who settled in Fond du Lac and Sheboygan counties, remained highly concentrated in the townships where they had originally settled. By the turn of the century, Dutch Protestants were clearly more "rural" than Dutch Catholics, despite the definite trend toward urban growth and rural depopulation in both two-county areas. The Dutch Catholics exhibited a much stronger tendency to migrate to urban areas and seek industrial employment, which resulted in their nearly equal representation in rural townships, small towns, and urban centers.

The cityward migration among Dutch Catholic immigrants was accompanied by increased opportunities for social mobility and group interaction. Intermarriage rates among Dutch Catholic immigrants in Outagamie and Brown counties in 1905 were highest among urban residents and much lower among township and village residents (Table 16.1). For Dutch Protestant immigrants in Fond du Lac County, intermarriage rates were much lower than those of the total number of Dutch Catholics who intermarried. Only 8 percent of first-generation Dutch Protestant immigrants in Fond du Lac County in

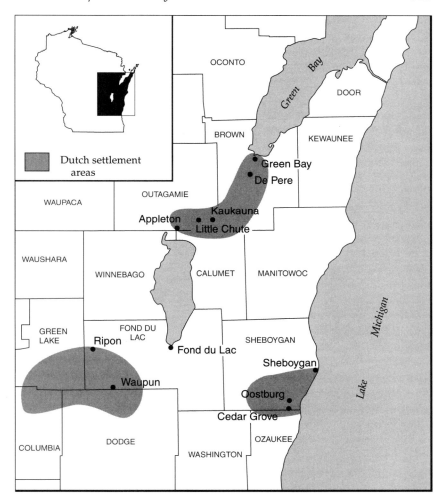

Figure 16.1. Map of Dutch immigrant settlement in Wisconsin. The shaded areas mark the Dutch Catholic immigrant settlements.

1905 had married with members of other groups. The difference between the Catholics and Protestants was especially striking for second-generation immigrants. Whereas, the intermarriage rate for second-generation immigrants among the Dutch Protestants of Fond du Lac County remained at a low 7.5 percent, the corresponding rate among Dutch Catholic second-generation immigrants in Brown and Outa-

Table 16.1. Numeric and percentage distributions of Dutch Catholics and Dutch Protestants inmarrying and intermarrying with other groups in Wisconsin, by locale,[a] 1905

	First generation				Second generation			
	Inmarriage		Inter-marriage		Inmarriage		Inter-marriage	
	n	%	n	%	n	%	n	%
	Dutch Catholics in Outagamie and Brown counties							
	(N = 194)				(N = 245)			
Freedom (t)	26	81.2	6	18.8	18	50.0	18	50.0
Vandenbroek (t)	24	100.0	0	0	40	80.0	10	20.0
Little Chute (v)	77	85.5	13	14.5	69	83.0	14	17.0
Kaukauna (c)	5	41.6	7	58.4	7	29.2	17	70.8
Appleton (c)	16	44.4	20	55.6	11	21.2	41	88.8
Total	148	76.3	46	23.7	145	59.2	100	40.8
	Dutch Protestants in Fond du Lac County							
	(N = 170)				(N = 93)			
Fond du Lac County	157	92.0	13	8.0	86	92.5	7	7.5

Source: Wisconsin State Census Manuscripts, 1905
[a] Township (t), village (v), city (c)

gamie counties was 40.8 percent.[8] Rural-urban migration by members of the Dutch Catholic immigrant group seems to have been a major factor in explaining their fairly rapid assimilation.

The Role of the Church

Why should intermarriage and rapid assimilation have been the case among Dutch Catholics but not Dutch Protestants? Sociologists have for a long time recognized the role of the Church as a factor in differential patterns of behavior among immigrant groups in the United States.[9] The Protestant denominational churches, for instance, often developed into highly distinctive community institutions. Lay leadership and the absence of an all-inclusive hierarchy allowed many Protestant churches to embark upon separate ways. Self-reliance further contributed to personal initiative among membership groups, creating an environment which was highly conducive to preserving group identity.[10] The Catholic church, on the other hand, remained in principle a universal church ·with a highly structured hierarchic organization. The hierarchy comprised archbishops, bishops, and priests, and higher-level clergy determined church policies and appointments.

In the mid-nineteenth century, nativist movements had an important impact on church policies. The influence of the Know Nothing movement was central in this respect.[11] The Know Nothing party was founded in 1853, attracted a large following among Protestant groups in eastern states, and often targeted Irish Catholics in its anti-immigrant campaigns. In turn, the Know Nothing movement sparked a major policy debate among the Catholic clergy. Some Catholic authorities, like Archbishop Hughes from New York, called for the strengthening of unity within the Catholic church, while others sought to relieve the impact of nativism by encouraging Catholics to organize Catholic colonization societies and disperse westward.[12] Proponents of greater centralization won the debate, and, as a consequence, the internal organization and structure of the Catholic church took on a distinctly defensive posture.[13] Fear of close contact with Protestants resulted in an emphasis on parochial schools and charitable organizations at the diocesan level. At the national level, Catholic synods and councils established a common body of laws strengthening the internal structure of the church. Loyalty to Rome and close alliance with international Catholicism were emphasized.

The Catholic church's centralized policy had a very profound and lasting influence on western Catholic immigrant settlements.[14] As a result of church policies, Catholic colonization on the western frontier remained sporadic. Little financial support was given to western pioneers, and settlement was discouraged. Consequently, there was often no parish priest to serve the isolated communities, and many who did migrate to the frontier frequently returned to the older settlements. Only after 1870 did Catholic colonization and settlement in rural areas of the Midwest develop.[15] However, in the period after the Civil War, which became increasingly known for heavy immigration of Catholics from southern and eastern Europe, the church became more and more an urban church. Faced with rapid growth and an increasing number of poor immigrants, the Catholic church officially designated the diocese, rather than the parish, as the main focus of its institutional structure and, in so doing, further promoted urban rather than rural Catholic settlement on the western frontier.[16]

In addition, the Catholic church in Wisconsin was marked by conflicts between the Irish and the German factions. Many Irish immigrants came to Wisconsin after having first settled in the northeastern states, where they frequently experienced discrimination in a predominantly Anglo-Saxon Protestant environment and where they often remained at the lower level of the socio-economic ladder. The impact of nativist movements like the Know Nothing movement in the 1850s left a clear mark on the political outlook of Irish Catholics in Wisconsin for the rest of the century. With respect to church policies, Irish priests and clergy usually favored unity among Catholic immigrants. They were sympathetic to Archbishop Hughes's views on integrating different Catholic immigrant groups within the Catholic church in the United States. Most of the German immigrants, on the other hand, came directly to Wisconsin from Europe and felt a need for the church to play a strong role in the preservation of their language and cultural traditions. As newcomers they favored a diocesan organization based on nationality rather than on geographic area.[17]

Initially, the Catholic church in Wisconsin struggled over these differences, partly because the first bishop to be appointed in 1844 was John M. Henni, who was German-speaking and of Swiss-German origin.[18] Although Henni favored rural settlement in the West and in fact promoted immigration to the western states, he opposed dispersal of immigrant groups. Ambiguously following church policies, Henni favored group integration at the level of the diocese, but promoted the idea of national-language parochial organizations at the parish level.[19] Henni often catered to the needs of German immigrants. By providing German clergy to help serve the German immigrant population, he established a strong German tradition in Wisconsin. Since he firmly believed that language was the best means of preserving the faith of the immigrant, he founded several foreign-language parochial schools. Although he was often criticized by Irish bishops for emphasizing the national origins of immigrant groups, he gained a high reputation among the non-English-speaking Catholic population.[20]

Division among the two major immigrant groups persisted. At the time of Henni's succession (in the late 1870s), the Irish Catholics did everything possible to prevent the appointment of another German-speaking bishop to the Milwaukee Archdiocese. Until that time, there had not ever been an English-speaking or Irish prelate, vicar-general, chancellor, or secretary in all Wisconsin. Opposition to German hegemony came mostly from the dioceses of La Crosse and Green Bay, where Irish and other non-German clergy were most strongly represented. The vicar-general of Green Bay was Eduard Daems, who was of Belgian origin and had been instrumental in attracting Dutch and Belgian priests to serve the Green Bay area Dutch and Belgian immigrant communities.[21] The Dutch and Belgian Catholics were less entangled in church politics and had a more immediate interest in preserving native language and traditions. Most of them had come directly to the Midwest from the Netherlands and Belgium and were like the German Catholics, less aware of the Know Nothing movement and its threat to the Catholic church. However, all non-English-speaking Catholic groups were exposed to the same, mostly Irish dominated, national church policy mandates. Consequently, Dutch and Belgian Catholics had no opportunity to establish their own separate immigrant church communities, although there were significant concentrations of both Dutch and Belgian Catholics in the Fox River valley and Door County, and several Dutch- and Belgian-born priests had been appointed to their respective settlements. Many Dutch and Belgian Catholics moved in subsequent decades to nearby urban centers rather than to the western frontier. This was the case especially after the Civil War, when emerging industrial centers like Green Bay, Kaukauna, De Pere and Appleton offered new employment opportunities.

Conclusion

The Catholic church did not encourage frontier colonization in the same way that Protestant churches did, and the established Catholic church hierarchy, based in the cities, ultimately gave Dutch Catholic immigrant settlement a distinctively urban character. The move to the city brought Dutch Catholic immigrants and their children in contact with members of other Catholic groups. Schools and charitable organizations were integrated at the diocese level, and Irish, German, Belgian, and Dutch Catholic immigrants, in most instances, shared the same parish church, making intermarriage a likely occurrence, especially for the second-generation immigrants who grew up speaking English and were associated with the American way of life.

Judging from the Wisconsin experience reviewed here, it seems that Dutch Protestant and Dutch Catholic immigrants exhibited a clearly different pattern of behavior with respect to settlement choice and interaction with members of other groups, as measured by urban versus rural settlement and intermarriage rates. In order to understand this differential pattern of behavior, we have considered the position of the immigrant group in the context of the larger society and looked in particular at the role of the Catholic church as an institution in explaining the urban focus and inclination of Dutch Catholic immigrants to marry with members of other Catholic immigrant groups. Unlike the Protestant church, the Catholic church in Wisconsin was clearly an assimilationist force. Urban- rather than frontier-oriented, the Catholic church gave a definite direction to settlement concentration and established conditions favorable to assimilation within the framework of the Catholic church. The role of the Protestant church was different. Both Lutheran church communities and Calvinist church communities were frontier-oriented and community-forming. It was the Dutch Reformed, or Christian Reformed, church which established and consolidated Dutch Protestant immigrant settlements in the Midwest.

Notes

1. Y. Schreuder, *Dutch Catholic Immigrant Settlement in Wisconsin, 1850–1905* (New York, 1989); R. P. Swierenga and Y. Schreuder, "Catholic and Protestant Emigration from the Netherlands in the 19th Century: A Comparative Social Structural Analysis," *Tijdschrift voor Economische en Sociale Geografie* 74 (1983): 25–40.

2. R. P. Swierenga, "Religion and Immigration Patterns: A Comparative Analysis of Dutch Protestants and Catholics, 1835–1880," *Journal of American Ethnic History* 5, 2 (Spring 1986): 23–45; R. P. Swierenga, "Religion and Immigration Behavior: The Dutch Experience," in R. P. Swierenga and P. R. Vander-Meer, eds., *Belief and Behavior: Essays in the New Religious History* (New Brunswick, 1991), 164–188.

3. H. S. Lucas, *Netherlanders in America: Dutch Immigration to the United States and Canada, 1789–1950* (Ann Arbor, 1955); J. van Hinte, *Nederlanders in Amerika: Een Studie over Landverhuizers en Volksplanters in the Negentiende en Twintigste Eeuw,* 2 vols. (Groningen, 1928), translated into English by R. P. Swierenga et al. under the title *Netherlanders in America: A Study of Emigration and Settlement in the 19th and 20th Centuries in the United States of America* (Grand Rapids, Mich., 1985).

4. R. M. Miller and T. I. Marzik, eds., *Immigrants and Religion in Urban America* (Philadelphia, 1977); J. Bodnar, *The Transplanted: A History of Immigrants in Urban America* (Bloomington, Ind., 1985); J. P. Dolan, *The Immigrant Church: New York's Irish and German Catholics, 1815–1865* (Baltimore, 1975); R. J. Vecoli, "Prelates and Peasants: Italian Immigrants and the Catholic Church," *Journal of Social History* 2 (1969): 228–251; D. A. Liptak, *European Immigrants and the Catholic Church in Connecticut* (New York, 1967); J. S. Olson, *Catholic Immigrants in America* (Chicago, 1987).

5. Farming occupations prevailed among all Dutch immigrants in both Outagamie and Brown counties and Fond du Lac and Sheboygan counties according to the 1860 census. In order to determine to what extent occupational and rural or urban backgrounds in the Netherlands had an impact on choice of settlement and socio-economic adaptation, linked sets of data from municipal records and emigration records in the Netherlands and census records in Wisconsin were used to determine similarities and differences between the two groups. Occupational backgrounds of members of the two groups were quite similar in that most were of a rural and farming background. Over 75 percent listed farming occupations in Wisconsin in the 1850 and 1860 censuses.

6.. For a discussion about the different phases of regional economic development, see E. K. Muller, "Regional Urbanization and the Selective Growth of Towns in American Regions," *Journal of Historical Geography* 3, 1 (January 1977): 21–39. Specifically related to the area of study, see E. E. Lampard, *The Rise of the Dairy Industry: A Study in Agricultural Change, 1820–1920* (Madison, 1963); and C. N. Glaab and L. H. Larsen, *Factories in the Valley: Neenah-Menasha, 1870–1915* (Madison, 1969).

7. A. G. Bogue, *Money at Interest: The Farm Mortgage on the Middle Border* (Ithaca, N.Y., 1955); and D. L. Winters, *Farmers without Farms: Agricultural Tenancy in Nineteenth Century Iowa* (Westport, Conn., 1978). For a general explanation of rural-urban migration in the United States, see C. W. Thornthwaite, *Internal Migration in the United States* (Philadelphia, 1934); and R. A. Easterlin, *Population, Labor Force, and Long Swings in Economic Growth: The American Experience* (New York, 1968). More specifically with a focus on the Midwest, see M. P. Conzen, "Local Migration Systems in Nineteenth Century Iowa," *Geographical Review* 64 (1974): 339–362.

8. Y. Schreuder, "Ethnic Solidarity and Assimilation among Dutch Protestant and Dutch Catholic Immigrant Groups in the State of Wisconsin, 1850–1905," in R. Kroes, ed., *The Dutch in North-America: Their Immigration and Cultural Continuity,* European Contributions to American Studies, 20 (Amsterdam, 1991), 195–218.

9. W. Herberg, *Protestant, Catholic, Jew: An Essay in American Religious Sociology* (Garden City, N.Y., 1960); G. Lenski, *The Religious Factor* (Garden City, N.Y., 1961); M. M. Gordon, *Assimilation in American Life: The Role of Race, Religion, and National Origin* (New York, 1964); A. M. Greeley, *The Denominational Society: A Sociological Approach to Religion in America* (Glenview, Ill., 1972).

10. R. H. Niebuhr, *The Social Sources of Denominationalism* (New York, 1929); T. S. Miyakawa, *Protestants and Pioneers: Individualism and Conformity on the American Frontier* (Chicago, 1964).

11. A. M. Greeley, *The Catholic Experience: An Interpretation of the History of American Catholicism* (New York, 1967).

12. H. J. Browne, "Archbishop Hughes and Western Colonization," *Catholic Historical Review* 35 (October 1950): 257–285; J. P. Dolan, "A Critical Period in American Catholicism," *Review of Politics* 35 (1973): 523–536; J. T. Ellis, *American Catholicism* (Chicago, 1955); J. P. Shannon, *Catholic Colonization on the Western Frontier* (New Haven, Conn., 1957).

13. J. P. Dolan, *The Immigrant Church: New York's Irish and German Catholics, 1815–1865* (Baltimore, 1975), 162–165.

14. J. Schafer, "Know-Nothingism in Wisconsin," *Wisconsin Magazine of History* 8, 1 (September 1924): 3–21.

15. Shannon, *Catholic Colonization,* 18–22.

16. Dolan, *Immigrant Church,* 163.

17. E. H. Rothan, *The German Catholic Immigrants in the United States, 1830–1860* (Washington, D.C., 1946).

18. P. L. Johnson, *Crosier on the Frontier: A Life of John Martin Henni* (Madison, 1959).

19.   Johnson, *Crosier*, 100–122; Rothan, *The German Cathoℇc Immigrants*, 61–76.

20.   Johnson, *Crosier*, 113, 125.

21.   For an account of Daems life and work in Wisconsin, see A. de Smet, *La Communaute Belge du Nord-Est du Wisconsin* (Wavre, Belgium, 1957). See also H. van Stekelenburg, *Landverhuizing als Regional Verschijnsel: Jan Noord-Brabant naar Noord-Amerika, 1820–1880* (Tilburg, Netherlands, 1991).

17

Four Worlds Without an Eden
Pre-Columbian Peoples and the Wisconsin Landscape

William Gustav Gartner

. . . Then [Trickster] said to the waterfall, "Remove yourself
to some other location for the people are going to inhabit this
place and you will annoy them." Then the Waterfall said, "I will
not go away. I chose this place and I'm going to stay here." . . .
[Then Trickster said,] "I am telling you that the earth was made
for man to live on and you will annoy him if you stay here. I
came to this earth to rearrange it. If you don't do what I tell
you, I will not use you very gently."
—Paul Radin, *The Trickster*

In the beginning of Winnebago times, Earthmaker sent several supernatural beings to rid the world of evil spirits and giants, as well as to
teach Indian peoples how they might secure a happy life. The first of
these world transformers, an amorphous fellow called Wakdjunkaga,
or Trickster, broke every known Winnebago taboo and thus found
himself apart from human society, ignorant of nature, and an aimless
wanderer. (Appropriately, *Wakdjunkaga* also means the "foolish one"
in Winnebago.) During the course of his travels, Trickster gradually
became a differentiated being and, in the process, altered portions of
Earthmaker's creation into the presently known world. Since Wak-

djunkaga was not entirely successful in his task, Earthmaker sent at least three other supernatural beings to finish the job of world transformation. Thus, Wisconsin was remolded a minimum of four times. Although each earthly creation was a terrestrial paradise, none can be called Eden, since the creative effort was never entirely completed by God (Earthmaker) alone.

Wakdjunkaga, both culture hero and buffoon, is partly responsible for shaping the Winnebago world. Trickster humorously illustrates that neither culture, nor nature, nor the human self can exist without the others. The Winnebago Trickster myth also exemplifies the main tenet of this chapter: the pre-European Wisconsin landscape has been humanized since the beginning of indigenous times.

Humanizing the Natural Landscape

Wisconsin Indians, whose pre-Columbian numbers may have approached 60,000–70,000, were decimated by diseases and disruptions to traditional life ways brought by the earliest missionaries, fur traders, and explorers.[1] When subsequent chroniclers scanned the landscape, then emptied by these maladies, they viewed it, incorrectly, as natural. They could not have known that centuries of living by the generations of people who first called this land home had, in fact, humanized the scene (Figure 17.1).

Fire

The widespread effects of Indian burning were apparent to some early observers. Lapham, in 1855, interpreted the presence of pre-Columbian mounds, presumably constructed in open vegetation, in areas of closed forest as evidence of a cessation of frequent fire: "Whether the greater extent of treeless country in former times was owing to natural or artificial causes, it is now difficult to determine; but the great extent of ancient works within the depths of the present forest, would seem to indicate that the country was at least kept clear from trees by the agency of man." Similarly, Owen, in 1852, saw the high density of mounds and the associated mosaic of vegetation in the northern Mississippi River valley as an indication of a humanized landscape: "The whole combination suggests the idea, not of an aboriginal wilderness inhabited by savage tribes, but of a country under a high state of cultivation and suddenly deserted by its inhabitants."[2]

Echoing these early impressionistic evaluations, scientists have argued that Indian burning was essential to the maintenance of prairies

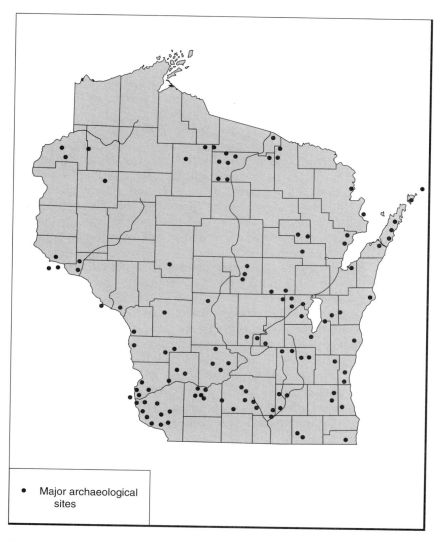

Figure 17.1. Major Wisconsin archaeological sites. The Wisconsin landscape has been widely inhabited by people for thousands of years. The ages vary from the Early Paleo-Indian stage (11,500–10,200 years B.P.) to historical times.

and oak openings in Wisconsin. Asa Gray, in 1878, commented that "the demarcation between our woods and our plains is not where it was drawn by nature."[3] Gleason identified anthropogenic fire as a primary cause of the distribution of grasslands and open-canopy forests in the Midwest and assembled numerous historical accounts of upper midwestern Indian fires to bolster his claim. Curtis, in his seminal book on Wisconsin vegetation, did the same.[4]

Indigenous peoples routinely cleared the Wisconsin landscape with fire. They burned to facilitate hunting and game drives, clear village and agricultural lands, assist in fuel-wood cutting, improve visibility and overland travel, manage pests, and facilitate warfare.[5] Frequent fires provided favorable habitats for deer, elk, and upland game birds such as wild turkeys and grouse by increasing the biomass of their preferred browse.[6] Native peoples used many herbs and grasses for food, medicine, and fiber that depended on frequent fires for habitat regeneration and reproduction.

Fire as a hunting tool—probably employed in the fall season after the first frost—is often described in seventeenth-century historical documents. La Salle described the Illinois River tall grass prairies as "scorched by the fires kindled in the dried grass by the Indian hunters and strewn with the [then hunted] carcasses and the bleached skulls of innumerable buffalo." Hennepin details the tall grass prairie fires lit by the Miami tribe for their annual hunts in both Illinois and Wisconsin.[7]

Controlled fires for agriculture and other purposes were lit in the spring, when tree sap began to flow. Preparing agricultural fields was a specialized process, involving tree girdling and the piling of branches, grasses, and the previous year's crop residues.[8] Such fires were essential for native agriculture, adding nutrients such as carbon and phosphorous to the soils, destroying plant and animal pests, enhancing microclimatic warming, and easing tilth preparation. Both the Sand Lake and Hulburt Creek ridged field sites demonstrate the aboriginal use of fire in agricultural settings.[9]

Native settlement patterns in general suggest frequent fires. Nearly all nineteenth-century Winnebago villages in eastern Wisconsin were located adjacently to natural fire barriers, such as on the east side of rivers or on peninsulas in large water bodies like Lake Koshkonong and Horicon Marsh. Plant ordinations from nineteenth-century records suggest an average return time of 112 years for fires in these locations, while 16 years was the norm outside these protected areas.[10]

Pre-Columbian Oneota and Mississippian villages are often located on or next to river bottoms and wetlands, presumably for the rich resources of such regions. However, these areas are also natural firebreaks. One of the few exceptions is Aztalan, which is located on the west side of the Crawfish River. Charred palisade posts and wattle and daub correlate with the last major occupation of Aztalan,[11] suggesting the site was prone to fire during warfare.

Taken together, then, native peoples had reasons to burn the early Wisconsin landscape and did so repeatedly. This humanizing imprint was nearly ubiquitous in the pre-Columbian scene.

Hunting and Trapping

Spectacular recent archaeological finds near Kenosha of mastodonts and mammoths butchered by Wisconsin's first human inhabitants will not resolve the debate concerning the cause—climate or humans—of the late Pleistocene extinctions of large mammals in North America.[12] Dating issues complicate the argument because ongoing research indicates that both the periods of human arrival in the New World and the timing of animal extinction lasted longer than once thought. The Kenosha finds do demonstrate that Paleo-Indians had at least some impact on Wisconsin's megafauna, even if only to accelerate the demise of an already doomed Pleistocene species.

Quite apart from the extreme event of the Pleistocene extinctions, however, Paleo-Indian descendants probably influenced Holocene animal populations for millennia. At the Raddatz Rockshelter, Archaic peoples collected several species of turtles and freshwater mussels, hunted or trapped nearly 20 species of birds, and hunted over 25 species of mammals.[13] Such broad-spectrum hunting suggests that animal products had specialized food, clothing, medicinal, ceremonial, ornamental, and artifactual uses. Faunal studies generally show that broad-spectrum hunting and trapping continued and expanded throughout the pre-Columbian period. Shell middens appear during Early Woodland times, and it is not uncommon to find 25 or more freshwater mussel species in a single deposit. Aquatic mammals, such as beavers, and once expiated mammals, such as the fisher, occur frequently in Middle Woodland midden contexts. Although both Archaic and Woodland Indians occasionally hunted bison and bear, deer and elk were the staples for most Archaic and Woodland tradition subsistence.

Woodland tradition Indians killed many deer and elk during the fall rutting and spring yarding seasons, times of minimal vegetative cover and high animal congregation.[14] The potential demand for deer and elk was great because each person required multiple animals for meat and clothing. Ethnohistorical records for the Huron indicate that an adult human male needed 3.5 deer hides per year for clothing and an additional 2–3 hides per year for moccasins. Women and children required fewer animals. Over 62,000 deer are estimated to have been harvested per year to clothe Huron families in southern Ontario. Many critics feel that these deer harvest estimates are conservative because of ecological heterogeneity (variability of animal size and population density by habitat) and underestimating human nutrition.[15]

If we accept that about 70,000 Indians lived in Wisconsin just before European contact, potentially hundreds of thousands of deer may have been taken in a single year. What effect this had on faunal populations is debatable. Animal populations often rebound slowly after hunting during reproductive periods. Yet, the harvest of deer, elk, and other animals may have been offset by the creation of preferred open-canopy habitats through anthropogenic fire and other Indian practices.

Fishing

Nearly all known aboriginal fishing technologies are found in archaeological and early ethnohistorical records of Wisconsin—spearing and torchlight fishing, net fishing with seines and set gill nets, fish weirs and traps, decoy fishing, fish hooks and toggles, and fish poisoning.[16] Up to 91 percent of all faunal remains recovered from spring- and fall-occupied Late Woodland villages around Lake Michigan are fish bones. The presence of fishing artifacts both extends back in time occurs widely over the state: A radiocarbon age of 4,210 years B.P. of a wooden shaft from inside a copper harpoon head found near Green Bay shows that fishing had great antiquity.[17] Even more substantial fishing impacts were likely in the more productive and accessible streams of the Mississippi River drainage and Wisconsin's inland waterways.

Some of this fishing activity, moreover, involved substantial landscape modification. The Big Bend fish weir, for example, spanned the Oconto River except for a small opening near the center. The rock-piled walls varied between 3.3 and 4.9 meters in thickness and spanned 41 meters of the channel.[18] In contrast, the Ehrke fish weir on the smaller Rock River, with its unique cantilevered construction, was only 1 meter thick. Stone fish weirs are also associated with the fortified village of Aztalan, suggesting that weirs may have regulated canoe as well as fish traffic.

Arboriculture

Nuts and fruit trees have been an important part of native diets since Archaic times. Archaeological and ethnohistorical evidence suggests that many specialized wood products were made from these trees, including bows from Osage orange trees, ceremonial posts from cedar, and medicines from willow and Kentucky coffee trees. Trees were important enough to the daily lives of native peoples that their over-exploitation was a factor in Cahokia's rapid decline after A.D. 1200.[19]

Early descriptions of upper midwestern "planted tree groves" are reminiscent of South America's anthropogenic forest islands in terms of species diversity, economic utility, and patchiness: "It is nothing but prairies as far as the eye can see, dotted here and there with small patches of woods, with orchards, and with avenues of trees. . . . These woods are full of chestnuts, locusts, oaks, ashes, basswoods, beeches, cottonwoods, maples, pecans, medlars, mulberries, chestnuts, and plums. All these trees are covered with a vine that bears a handsome grape and which has large seeds."[20] Raudot goes on to describe crabapple, papaw, and Kentucky coffee trees, as well as medicinal roots and other miscellaneous plants, in groves associated with the Illinois tribe.

Marquette describes planted tree groves containing plums and grapes interspersed between the prairies and cornfields of eastern Wisconsin in 1673. Hennepin also believed that Indians planted trees in the prairies of Wisconsin and Illinois. Allouez refers to wood patches in southern Wisconsin as "planted designedly to form alleys more agreeable to the sight than orchards." Father Membre goes so far as to compare indigenous orchards of the southern Great Lakes with Roman villas.[21] Fruits and nuts from the tree species described by Raudot and others have been found in Wisconsin archaeological contexts for millennia.

Early chroniclers may have exaggerated the spatial organization of Indian "orchards." Yet, centuries of selective tree felling and planting in fire-protected areas, combined with topographic, edaphic, and microclimatic factors, could have resulted in the "forest islands"—a humanized landscape element—once described for Wisconsin's prairies.

Foraging

The gathering of wild plants for food, medicinal, fiber, and dye uses also dates to Archaic times in Wisconsin and may have influenced plant communities in several ways. Collected plants tend to become overrepresented in biotic communities because of Indian dispersal. Native peoples may also extend the geographic range of useful plant species.

Wild rice is probably the best-known nondomesticated plant whose range and density were influenced by pre-Columbian peoples. Most Woodland tradition peoples used wild rice in Wisconsin. In northern contexts, Middle and Late Woodland settlement patterns may actually have reflected the density of wild rice stands. Its human dispersal is well documented historically, for as the Menominee elders say, "whenever the Menomini enter a region the wild rice spreads ahead, whenever they leave it the wild rice passes."[22]

Pre-Columbian peoples extended the geographic range of many other plants. The Fred Edwards site, a Late Woodland community involved in Mississippian trade, contains the northernmost occurrence of maygrass (*Phalaris caroliniana*). Some species of amaranth (*Amaranthus* sp.) that are common in the archaeobotanical record may also occur north of their natural range. Indians also extended the geographic range of medicinal plants such as sweet flag and the Kentucky coffee tree.[23]

Agriculture

Plant domestication was a gradual process in North America. Middle and Late Archaic stage peoples domesticated cucurbit (*Cucurbita pepo*), bottle gourd (*Lagineria siceraria*), sumpweed (*Iva annua*), and sunflower (*Helianthus annuus*) between 5,000 and 3,000 years B.P. in the central midwestern United States. Disturbed earth seed plants such as chenopods (*Chenopodium berlandieri*), knotweed (*Polygonum erectum*), and little barley (*Hordeum pusillum*) later complemented these domesticates and formed the basis of Late Archaic–Early Woodland cultivation. Although Middle Woodland Indians knew about tropical cultigens such as maize, at least 600 years passed before it was grown in any quantity. By the time of Columbus, Indians had domesticated or quasi-domesticated about a dozen native North American species.[24]

Unfortunately, there are few studies of plant domestication in Wisconsin. The earliest firm evidence for plant domestication comes from Millville phase sites, and cultivated sumpweed and squash remains date to around A.D. 200 at the Hunters Channel IIa and Mill Coulee sites. Millville peoples also utilized a variety of wild plants for seeds and greens, including little barley, chenopods, knotweed, and possibly tobacco.[25]

Maize is present in many classic Effigy Mound contexts, with a radiocarbon age of A.D. 920 ± 80. Maize is also found at Kekoskee phase Effigy Mound sites, with radiocarbon ages ranging from A.D. 800–1200.

Ridged Fields and Other Landforms of Cultivation

Indigenous landscapes of cultivation came well after domestication, as cultigens contributed only a small portion of the upper midwestern diet until around A.D. 800. Native Wisconsin peoples engaged in agricultural practices that varied from selective weeding to polyculture, using tools that ranged from the simple digging sticks to finely crafted chert and bison scapula hoes, and with production modes as different as household gardens and extensive field agriculture.

The most important agricultural innovation in Wisconsin was ridged fields, which are often called garden beds (Figure 17.2). Ridged fields improve cultivating conditions by draining water at the planting surface, storing water at the base of the bed, minimizing temperature fluctuations, draining radiation frost, enhancing nutrient cycling, slowing disease and pest vectors, and improving tilth.[26] Ridged fields permitted aboriginal cultivation in a variety of settings ranging from wetlands to uplands, saturated mucks to well-drained loamy sands, grasslands to closed-canopy forests, and areas that today range from fewer than 90 to more than 160 frost-free days. Extant plots have many relief patterns and range in size from one to more than several hundred acres. More than 175 garden bed sites have been reported in Wisconsin.

More than 120 corn hill sites, small circular heaps of soil as opposed to large linear ridges, are also reported in Wisconsin. Keokuck's village, consisting of Sauk and Fox Indians, had 5,000 acres under cultivation along the mouth of the Rock River. Many early explorers, such as Jonathan Carver, and military campaigns, including the massive British 1779–1780 campaign to suppress the St. Louis rebellion, were dependent on "the endless cornfields of the Wisconsin River" for supplies. Such accounts reinforce the admittedly fragmentary character of pre-Columbian agricultural landscapes that survive to the present day.[27]

Mining

Native peoples exploited Wisconsin's lithic resources soon after entering Wisconsin. Although Clovis peoples all collected local chert materials, subsequent mining was both major in volume and supportive of a widespread trade. Hundreds of prehistoric quarry pits and several workshops are still visible at Silver Mound, in Jackson County, and

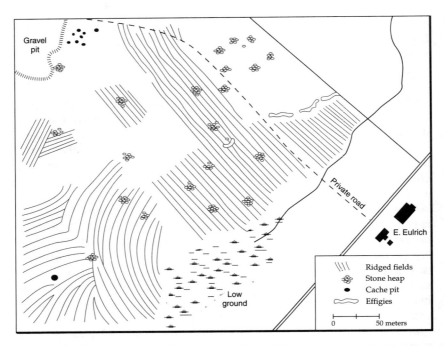

Figure 17.2. Plan view of the Eulrich farm ridged fields, near present-day Appleton (A.D. 1000–1200) (Adapted from G. R. Fox, "Stoneworks and Garden Beds in Winnebago County," The Wisconsin Archeologist 1 (1922): 55.)

the surrounding fields are literally choked with debitage. Its materials have been found in Paleo-Indian contexts as far away as Mammoth Cave, Kentucky, and it maintains a wide distribution during many subsequent archaeological stages.[28] Similarly, at the Bass site in southwestern Wisconsin debitage averages nearly 400 items per 100 square meters over an area of 43 hectares.[29]

Copper artifacts, manufactured by cold hammering and annealing, proliferated in Wisconsin during the Middle Archaic manifestation in what was referred to as the Old Copper culture in the older literature. Trace element analyses demonstrate that many copper artifacts in eastern North America originated from the Lake Superior region. Some copper deposits could be collected at the surface, but there is ample evidence of rock having been removed by fire cracking and hammering to depths of 10 meters and more. The McCargoe Cove copper pits near Minong, Michigan, have such "ancient workings that almost touch each other for miles."[30]

Anthropic Soils and Sediments

Human occupation of land surfaces often inadvertently changes the color, texture, structure, and chemistry of soils and sediments. If such changes are significant and mappable, the term *anthropic* is often used to distinguish these deposits from natural soils and sediments.

Pre-Columbian peoples significantly altered soils and sediments in a variety of ways. Fire, cooking, and construction activities add substantial amounts of ash, charcoal, wood, grass, and mud to preexisting deposits. Artifact manufacture may contribute lithic and other raw materials. As trash and human waste decay, organic matter, fine particulites, phosphorous, calcium, and potassium are released to natural deposits and alter their pH, chemistry, and texture. Intentional land clearance and pedestrian traffic devegetate sites, causing erosion and mass wasting of earthen deposits. These occurrences are so common that natural scientists have considered adding humans to the list of soil-forming factors.[31]

Anthropic soils are common in Wisconsin. Human-induced slope wash and mass-wasting transported artifacts and earthen deposits from a living area to a swale at Aztalan in Jefferson County. This swale was also used to dump trash. Calcium, potassium, phosphate, and pH values are all higher in soils from the swale area than in control samples taken from nonarchaeological contexts.[32] Also, several Wisconsin Indian villages have been located and mapped on the basis of elevated phosphate levels in anthropic soils.[33]

Pre-Columbian subsistence and settlement also changed entire landforms. Land clearance associated with agriculture and village construction in coulees near La Crosse caused erosion and the subsequent accumulation of deposits downstream. Thus, alluvial fans in this densely occupied area are much larger than their nearby counterparts along the Mississippi Valley. Some depositional strata within the La Crosse fans are so distinctive that they are termed Oneota settlement alluvium.[34]

Dark-colored deposits, reflecting anthropic accumulations of organic matter, outline house basins and plowed-down mounds in aerial photographs of the Red Wing and Diamond Bluff areas.[35] Many plowed-down mounds are also visible in aerial photographs of the Cranberry Creek Archaeological District.[36] Soil changes visible in aerial photographs show that archaeological inventories underrepresent the number of mounds and sites that once existed before nineteenth-century plowing (Figure 17.3).

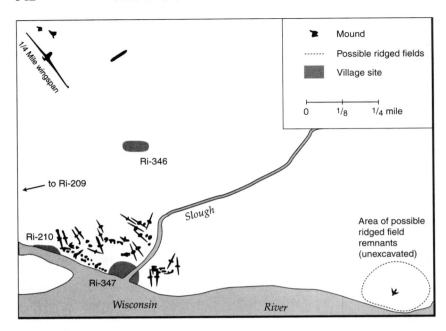

Figure 17.3. Plan view of the western portion of the Muscoda Mounds of the Late Woodland stage (1,500–800 years B.P.) (Adapted from an 1886 survey by T. H. Lewis and a 1993 base map created by Jan Beaver, James P. Scherz, and the Wisconsin Winnebago Nation GIS Department. On file at the Office of the State Archaeologist, Wisconsin State Historical Society.)

Anthrosols and Anthroseds

Pre-Columbian peoples not only inadvertently but also intentionally modified soils and sediments; such purposefully created soils are termed Anthrosols, and the sediments are called Anthroseds. These artifacts provide valuable information about past peoples, nature-society relations, and culture change.

Anthropogenic soils, or Anthrosols, are common on agricultural sites, where indigenous management practices often altered the physical and chemical properties of soil. At the Hulburt Creek site, indigenous agriculturalists positioned ditches to collect organic matter and water. They used the distinctive organic-rich mucks that developed in the ditches to rebuild the fields and maintain the planting surfaces. At the Sand Lake site, Oneota peoples used household garbage and fire to replenish agricultural soils. The proximity of garden and household wastes at many archaeological sites suggests periodic rotation of houses and gardens to manage soils.[37]

Anthropogenic sediments, or Anthroseds, are manufactured sediments that often contain disparate materials combined in a prescribed manner. Anthroseds are common in such earthen architectural features as mounds, burial pits, plazas, and certain houses. At the Gottschall Rockshelter, the same "recipe" for Anthroseds was followed for nearly 700 years, and contained such exotic materials as limestone, clamshell, bone, hornblende, and galena mixed within an ashy matrix of cedar and various grasses.[38] Although they were formally defined in the Woodland levels at the Gottschall Rockshelter, Anthroseds date to much earlier times in Wisconsin and elsewhere.

The Built Landscape

Native influences on the Wisconsin landscape are not only the result of subsistence pursuits and technological activities. The indigenous landscape legacy also includes the vestiges of a built world—the roads, villages, and earthworks that represent the quintessential transformation of the earth into a home (Figure 17.1).

Roads

Hinsdale described the overland trail network in aboriginal America as "a prodigious spider web" that linked contemporary villages and permitted access to many environments and waterways for purposes of food and raw material procurement, commerce, warfare, and general ease of travel. Mound groups often occur along ancient pathways, such as those along Military Ridge and around Lake Koshkonong. Native peoples tied down young hardwood trees at the junction of overland travel ways to serve as signposts. These "trail marker trees" are widely reported in eastern Wisconsin. Many ancient Indian roads formed the base for our own modern highway system, including Route 18 along Military Ridge and Route 141 from Milwaukee to Green Bay.[39]

Villages

Wisconsin's native settlements ranged from single households to planned fortified villages with public spaces and monumental earthen architecture (Figure 17.1). Middle Woodland Stage peoples built a single conical pit house at the Robinson site and a planned village of 14 circular and oval houses organized around a central open space at the Millville site. House basins shaped like keyholes have been found at the Late Woodland Stage Statz, Weisner III, and Elmwood Island sites.

Weisner III and the nearby Weisner IV sites were both stockaded, as was the contemporary Camp Indianola site.[40]

House architecture diversified during the Developmental Oneota Horizon between A.D. 900 and 1300. Rectangular semisubterranean house basins have been found at the Old Spring, Diamond Bluff, and Armstrong sites, while ground-level, oval-shaped houses were built at the Walker-Hooper and Pipe sites. The Mississippian Aztalan site and the Oneota settlement at Carcajou Point are both fortified and have rectangular houses with post-in-wall trench structures, as well as circular and rectangular shaped houses with singly set posts. The Walker-Hopper stockade was completed shortly after Aztalan's demise ca. A.D. 1200.

Pit and midden features, ridged fields, and surface concentrations of Classic Oneota Horizon pottery (A.D. 1300–1650) are nearly contiguous throughout the North Asylum Bay area, while occupational debris lines the shores of Lake Winnebago and the Fox River (Figures 17.1 and 17.2). The Karow site has a rectangular house with individually set posts. More than 13 Classic Oneota site complexes have been identified around La Crosse. The Tremaine site consists of three adjoining villages covering more than 51 hectares. Similar-sized villages occur at Greens and Sand Lake coulees. Longhouses at the Tremaine site have floor areas of no less than 190 square meters and some are in excess of 360 square meters. An oval structure at the Overhead site may represent a wigwam. The Valley View, Midway, and Lane sites were all fortified.[41]

Mounds

An estimated 20,000 conical, linear, and effigy mounds had been reported to Wisconsin historical societies by World War I. In a more recent review of the literature, over 3,250 mounds were documented in just six Wisconsin counties. Sadly, more than 80 percent of these mounds have been destroyed by twentieth-century "progress."

The scale of earth moving for Wisconsin mound construction is truly astonishing. Approximately 2,000 mounds have been identified in the Lake Pepin area alone. Effigy Mound peoples constructed upwards of 500 mounds in the Cranberry Creek Archaeological District. The wing span of a single bird effigy near Muscoda, Wisconsin, is over 0.4 kilometer long (Figure 17.3). In what is now the modern town of Trempealeau, Mississippian peoples constructed a 5.5-meter-high, flat-topped pyramid mound that contained nearly 4,500 cubic meters of deposits.[42]

Conclusions

The Wisconsin landscape in pre-European times was anything but a "pristine" wilderness, an "Eden" untouched by human hands. The imprints of indigenous peoples who called this land home were varied: fire maintained prairies and oak openings; gathering, hunting, and fishing altered plant and animal numbers; selective cutting and planting influenced plant ranges; agricultural clearing replaced the wild vegetation with domesticated crops; the construction of ridged fields modified the surface of the earth; mining and changing vegetation cover altered soils and sediments.[43] An elaborate cultural landscape consisting of roads, villages, and mounds was also present. Natural processes and cultural activities are interwoven in the creation of landscape; nature and people cannot be separated from one another. The native peoples of Wisconsin clearly continued the work of Wakdjunkaga in transforming the world.

Notes

1. Nicolet estimated between 3,000 and 5,000 Winnebago and Menominee around Green Bay in 1634. Marquette and other Jesuit missionaries estimated 3,800 Winnebago and 3,000 Menominee in easternmost Wisconsin in the 1670s. William Green compiles evidence supporting native depopulation in Wisconsin prior to sustained European contact, in "Examining Protohistoric Depopulation in the Upper Midwest," *Wisconsin Archeologist* 74 (1993). An average depopulation rate of 72 to 90 percent is postulated for most of the Americas. See J. Verano and D. Ubelaker, *Disease and Demography in the Americas* (Washington, D.C., 1992); and W. M. Denevan, ed., *The Native Population of the Americas in 1492* (Madison, 1992). However, sedentary agricultural populations in major drainage basin areas often had the higher rate of decline. See A. Ramenofsky, *Vectors of Death* (Albuquerque, 1987); and Denevan, *Native Populations of the Americas.* If we apply a 72 percent depopulation rate to the Jesuit numbers and then multiply by an areal projection factor, we arrive at an estimate between 60,000 and 70,000 Indians prior to European contact. Since many Wisconsin Indians were agricultural and occupied a region that included the Mississippi and Great Lakes watersheds, a density in excess of 0.45 person per square kilometer is likely in many areas.

2. I. A. Lapham, *The Antiquities of Wisconsin, as Surveyed and Described* (Washington, D.C., 1855), 91; D. D. Owen, *Report of a Geological Survey of Wisconsin, Iowa, and Minnesota; and Incidentally of a Portion of Nebraska Territory* (Philadelphia, 1852), 66.

3. A. Gray, "Forest Geography and Archaeology," in R. L. Stuckey, ed., *Essays on North American Plant Geography from the Nineteenth Century* (New York, 1978), 94.

4. H. A. Gleason, "The Relation of Forest Distribution and Prairie Fires in the Middle West," *Torreya* 13 (1913): 173–181; J. Curtis, *The Vegetation of Wisconsin* (Madison, 1971), 296–301, 461–462.

5. C. Sauer, "Grassland Climax, Fire, and Man," *Journal of Range Management* 3 (1950): 16–21; O. Stewart, "Burning and Natural Vegetation in the United States," *Geographical Review* 41 (1951); G. M. Day, "The Indian as an Ecological Factor in the Northeastern Forest," *Ecology* 34 (1953); H. T. Lewis, "Indian Fires of Spring," *Natural History* 89 (1980).

6. For a study listing the natural variables that influence fire impacts on browse, see L. C. Hulbert, "Causes of Fire Effects in Tall Grass Prairie," *Ecology* 69 (1988): 46–58. For case studies on fire effects on wildlife, see G. W. Wood, ed., *Prescribed Fire and Wildlife in Southern Forests* (Georgetown, 1981). For an overview of indigenous fire ecologies, see H. T. Lewis, *A Time For Burning* (Boreal Institute for Northern Studies, University of Alberta, No. 17, (Alberta, 1982). For methodological and empirical studies concerning fire and past soil and vegetation landscapes, see the articles in *The Ecological Role of Fire in Natural Conifer Forests of Western and Northern North America*, M. L. Heinselman and H. E. Wright, eds., special issue of *Quaternary Research* 3 (1973).

7. See La Salle's letter to Minister Colbert in F. Parkman, *France and England in North America*, vol. 1, ed. D. Levin (New York, 1983), 833. Hennepin describes native-lit fires during conflict in "Narrative of the Voyage to the Upper Mississippi," in J. G. Shea, ed., *Discovery and Exploration of the Mississippi Valley* (New York, 1853), 123. Both O. Anderson, "The Phytosociology of Dry Lime Prairies of Wisconsin," doctoral dissertation, University of Wisconsin–Madison, 1954, 15–22; and H. A. Gleason, "The Relation of Forest Distribution and Prairie Fires in the Middle West," *Torreya* 13 (1913): 173–181, compile nineteenth-century eyewitness accounts of Indian fires and seventeenth-century descriptions of prairies in areas later mapped as forests by the United States General Land Office surveys.

8. G. Imlay, *A Topographical Description of the Western Territory of North America* (1797; reprint, New York, 1969), 236; G. L. Wilson, *Agriculture of the Hidasta Indians, An Indian Interpretation*, Bulletin of the University of Minnesota (Minneapolis, 1917), 15.

9. R. F. Sasso and W. G. Gartner, "Garden Beds, Corn Hills, and Corn Fields: Current Investigations into Agricultural Sites in Wisconsin," in *Program and Abstracts*, Midwest Archaeological Conference (Milwaukee, 1993), 23; W. G. Gartner, "The Hulbert Creek Fields," in *Program and Abstracts*, Midwest Archaeological Conference (La Crosse, 1991), 31.

10. J. R. Dorney, "The Impact of Native Americans on Presettlement Vegetation in Southeastern Wisconsin," *Wisconsin Academy of Sciences, Arts and Letters* 69 (1991).

11. J. E. Freeman, "Aztalan: A Middle Mississippian Village," *Wisconsin Archeologist* 67 (1986): 341–342.

12. Arguments supporting each position are detailed in P. S. Martin and R. G. Klein, eds., *Quaternary Extinctions* (Tucson, 1984).

13. P. W. Parmalee, "Animal Remains from the Raddatz Rockshelter," *Wisconsin Archeologist* 40 (1959): 83–90; C. E. Cleland, *The Prehistoric Animal Ecology and Ethnozoology of the Upper Great Lakes Region*, Museum of Anthropology, University of Michigan, No. 29 (Ann Arbor, 1966), 102–106.

14. J. Theler, *Woodland Tradition Economic Strategies: Animal Resource Utilization in Southwestern Wisconsin and Northeastern Iowa*, Office of the State Archaeologist, Report 17 (Iowa City, 1987); L. K. Lippold, *Aboriginal Animal Resource Utilization in Woodland Wisconsin*, doctoral diss., Department of Anthropology, University of Wisconsin–Madison, 1971; W. A. Starna and J. H. Relethford, "Deer Densities and Population Dynamics: A Cautionary Note," *American Antiquity* 44 (1985).

15. R. M. Gramley, "Deerskins and Hunting Territories: Competition for a Scarce Resource of the Northeastern Woodlands," *American Antiquity* 42 (1977): 601–605; E. R. Turner and R. Santley, "Deer Skins and Hunting Territories Reconsidered," *American Antiquity* 44 (1979); G. S. Webster, "Deer Hides and Tribal Confederacies: Gramley's Hypothesis Reconsidered," *American Antiquity* 44 (1979); W. A. Starna and J. H. Relthford, "Deer Densities and Population Dynamics: A Cautionary Note," *American Antiquity* 44 (1985).

16. E. Rostlund, *Freshwater Fish and Fishing in Native North America* (Berkeley, 1952), 81–133, 291–298.

17. Cleland, *The Prehistoric Animal Ecology and Ethnozoology of the Upper Great Lakes Region*; T. Pleger, "A Functional and Temporal Analysis of Copper Implements from the Chautauqua Grounds Site (47-Mt-71)," *Wisconsin Archeologist* 73 (1992): 160–176.

18. P. Stark and D. Denor, "Two Fish Weirs on the Oconto River," *Fox Valley Archeology* 16 (1990): 29–42; John Richards, "Prehistoric Stone Fish Weirs," in David Overstreet, ed., *Phase II Archaeological Investigations at 47Je932, 47Je933, and 47Wk445* (Milwaukee, 1992).

19. N. Lopinot and W. Woods, "Wood Overexploitation and the Collapse of Cahokia," in C. M. Scarry, ed., *Foraging and Farming in the Eastern Woodlands* (Gainesville, 1993), 206–231.

20. Antoine Denis Raudot, in a letter to his superior, 1710, quoted in W. V. Kinietz, *The Indians of the Western Great Lakes: 1615–1760* (Ann Arbor, 1991), 384–385.

21. Marquette in Claudius Dablon's "Relation of the Voyages, Discoveries, and Death of Father James Marquette," in Shea, ed., *Discovery and Exploration of the Mississippi Valley*, 14; Hennepin, "Narrative of the Voyage to the Upper Midwest," 108; C. Allouez, "Narrative of a Voyage Made to the Illinois," in Shea, ed., *Discovery and Exploration of the Mississippi Valley*, 72; La Salle in the "Narrative of Father Membre" in Shea, ed., *Discovery and Exploration of the Mississippi Valley*, 91.

22. G. Rajonovich, "A Study of Possible Prehistoric Wild Rice Gathering on Lake of the Woods, Ontario," *North American Archaeologist* 5 (1984); F. M. Keesing, *The Menomini Indians of Wisconsin: A Study of Three Centuries of Cultural Contact and Change* (Madison, 1987), 22.

23. C. Arzigian, "The Emergence of Horticultural Economies in Southwestern Wisconsin," in W. F. Keegan, ed., *Emergent Horticultural Economics of the Eastern Woodlands,* Center for Archaeological Investigations, Southern Illinois University at Carbondale, Occasional Paper no. 7 (Carbondale, 1981), 235; M. J. Black, "Plant Dispersal by Native North Americans in the Canadian Subarctic," in R. I. Ford, ed., *The Nature and Status of Ethnobotany,* Museum of Anthropology, University of Michigan, No. 67 (Ann Arbor, 1978), 255–262; C. Bloom, "Pathologies, Plants, and Healers: Medicinal Ethnobotany of the Winnebago and Other Eastern North American Tribes," manuscript on file at the Geography Library, University of Wisconsin–Madison, 1994.

24. Recent summaries of native agriculture in eastern North America include: B. D. Smith, "Prehistoric Plant Husbandry in Eastern North America," in C. W. Cowan and P. J. Watson, eds., *The Origins of Agriculture* (Washington, D.C., 1992); G. J. Fritz, "Multiple Pathways to Farming in Precontact Eastern North America," *Journal of World Prehistory* 4 (1990); W. I. Woods, ed., *Late Prehistoric Agriculture: Observations from the Midwest* (Springfield, Ill., 1992); and Scarry, ed., *Foraging and Farming in the Eastern Woodlands.*

25. C. Arzigian, "The Emergence of Horticultural Economies in Southwestern Wisconsin"; L. A. Zalucha, "Floral Analyses of the DEET Thinker and Cipra sites, Crawford County, Wisconsin," in J. D. Richards, ed., *Archaeological Recovery at the DEET Thinker (47 CR 467), Cipra (47 CR 414), and McCarthy (47 CR 108) Sites, Crawford County, Wisconsin,* Great Lakes Archaeological Research, Reports of Investigations no. 294 (Milwaukee, 1993).

26. Gartner, "The Hulburt Creek Fields"; Sasso and Gartner, "Garden Beds, Corn Hills, and Corn Fields: Current Investigations into Agricultural Sites in Wisconsin"; S. C. Lensink and W. G. Gartner, "Early Agricultural Field Systems from the Upper Midwest and Eastern Plains," in *Program and Abstracts,* Fifty-first Southeastern Archaeological Conference and the Thirty-ninth Midwest Archaeological Conference (Lexington, Ky., 1994); J. P. Gallagher, "Agricultural Intensification and Ridged Field Cultivation in the Prehistoric Upper Midwest of North America," in D. R. Harris and G. C. Hillman, eds., *Foraging and Farming: The Evolution of Plant Exploitation* (London, 1989); J. P. Gallagher and R. F. Sasso, "Investigations into Oneota Ridged Field Agriculture on the Northern Margin of the Prairie Peninsula," *Plains Anthropologist* 32 (1987).

27. J. Shaw, "Indian Chiefs and Pioneers of Northwest," *Wisconsin Historical Collections,* vol. 10 (1885; reprint, Madison, 1909), 220; B. H. Hibbard, "Indian Agriculture in Southern Wisconsin," *Proceedings of the State Historical Society of Wisconsin,* 52d Annual Meeting (Madison, 1904).

28. M. G. Hill, "PaleoIndian Projectile Points from the Vicinity of Silver Mound (47Ja21), Jackson County, Wisconsin," *Midcontinental Journal of Archaeology* 19 (1994); J. A. Behm, "Comments on Brown's Research at Silver Mound," *Wisconsin Archeologist* 55 (1984): 169–173; K. B. Tankersley, "The Exploitation Frontier of Hixton Quartzite," *Current Research in the Pleistocene* 5 (1988): 34–35.

29. J. B. Stoltman, "The Bass Site: A Hardin Barbed Quarry/Workshop in Southwestern Wisconsin," in B. M. Butler and E. E. May, eds., *Prehistoric Chert*

Exploitation: Studies from the Midcontinent, Center for Archaeological Investigations, Southern Illinois University, No. 2, (Carbondale, 1984).

30. G. A. West, *Copper: Its Mining and Use by the Aborigines of the Lake Superior Region,* Bulletin of the Public Museum of the City of Milwaukee 10 (1929): 32. See also J. J. Houghton, "Ancient Copper Mines of Lake Superior," *Wisconsin Historical Collections,* vol. 8 (Madison, 1879); and W. H. Holmes, "Aboriginal Copper Mines of Isle Royale," *American Anthropologist* 3 (1901).

31. O. W. Bidwell and F. D. Hole, "Man as a Factor in Soil Formation," *Soil Science* 99 (1965); R. Amundson and H. Jenny, "The Place of Humans in the State Factor Theory of Ecosystems and Their Soils," *Soil Science* 151 (1991).

32. M. F. Kolb, N. P. Lasca, and L. G. Goldstein, "A Soil-Geomorphic Analysis of the Midden Deposits at the Aztalan Site, Wisconsin," in N. P. Lasca and J. Donahue, eds., *Archaeological Geology of North America* (Boulder, 1990), 199–218.

33. E. F. Dietz, "Phosphorous Accumulation in a Soil of an Indian Habitation Site," *American Antiquity* 22 (1957); D. F. Overstreet, "A Rapid Chemical Field Test for Archaeological Site Surveying: An Application and Evaluation," *Wisconsin Archeologist* 55 (1974).

34. K. Stevenson and R. F. Boszhardt, *The Current Status of Oneota Sites and Research in Western Wisconsin,* Mississippi Valley Archaeology Center, Reports of Investigations no. 163 (La Crosse, 1993).

35. C. A. Dobbs, "The Application of Remote Sensing Techniques to Settlement Pattern Analysis at the Red Wing Locality," in *Program and Abstracts,* 56th Annual Society of American Archaeology Meeting (New Orleans, 1991).

36. R. F. Boszhardt, *Mound Mapping and Related Investigations at the Cranberry Creek Archaeological District,* Mississippi Valley Archaeology Center, Reports of Investigations no. 75 (La Crosse, 1988).

37. Gartner, "The Hulburt Creek Fields"; Sasso and Gartner, "Garden Beds, Corn Hills, and Corn Fields: Current Investigations into Agricultural Sites in Wisconsin"; Lensink and Gartner, "Early Agricultural Field Systems from the Upper Midwest and Eastern Plains."

38. W. Gartner, "The Geoarchaeology of Sediment Renewal Ceremonies at the Gottschall Rockshelter, Wisconsin," master's thesis, University of Wisconsin–Madison, 1993.

39. See W. B. Hindsdale, "Indian Overland Travel Ways," *Wisconsin Archeologist* 9 (1930): 118–119; L. P. Kellogg, "The Chicago-Milwaukee-Green Bay Trail," *Wisconsin Archeologist* 9 (1930); H. E. Neuenschwander, "Indian Trails and Villages of Dodge County," *Wisconsin Archeologist* 39 (1958); A. B. Stout and H. L. Skavlem, "The Archaeology of the Lake Koshkonong Region," *Wisconsin Archeologist* 7 (1908): 102; C. E. Brown, "The Brule–St. Croix Portage Trail," *Wisconsin Archeologist* 14 (1933); A. O. Barton, "Black Hawk Retreat in Dane County," *Wisconsin Archeologist* 24 (1943); R. E. Ritzenthaler, "Trail Marker Trees," *Wisconsin Archeologist* 46 (1965); and P. Sander, "A Trail Marker Tree at Twin Lakes," *Wisconsin Archeologist* 46 (1965).

40. For Middle Woodland Stage sites, see R. J. Salzer, "The Middle Wood-

land Stage," *Wisconsin Archeologist* 67 (1986): 275–277; and J. E. Freeman, "The Millville Site, A Middle Woodland Village in Grant County, Wisconsin," *Wisconsin Archeologist* 50 (1969): 37–87. For Late Woodland Stage sites, see P. H. Salkin, *Mitigation Excavations in the Kekoskee Archaeological District in Dodge County, Wisconsin*, Archaeological Consulting Services, Report no. 700 (Verona, 1993), 205–220; P. H. Salkin, *Archaeological Mitigation Excavations at Fox Lake, Dodge County, Wisconsin*, (Verona: Archaeological Consulting Services, Report no. 500 (Verona, 1989), 326–335. Norm Meinholz, in press, "A Late Woodland Community at the Statz Site," Museum Archaeology Program, Wisconsin State Historical Society, (Madison, in press); V. Dirst, *Research in Pursuit of the Past at Governor Nelson State Park, Dane County, Wisconsin*, Bureau of Parks and Recreation, Department of Natural Resources, No. PR502 (Madison,), 85–98, 186.

41.   A compilation of Oneota houses from the upper Midwest may be found in R. E. Hollinger, "Investigating Oneota Residence through Domestic Architecture," master's thesis, University of Missouri at Columbia, 1993. For the more recently discovered Oneota houses and settlement patterns in western Wisconsin, see K. Stevenson and R. F. Boszhardt, *The Current Status of Oneota Sites and Research in Western Wisconsin*, Mississippi Valley Archaeology Center, Reports of Investigations no. 163 (La Crosse, 1993); J. O'Gorman, *The OT Site*, State Historical Society of Wisconsin, Museum Archaeology Program (Madison, 1993); J. O'Gorman, *The Filler Site*, State Historical Society of Wisconsin, Museum Archaeology Program (Madison, 1994).

42.   A. B. Stout, "Prehistoric Earthworks in Wisconsin," *Ohio Archaeological and Historical Publications* 20 (1911); R. W. Peterson, "A Survey of Destruction of Effigy Mounds in Wisconsin and Iowa," *Wisconsin Archeologist* 65 (1984); R. F. Boszhardt, *Mapping Mounds in the Cranberry Creek Archaeological District, Year 1*, Mississippi Valley Archaeology Center, Report of Investigations no. 58 (La Crosse, 1987); and Boszhardt, *Mound Mapping and Related Investigations at the Cranberry Creek Archaeological District, Year 2*; W. Green and R. L. Rodell, "The Mississippian Presence and Cahokian Interaction at Trempeleau, Wisconsin," *American Antiquity* 59 (1994).

43.   Soil and vegetation surveys suggest that nearly half the vegetated land surface in Wisconsin was directly affected by native peoples prior to European settlement. See Curtis, *Vegetation of Wisconsin*, 464; F. D. Hole, *Soils of Wisconsin* (Madison, 1976), 28. See also W. Denevan, "The Pristine Myth: The Landscape of the Americas in 1492," *Annals of the Association of American Geographers* 82 (1992).

PART THREE

REGIONAL ECONOMIES AND LANDSCAPES

Over the last several centuries, human activities have increasingly modified the natural world of Wisconsin. These activities have interacted with the forces and features of nature to create ways of living and landscapes in which the dichotomous realms of people and nature are blended and muted. It becomes difficult to know where the natural world ends and the human begins.

The fusing interactions often have strong spatial expressions, helping to make Wisconsin look different from, say, central Illinois or western Minnesota. The effects of other linkages may be more local, creating those distinctive landscapes that we Wisconsinites readily appreciate as the familiar regional places of our home state—the forests of dense aspen and young pine surrounding boat-speckled lakes in the northwoods, or the deep valleys with wooded slopes and small ridgetop or valley-bottom farms in the Driftless Area.

The same blending of nature and people has created distinctive regional economies within Wisconsin. We all readily accept this fact: we think of the small dairy farms in the southwest, the urban commercial centers in the southeast, the large commercial farming operations in the center, and the wood and tourist-based activities in the north. These economic activities, in turn, are often tied to the appearance of the landscape; for example, German immigration has influenced the appearance of Milwaukee neighborhoods, and the harvesting of both fish and wild rice continues to stamp the environment of northern Wisconsin.

The chapters in this section of the book celebrate the creation of such regional landscapes and economies in Wisconsin. Some treat elements of the visual scene; others explore the historical creation of regional identities; still others analyze forces that tend to create those spatially expressed personalities. A more distant past may be stressed in some chapters, whereas others may be focused on the present. Like the chapters in other sections of the book, those in this section illustrate what we, as curious members of the Wisconsin community, might learn about a myriad of topics that link people and nature in our home state.

18

Changing Technology, Values, and Rural Landscapes

Clarence W. Olmstead

John Fraser Hart, in his book *The Look of the Land* (on which this chapter leans), declares that the Kentucky Bluegrass country is "the loveliest rural area in the United States."[1] Indeed it is lovely, as are parts of rural New England, New York, and Pennsylvania. Much of the charm of these four areas, however, is due to personal wealth and aesthetic ideas introduced from outside, whereas still another rural Eden, the larger southern Wisconsin area, retains a loveliness derived from internal primary economic activity, with its productive farms of dairy herds and green pastures, of contoured fields, and dense woodlots. This chapter supports the proposition that southern Wisconsin is an area of unusually attractive rural farm landscapes. It will attempt to explore the following questions: What is the mix of ingredients that makes them so attractive? How were the landscapes created? How have they changed and how are they changing? What are the prospects for the future? We will confine our exploration and examples to southwestern Wisconsin, although the processes and patterns have broader application to much of the southern part of the state.

The Concept

Like many things studied by geographers, landscapes are part of the everyday lives of people, yet they challenge scholars both to define exactly what they are and to understand how they develop and change. The term is used variously, often confusingly, as a synonym for nature, environment, habitat, place, region, scenery, view, or even a painting.[2] We will use the term to represent the integrated ensemble of all that the human eye can observe from a single vantage point. Further, it is an *informed* view from a *selected* vantage point. It is informed because what one sees depends upon what one knows. That is, to *read* a landscape requires informed curiosity and the skills to derive meaning from what one sees. The vantage point is selected so that the landscape observed is representative of a significant area and set of processes. The rewards can be exciting: Such a landscape is not only a meaningful piece of our Earth Home, but also a record of ourselves—who and what we are, what we have created, what we value.

A landscape is an ensemble—a unity. But to understand and appreciate a unity, one must identify and assess the individual features which are unified. These particular features are derived from earth processes, human processes, or combinations of both. We will explore how the features are meaningfully interrelated to form the rural landscapes of southwestern Wisconsin. In doing so, I propose a corollary to our original proposition: The unusual attractiveness of Wisconsin rural landscapes results from the diversity and balance of their ingredients.

The Ingredients of Landscapes: Earth Processes

The most basic structural ingredients of landscapes are the landforms on, or over, which all else rests. The landforms of southwestern Wisconsin are neither spectacular nor monotonous. They lie within the Driftless Area, the only part of the state not overrun by the continental ice sheets, which modified landforms, generally softening their features, elsewhere. The basic geological structure is one of layers of sedimentary rock, strata of relatively weak sandstone and shale alternating with more resistant strata of dolomitic limestone, all sloping gently toward the southwest. These rocks have been incised by streams that flow directly to the Mississippi River or to its tributaries, especially the east-to-west-flowing section of the lower Wisconsin or, farther south, the Pecatonica. The resulting valleys are particularly deep near the Mississippi and Wisconsin rivers, where the local relief may exceed 400 feet. The uplands comprise fingerlike ridges, the broadest and most extensive being Military Ridge, a cuesta formation lying just south of and parallel to the lower Wisconsin River (Figure 18.1).

Prairie and oak savannas, or oak openings, covered most of the

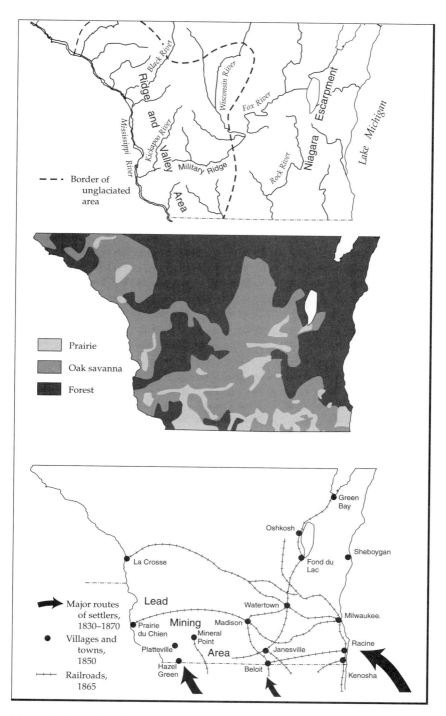

Figure 18.1. Map of Southern Wisconsin showing selected features in the creation of rural landscapes, and the major routes of the settlers during the pioneer period, 1830–1870.

Figure 18.2. Views north and south from East Blue Mound. West and East Blue Mounds are remnants of the Silurian and Maquoketa geological formations, which have been removed elsewhere from all but the eastern border of Wisconsin. Sitting atop the eastern end of Military Ridge, they afford wide views of landscapes. The foreground, similar in each view, is the sloping edge of East Blue Mound. But beyond, the contrast between predominantly woodland to the north and cropland to the south is sharp. In both cases, trees occupy steep slopes. To the south *(top)*, streams draining the long, gentle dip slope of the Military Ridge cuesta flow first over its surface, their valleys deepening gradually

ridge and valley topography in pre-European times, with woodland more prominent northward and on steep slopes, prairie more dominant southward and on uplands (Figure 18.2). Today, the vegetation has been widely replaced by crop plants and introduced weeds, but on areas of steep slope or thin soil, or on unplowed strips bordering railroads, remnants of the wild vegetation remind us of past conditions. Rich soils developed beneath the prairie grasses and oak groves, and, although visible only during the brief planting season when they are exposed by tillage, their role in the use of land, and thus in the appearance of the landscapes, is fundamental. Finally, much of the natural landscape owes its character to the central importance of weather and climate. The various forms of precipitation, their amounts and seasonal occurrences, the presence or absence of clouds, the humidity and other qualities of the air, the nature of the winds—all influence the appearance of the landscape on any given day. Even more important, weather and climate play a fundamental role in the *making* of the landscape: in the erosion of its landforms, the growth of its plant cover, the development of its soils, and the various human processes that build upon the natural scene.

The Ingredients of Landscapes: Human Processes

Native Americans and their ancestors have lived in southwestern Wisconsin for thousands of years. They probably affected, more than has been generally recognized, the landscape ingredients briefly reviewed above as having resulted from earth processes. Nevertheless, it remained for European Americans, mainly after 1830, to settle the land more completely and create landscapes dominated by people and their activities. Among the dominant human works are the system of land survey and division, the systems for transport and communication, the farms and farm structures, the service centers (hamlets, villages, towns, and cities), rural economic and social establishments (mines, mills, factories, stores, taverns, inns, schools, churches, parks, government buildings), and rural nonfarm residences.

---

southward. Most of the land there remains in upland with fertile prairie soils developed on loess (silt carried by wind from the Mississippi Valley during glacial time). To the north *(bottom)*, the many small streams plunging down the escarpment edge of the cuesta to the Wisconsin River, only 10–16 miles distant, have cut the land into deep, steep-sided valleys with narrow ridges between. Some farms and fields on the ridges can be seen; those in the valleys are hidden. The Baraboo Range, 25 miles distant, marks the horizon in the northward view. (Photographs by the author)

The Creation of Rural Farm Landscapes

The settlement and transformation of southwestern Wisconsin by Europeans are examined elsewhere in this book. We will confine our commentary here to the creation and modification of rural landscapes. That has been and is a continuous process, but one which proceeds at different rates and in different ways as new ideas are introduced to and diffused across an area over time. To aid our review of the process, I identify four overlapping stages whose time spans are only approximate.

The Pioneer Period, 1830–1870

Wisconsin lands were taken from the Native American peoples by the federal government mainly between 1827 and 1848, when a state government was established. Offices for sale of land to settlers were opened at Green Bay and Mineral Point in 1834 and at Milwaukee in 1836. Settlers were coming even earlier, primarily from two directions: up the Mississippi Valley into the lead mining area in the southwest, and through the ports of Chicago, Southport (Kenosha), Racine, and Milwaukee into the southeast (see Figure 18.1). State population reached 30,000 by 1840, and 300,000 by 1850; it was mostly rural and south of a line from Prairie du Chien to Green Bay.

Fundamental in the making of most midwestern landscapes is the rectangular system of land survey and division adopted by the Ordinance of 1785.[3] As adapted, it created 6-mile-square townships, each divided into 36 numbered, square-mile sections of 640 acres. Successive quartering of the sections created convenient square blocks of 160, 40, and 10 acres. The rectangular pattern is reflected in most kinds of land use, such as farms, fields, roads, street patterns, utility lines, and government administrative units.

John R. Stilgoe has pointed out some basic dichotomies in the making of landscapes.[4] One is natural versus cultural, or wilderness versus domestication. In 1830, southwestern Wisconsin was dominated by wilderness, except for the beginnings of commercial lead mining. By 1870 there was a greater balance, with dominance swinging toward the side of human modification. A second dichotomy is husbandry (cultivating but nurturing the earth) versus artifice and industry (using earth materials to manufacture desired articles). During the earliest pioneer stage in southwestern Wisconsin, artifice was dominant in the form of lead mining, but after a decade or two, that dominance was lost to farming.

Although speculators in land and founders of railroads, villages, and industries often established important patterns upon the land, most of the settlers were individual farmers. A very few of them could purchase land already partly improved, or could afford to hire the land cleared and to build substantial initial houses and farm buildings. The majority, however, especially immigrants from Europe, had to start from scratch, with very little capital. The ideal was to arrive in spring so as to plow enough prairie sod or clear enough forest to plant and harvest a crop of wheat, corn, or potatoes sufficient to survive the winter. But to break the prairie sod properly required teams of four or five yokes of oxen and a heavy wheeled plow. Only three to five acres of heavy timber could be cleared by a man in a winter-spring season, with the stumps left to rot. The family would live in a tent, lean-to, or dugout until a log cabin could be ready, preferably before winter. Cattle, and especially hogs and chickens, were often left to forage for themselves. Fences, usually of split rails laid up in zig-zag fashion, were needed to protect crops from livestock. Contrary to stories, most settlers, especially those from Europe, were not, initially, accomplished hunters or fishermen, nor had they time for such activities except perhaps in winter. Neither are the stories of self-sufficient farms representative. To be sure, the first priority was to grow enough food to sustain family and livestock, but a goal from the beginning was to produce one or more products for sale so as to be able to pay for land, tools, and other necessities. The dominant commercial crop was wheat. It could be hand sown on plowed soil, even among stumps, and largely neglected until harvested by scythe and cradle, hand tied into bundles and shocks, and threshed by trampling oxen. The major problem was getting it to market.

Roads were built to align with the rectangular survey, the eventual goal being a road along the sides of each mile-square section, or even quarter-section. But the goal was long-delayed and incompletely attained. Landforms, especially the steep-sided ridges and deep valleys of western Wisconsin, forced major irregularities. Second, almost no roads were surfaced, even with gravel, during the pioneer stage. It was the completion of railroads, from Lake Michigan port cities to Prairie du Chien and La Crosse in the late 1850s, that provided farmers with a means of marketing wheat, provided they could get it from their remote fields to the rail lines.

Besides access to the world, a farm settler needed water, tillable soil, pasture for livestock, wood for buildings, tools, fuel and shade, and some protection from winter winds. A site near the beginning of a stream at the head of a valley could provide all: water from natural springs, pasture along the stream, trees on the valley sides, fertile soils on the prairie upland, and even some shelter from wind. But the number of such sites is limited. Most settlers had to choose between a ridge farm and a valley farm, either one often including parts of the steep wooded slopes between.

In brief summary, a typical landscape of the period included scattered, partly improved farms of perhaps 40–100 acres. Small, irregular fields, enclosed by rail fences and worked by oxen, men, women, and children, produced wheat, corn, potatoes, and subsidiary crops such as barley, rye, or hemp. Some original log houses had been enclosed within, or replaced by, larger wood-frame structures, only a few by brick or stone. A replaced log dwelling might have been converted to a summer kitchen, a livestock shelter, or other use. On older or better farms, there would be a log or frame stable, corncrib, chicken coop, springhouse, or other structures. An outdoor privy, a small garden, and perhaps a patch of tobacco and a few fruit trees completed the farmstead. Every few miles along the muddy roads would be a one-room schoolhouse, and perhaps a log or frame church with a steeple but no bell. Each six-mile-square town would have a small frame town hall, barely distinguishable from the schools. The service centers, many located along the few railroads, would be small, with a grist mill, sawmill, blacksmith, general store (probably including a post office), school, and church, all along an unpaved street. It was a landscape in transition—unfinished, unkempt, and entirely utilitarian, with no frills or embellishments except on the rare holdings of the wealthy, and those mainly in developing towns.

Consolidation, 1870–1920

During the next half-century, the process of creating a human-dominant landscape out of wilderness was essentially completed. In 1870 about half the land was in farms; by 1920 virtually all was so transformed. Not only did people bring most of the land into use; they also changed the kinds of use, especially the system of farming, thereby creating new ensembles of features—new landscapes.

The pioneer system of farming provided subsistence for the family and a wheat crop for cash while the land was being cleared, and while structures, tools, and livestock were being added or improved. Output of wheat in Wisconsin reached a peak in 1860, but then began to decline in the face of disease, pests, diminishing yields, and competition from cheaper or better land farther west. As alternatives were sought during the following decades, dairying emerged as the new and dominant system. The dairy cow, which in earlier times was a dual-purpose animal often left to forage for herself or even subjected to the yoke with oxen, was now selected for improved breeding and production. Homemade butter, credited at the local general store, had commonly been an auxiliary earner of cash during flush milk season in spring and early summer. As butter production increased, its manufacture was taken over by creameries located in villages or towns. Cream was separated on farms with new hand-cranked centrifugal machines, allowing skim milk to be retained for animal feed. Five-gallon cream cans, sitting on small wooden platforms at the ends of farm driveways awaiting pickup by haulers, were a common feature of rural landscapes.

Making quality cheese, a biochemical process which involves proper curing, was not easily accomplished on farms. In 1851 in New York, and then in 1864 in Wisconsin, the first examples of a new factory system were established. Small hillside factories were built at country crossroads to which farmers within two or three miles could bring fresh whole milk in cans. The cheese was made by a skilled artisan using appropriate vats and equipment, and placed in a cool curing room carved into the limestone or sandstone hillside. Early factories were usually owned by individual cheese makers or by investors. Later, many became cooperatives, owned by the farmer-suppliers who hired the cheese maker. By 1910 there were 1,000 creameries and probably twice that number of crossroads cheese factories in Wisconsin. The development of dairy farming into a productive economic system was a complex process involving the input of many people, from progressive farmers to scientists. The University of Wisconsin College of Agriculture played a leading role, not only in developing basic technology such as the Babcock butterfat test, but also in its "Wisconsin Idea" of transferring the results of research on crops, livestock, products, and methods directly to farmers through extension programs, bulletins, institutes, demonstrations, and winter short courses for farm boys.

Moderately cool Wisconsin summers with usually ample rain are well-suited for oats, hay, and pasture growth, and for animal health. But to provide feed and shelter during the relatively long, cold winters requires capital, effort, and ingenuity. Two key adaptations during this period, both distinctive features of the landscape, were the large Wisconsin dairy barn and the silo. The way barns have evolved and diversified in their diffusion from Europe to and across America is a complex and fascinating subject, a Wisconsin facet of which is treated in another chapter.[5] Suffice it to say here that large barns designed to provide stanchions for up to 30 milk cows at ground level, as well as a huge superstructure for storing loose hay, were becoming a distinguishing feature of Wisconsin farm landscapes by the close of the consolidation period.

Likewise fascinating are the role and forms of the silo. Corn was important for the feeding of people, and especially livestock, during the pioneer period. But the warmth and length of Wisconsin summers were often insufficient for the grain to mature. The silo is designed to preserve green fodder crops, especially corn, in a succulent condition by finely chopping the whole plant and storing it, tightly packed so as to reduce or prevent exposure to air. Experimentation occurred in France, New York, and then Wisconsin (again with leadership from the university's College of Agriculture) after 1870, first with excavated pits, then upright structures of various building materials, shapes, sizes, and roof forms. Emerging as most successful were cylindrical towers of vertical tongue-in-groove staves, first of wood, then of concrete, reinforced by encircling iron or steel bands. By 1924, there were over 100,000 such silos in Wisconsin, each situated next to a dairy barn so that the ensilage (silage) could be forked off the top and dropped daily into the feeding alley of the stable.

By early twentieth century the rural landscapes of southwestern Wisconsin had become more finished and diversified and more balanced between "natural" and human features. Although most of the land was now in farms, up to half or more, mainly on the steep slopes separating upland from valley, remained in woodland pasture.

The oxen-drawn plow and much of the hand labor in fields had been replaced by gradually improved horsedrawn implements—drills for planting, mowers and rakes for haymaking, reapers and steam-powered threshers for harvesting small grains. Horses were the prime means, also, for hauling milk, cream, or produce to factories, service centers, or rail stations, and for travel on the still little-improved, but extended, system of roads.

The fields of hay, corn, and oats were still relatively small, separated by fencerows, and dotted in summer and fall by handmade haycocks and shocks of small grains or corn. The hay was cut and cured two or three times each summer. Part of the corn was cut and chopped while green and stored in silos; part was left to ripen for grain, the ears husked by hand, the fodder used for feed and bedding. Oats served as feed, especially for horses, as a "nurse" crop for spring-planted hay crops and as straw for bedding. Although dairying and its three associated feed crops had become dominant, most farms retained auxiliary enterprises. Surplus dairy animals, especially young males and retired milkers, were butchered for home use or sold. Whey was hauled home from cheese factories as supplementary feed for pigs, which were also butchered or sold. Almost every farm had flocks of poultry, usually tended by farm wives and children. Some areas developed more specialized enterprises, but most of these, except for tobacco in the Viroqua area, were outside southwestern Wisconsin. In the extreme southwest, where a longer growing season made the maturing of corn for grain a more dependable occurrence, it was fed to beef cattle and hogs. Such "Corn Belt" farms tended to be larger than the dairy farms elsewhere, but with smaller barns and, often, no silo.

Farmsteads were now more finished and elaborate, with modest, usually two-story, white frame dwellings, large dairy barns with attendant cylindrical silos, windmills drawing up groundwater (now the chief means of supplying increased needs), and auxiliary buildings, yards, and gardens. Shade trees, flowers, shrubs, or even a front lawn suggested permanence and, perhaps, relief from the earlier struggle to subsist. The basic rectangular pattern designed to provide a road along each section line was highly modified, with roads often following the sinuous ridges and valleys. White-painted, one-room schools and cheese factories, the latter tucked into hillsides, were arranged along the still-unimproved roads so as to require no more than a two-mile walk or three-mile haul, respectively, to be reached from each farm. A somewhat less regular and dense arrangement of white frame or stone churches, now usually with bells, added charm to the countryside.

Mechanization, 1920–1970

Few, if any, innovations have affected American life and landscapes more than the internal combustion engine and electricity. Both were being developed during the decades just before and after 1900, but began to change farming systems and rural landscapes in Wisconsin slowly after 1920, rapidly after 1940.

The gasoline engine was first used on farms to provide belt power for tasks such as chopping, grinding, pumping, and elevating. It was the use of engines in tractors, motor trucks, and automobiles, and the concomitant improvement of roads, that sparked a revolution. Just as tractors replaced horses in fields and roads, electricity relieved human power in barns and farmsteads. Lighting, milking machines, pumps, pipelines, cooling tanks, fans, silo unloaders, grain dryers, elevators, forklifts, electric fences, power tools—all tended to change farm operators from laborers to managers and to increase labor productivity.

The new off-farm sources of farm energy were applied through an increasing variety of machinery and farm implements and accompa-

Figure 18.3. Upland landscape, west of Westby, Vernon County. The broadest ridge in the ridge and valley country north of the lower Wisconsin River is that extending for more than 50 miles, north to south, between the Mississippi and Kickapoo rivers. The rolling upland supports large dairy farms, some milking more than 100 cows, some in the third or fourth generation of ownership. The neat farmsteads and contoured fields of corn, alfalfa, and some oats suggest respectful care of the land. (Photograph by the author)

nied by comparable developments in plant and animal breeding, fertilization, control of disease and pests, and processing and marketing of the increasing crop and animal yields. A primary response of management was to enlarge the scale of operations so as to pay for and use the new technologies efficiently. And the landscape changed.

Fields were enlarged by removal of fencerows. Use of land, even steep slopes, for pasture declined; the new machines made it possible to bring feed to the dairy cows, often distributing it mechanically, ready to eat, in feedlot or barn troughs. The hay-corn-oats trilogy of major crops persisted but in different proportions and forms. Oats, no longer needed to feed horses, declined. Larger acreages of corn for grain, the yield vastly increased by new earlier-maturing varieties of hybrid seed, spread northward from their former concentration in the southwest corner of the state. The picturesque shocks of corn and small grains and the cocks of curing hay were no longer to be seen; the crops now disappeared into voracious machines.

Farmsteads changed (Figures 18.3 and 18.4). The already-large dairy

Figure 18.4. Valley farms. This former farm *(top)*, probably once prosperous and considered ideal, is near the head of a small valley, tributary to Vermont Creek, Iowa County. It is now a nonfarm residence, with a fine old three-story restored farmhouse surrounded by white fence, lawn, and flowers beyond the well-maintained farm buildings. The narrow, irregular fields and the old tile silo are inadequate for modern, large-scale farming. The foreground field is mowed for appearance; larger fields beyond the buildings are incorporated into a larger farm farther down the widening valley. Large farm *(bottom)*, Mill Creek valley, Iowa County. Here the valley is broad enough for large, productive fields. The large brick house, center, is dwarfed by the elaborate farmstead, most of whose steel grain bins, low pole barns, and giant concrete and steel silos were constructed after 1950, as the farm grew in scale. Wooded valley sides, formerly pastured, separate the valley farm from upland farms. (Photographs by the author)

barn now showed one or several additions; the first was commonly an extension of the same height, roof form, and color, but the second and later incremental sections were often low additions constructed of different materials, a compromise with cost and utility. What farmers needed was larger ground-level stable for more milk-cow stanchions, not the huge haylofts above; bales of hay, now compacted by a field implement, could more easily be stored in an adjacent low building. The dominance among farmstead structures was shifting from dairy barn to silos. The original silo of wood, tile, or concrete, standing adjacent to the barn, was now dwarfed by a series of cylindrical structures, each newer one larger than the last, and often with auger-fed troughs leading into feed yards or low steel buildings for loose housing of cattle. Most conspicuous (and expensive) are the towering dark blue steel structures prominently labeled Harvestore. Their glass lining, air pressure bag at the top, and power unloader at the bottom provide air-sealed preservation and mechanized delivery of green chopped corn, or sometimes alfalfa, oats, or mixtures of these crops. The climb to the top of an earlier-style silo to fork down ensilage was both difficult and dangerous, especially in the dark and cold of winter. Outmoded structures (springhouse, privy, windmill, poultry or hog houses) were converted to or replaced by new buildings (garage, workshop, steel grain bins, pole barns for implements or hay) as subsidiary enterprises were dropped in order to concentrate on the larger scale and technology of the main activity.

Because most rural land was in farms by 1920, farm enlargement could occur only through consolidation. Larger, more progressive farms tended to take over smaller, poorer ones, often from older operators. Ironically, such a smaller farm was often one that was among the first settled in the area, near the head of a small valley with the combination of resources considered ideal at the time. But the necessarily narrow, irregular fields along a tiny, winding stream or ridgetop, or the steep wooded slopes between, are not ideal for large-scale, mechanized farming. Consequently, in the ridge and valley country, it was the small farms uppermost in the valleys, or outermost on the narrowing ridges, that were likely the first to become "abandoned." If the fields were indeed too small and irregular, or the farmstead too outdated to be useful in modern farming, such a farm was likely to become a rural home or retreat for an urban dweller. It was the next, slightly better or bigger, farms along the widening valley or ridge which were likely to be absorbed into still larger operations.

The two, three, or more farms that become consolidated into a single operational unit were seldom contiguous. Consequently, tractor-drawn wagons and field implements became regular users of the newly hard-surfaced roads.[6] So, too, did commuters to towns and cities, now living in the extra farmhouses of the merged farms. And so did yellow school buses bearing children to newly consolidated schools in villages, and gleaming stainless steel tank trucks carrying milk from new milk houses attached to every dairy barn to new or enlarged factories, usually in towns or villages. Even little country churches might lose their parishioners to consolidated village parishes. Blacksmith shops became converted to service or repair stations for motor vehicles. The quiet rural landscape had become part of a bigger-scale, faster-moving world.

Hybridization, 1970–1995

All the processes which characterized the mechanization stage continue: the development and application of technology, the use of off-farm or other-farm inputs (notably, energy, machinery, agricultural chemicals, hybrid seed, supplemental feed), and the resultant consolidation of farms and change in farming systems and landscapes. But a new process marks the late twentieth century—the hybridization of rural and suburban life and landscapes (Figure 18.5).

First, the country, including its farms, has adopted the city or city ways. The farmer has become manager of a large-scale business. A dairy farm that milked a dozen cows by hand in 1900, three dozen by machine in 1950, may now milk 70 in a stanchion barn with pipeline, or hundreds in a milking parlor with new low buildings for loose or free-stall housing. Detailed records of input, output, cost, and return are kept by computer. Many farms, especially on uplands with fertile prairie soils, now produce only grain corn and soybeans for the world market (Figure 18.6). They represent an extension of the Corn Belt into the dairy region. The larger farms are often family corporations involving parents and children, siblings, or other relatives. Such an arrangement facilitates not only largeness of scale and individual specializations in management, but also working hours and vacation times more comparable to those of city folk.

Other, often smaller, farms have become suppliers to the city. They may vend fruits and vegetables, some "organically" grown, at newly revived farm markets, or invite city folk to "pick your own" or even to contract in advance for garden produce. They may be recreational farms providing horses or trails for riding, hiking, skiing, or snowmobiling, perhaps with bed and breakfast. Still others have been converted to golf courses, parks, restored prairies, or nature preserves.

Figure 18.5. Landscape change. The two views, recorded about 1900 and 1994, look west-ward from the terminal moraine of the Wisconsin glaciation, and along Mineral Point Road, about 10 miles west of the state capitol at Madison. The road was, locally, part of Military Road, built 1835–1837, to connect Fort Howard (Green Bay) and Fort Winne-bago (Portage) to Fort Crawford (Prairie du Chien) and the lead mining region. In 1900, roads were little more than cleared rights-of-way. Since 1920, their improvements have matched those of vehicles. Note the broad paved surface with turning lanes, the land fill across the old glacial drainageway, and the deep cut in the bedrock dolomite upland beyond. The farmstead on the right, one of the earliest in the area, was removed after 1990, when the farm was converted to a federal wildlife refuge with restored pond. In the farmstead at left, the exterior of the farmhouse has been little changed. But all other structures, except the windmill frame, have been replaced. The 50-stanchion barn, with haylofts, was built in 1959, the silos and other structures since. The air-sealed silo, at

Figure 18.6. Arlington (Empire) Prairie, about 20 miles north of Madison, within the glaciated area. This was one of the largest islands of prairie in Wisconsin. The fertile prairie soils on near-level terrain support large farms, many of which produce corn and soybeans for the world market. Some have been incorporated into the University of Wisconsin Experimental Farms. The few trees, mainly in farmsteads, have been planted (Photograph by the author)

Former railroad beds are now bike trails. Many farm wives, older children, and part-time farmers commute daily to jobs in cities. Farm families now live like city families. They live in similar houses with similar conveniences, lawns, and landscaping. They shop at the same malls and markets and purchase similar items, including packaged milk, meat, and eggs. They go to the same schools, churches, clubs, and restaurants.

Second, the city has adopted the country. City residents have moved into extra farmhouses of merged farms, refurbished schoolhouses, and

right, is used to store high-moisture shelled corn. The ranch-type dwelling on the upland above the old farmhouse is a "grandparent house"; a feature seen on many farms, it houses the retired generation of the farm family. The electric power line, right, reached the two farms from the Madison area in 1900. The telephone line, along the left in the older photo, is now underground. The low, partly wooded ridge of the terminal moraine, from which the modern photo was taken, is now covered by the expensive homes of unsewered residential suburbs of Madison, which have leap-frogged farm land in order to lie beyond the reach of Madison's planning powers. A part of city-in-the-country, they add to the flood of morning commuters seen in the modern photo, in contrast with the single, local, horse-drawn vehicle in the other. (The old photograph is courtesy of Maurice Coyle and Jacob Dresen; contemporary photograph by the author)

Figure 18.7. Values, pride, and wealth play important roles in the creation and modification of landscapes. This former farmstead is about 10 miles from an urban center. The new small residence occupies the site of the former farmhouse. Although the barn and small silo, on the right, are falling down, two riding horses are enclosed in a small fenced lot attached to the stable. (Photograph by the author)

Figure 18.8. Many former crossroads cheese factories have been converted to residences. Such country factories were built into a hillside, with underground curing rooms, and near a crossroads so as to serve a maximum number of farms (usually 8–15) within minimum hauling distance. The wide, hard-surfaced road, garage addition with deck and TV satellite dish, and recreational vehicle, all suggest marked change in life and landscape from the early days of the factory. Western Dane County. (Photograph by the author)

crossroads cheese factories, even remodeled barns or log houses. They have built new houses in the country, sometimes on sufficient acreage for gardens or horses, sometimes in large suburbs in the countryside (Figures 18.7 and 18.8). Many have purchased as second homes the small but scenic valley, ridge, or hilly wooded farms no longer suited for large-scale agriculture. Roads originally trod by farm oxen and country school children are now commuter routes for both farm and city folk, indistinguishable.

Reflections

Questions arise concerning the new, hybridized landscapes. Are they attractive? Are they good? What of the future? In reflecting on such questions one need recognize two points. First, although we have directed our attention in this chapter to one small area for a period of less than two centuries, the process of landscape creation and change occurs in all places during all human time, and with many of the same forces and in many of the same ways as have been described for southwestern Wisconsin. Second, these are questions of values which all societies face but upon which individuals differ. They involve the basic dichotomies within landscapes: natural versus cultural, preservation versus development, husbandry versus artifice/industry, aesthetics versus utility or material gain, vernacular versus professional design.

During the pioneer stage of landscape creation, the highest value was to subsist and survive. Only gradually were people able to express other values in the care given to fields and livestock, the neatness and embellishment of homes and farmsteads, the design of schools, churches, parks, and roadside. As change became more rapid with mechanization and new technology, novel values became more critical in landscape decisions. Should a residential suburb or other fragment of city be set down in the midst of farms? Should an archway of trees or delightful bends in a sylvan road be sacrificed in order to speed travel? Should utility or financial gain erase beauty?

In the application of values, extremes are seldom good, even if possible. Either complete wilderness or complete urbanization is probably not in southwestern Wisconsin's future. Extremes of monocrop agriculture uninterrupted by farmsteads, such as are found in parts of southern Florida or coastal Texas—productive agriculturally but deserts aesthetically—are unlikely in Wisconsin, where diversity of natural features encourages diversity and balance in cultural characteristics, and where most farmers continue to live on the land they farm (Figure 18.9).

Figure 18.9. An attractive landscape of diverse natural and cultural features: upland and valley; unplanted woodland and carefully tended, contoured fields of alfalfa and corn; neat farms with old and newer structures. The nearest farmstead, now a nonfarm residence, retains the windmill and small old silo, although the barn has been replaced. The two farmsteads in the middle ground have added newer pole barns to the old-style ones, and newer, huge silos, which dwarf the old one at the third farmstead. (The newest chapter in the 100-year sequence of silo technology: Many farms are now using long tubes of tough plastic to enclose and preserve ensilage. Supported on the ground, filled and emptied by tractor forklift, and disposable—by what means?—they are less convenient but much less expensive than huge cylindrical, upright silos. The landscape ever changes.) The view, in southwestern Dane County, is eastward into the valley of the West Branch, Sugar River. (Photograph by the author)

Prediction of future technology and values and their effect on landscapes, however, is not possible. American society seems currently to have fallen into a kind of individual selfishness and cynical mistrust of governments (i.e., individuals acting cooperatively through chosen representatives) to make wise decisions, including those about land use and landscapes. But, in longer view, there seems also to have been some retreat from earlier, sometimes intemperate use of technology in the exploitation of land and landscapes for material benefit. Drained marshes (in eastern glaciated Wisconsin) have been withdrawn from cropping and allowed to play their more important roles in natural hydrology and support of wildlife. Dam construction has been curtailed, the anticipated material gains recognized as being outweighed by natural and aesthetic losses. Waste dumping has been slowed, stream banks protected, prairies and woodland restored, unsewered developments vetoed, parklands preserved, excessive use of agricultural chemicals curtailed.

Perhaps the most important questions concerning agricultural land are: Who owns it? Who manages it? Who cultivates it? Who lives on it? Who loves it? In the extreme case of parts of the Texas coast, the owner may live in Houston, the manager in Corpus Christi, the cultivator in a small town, and no one lives on the land, or loves it. In the ideal case in Wisconsin, the answer to all five questions is, "the Jones family." Furthermore, the Jones family has lived on the land for several generations, and they consider it a permanent Earth Home and themselves its current caretakers. The basic needs of land for expansion of cities, for production of food and fiber, for restoration of body and spirit, will long continue. With wisdom, and the values of the Jones family, they can long be met. The diversity and balance which make Wisconsin landscapes both productive and attractive today provide a key for the future.[7]

Notes

1.   J. F. Hart, *The Look of the Land* (Englewood Cliffs, N.J., 1975), 186.

2.   D. W. Meinig, ed., *The Interpretation of Ordinary Landscapes* (New York, 1979), 1–7.

3.   H. B. Johnson, *Order upon the Land* (New York, 1976).

4.   J. R. Stilgoe, *Common Landscapes of America, 1580 to 1845* (New Haven, Conn., 1982), 3–29.

5.   See also A. G. Noble, *Wood, Brick and Stone, The North American Landscape*, vol. 2: *Barns and Farm Structures* (Amherst, Mass., 1984); Hart, *The Look of the Land*, chap. 8; J. Apps and A. Strong, *Barns of Wisconsin* (Madison, 1977).

6.   Wisconsin was one of the first areas on the continent to pave most of its rural farm roads. A major incentive was to facilitate prompt and regular daily delivery of fresh milk to the crossroads cheese factories, increasingly, after 1920, by motorized vehicles.

7.   Such values, encompassed in a land ethic, were best expressed by Wisconsin's own Aldo Leopold in *A Sand County Almanac*, first published in 1949 by Oxford University Press. But this is the present. It is suggested that the reader acquire appropriate maps, a guide (even this book), a pair of binoculars, and perhaps some bread, wine, and cheese, and venture leisurely into the Wisconsin countryside to read, ponder, and enjoy the landscape. An ideal guide for part of the area is C. Mather, J. F. Hart, H. B. Johnson, and R. Matros, *Upper Coulee Country* (Prescott, Wis., 1975). A beginning source of maps is the University of Wisconsin Geological and Natural History Survey, Madison.

19

Milwaukee's German Renaissance Twice-Told
Inventing and Recycling Landscape in America's German Athens

Steven Hoelscher, Jeffrey Zimmerman, and Timothy Bawden

As a way to spruce up its languishing central business district, the Milwaukee Redevelopment Corporation commissioned world-renowned painter Richard Haas in 1977 to design and execute an expansive mural on one of Milwaukee's downtown theater buildings. The large-scale mural was to be painted on the art deco–style theater's northeast exterior wall, which had been conveniently opened to view with the razing of its neighboring building, leaving behind an immense concrete "canvas." Haas, a Wisconsin native and a great admirer of the city's German-inspired nineteenth-century landscape, chose the first and one of the grandest of the city's "golden age" buildings as his subject—the 13-story Pabst Building of 1892, which once towered above Milwaukee's generally level skyline as its first skyscraper and, more important, as its first major civic structure to employ a distinctively German-influenced style (Figure 19.1). With its elaborate curved gables and crowning tower, the Pabst Building reminded visitors and residents more of what they had seen in Hamburg or Munich than in Chicago, St. Louis, or Boston.

By the late 1970s, when Haas began his mural, the Pabst Building had fallen on hard times. Giving the building "a good shave with a cross-cut saw," its owners had stripped the Pabst of its tower and seriously altered its roofline, leaving behind a grim monument to the forces of modernization. For Haas, such destruction signaled a grave injustice to the character and life of a place he had known as a child. Thus, when he came to the task of designing the city's downtown mural, he chose to create an ironic statement of loss and regret (Figure 19.2). Moved, as he said, "to correct a crime," his mural depicts a mirrored window that reflects a historic view of the Pabst Building before its "insensitive modernization." With near photographic realism, the 10-story-tall Pabst mural mesmerizes. It forces the viewer to look over his or her shoulder to see the landscape that is reflected so beautifully. When standing halfway between the two buildings, separated by two city blocks and the river, Haas intended that one should see two strikingly different versions of the Pabst Building: to the east would stand the "real" Pabst, a mere shell of its former self; and to the west would be the "illusionary" Pabst, restored to its former nineteenth-century glory. Haas's mural would seem to beg the question to the downtown pedestrian: These are two versions of this remarkable building; which, do you think, is the better of the two?[1]

Richard Haas never had the chance to pose this landscape question, for unintentionally he created an even greater irony than he had first planned. Because of budgetary difficulties, Haas took several years longer than anticipated to complete the mural, and when it was finally finished in 1981, the building that he sought to mirror was no longer standing! With the Pabst Building's demolition, Haas revealed that a new and larger crime had occurred. Now only an echo of the past remained, and now the question would seem to become a statement: Milwaukee, this is your history and you are foolish to tear it down. The mural today achieves its power through the ironic juxtaposition of a historic landscape that no longer exists, set in a place that, seemingly, has little regard for that history. As if to correct, symbolically at least, a landscape injustice, Haas's Pabst mural forces us to think about Milwaukee's historic German landscape and the way in which that past influences our conception of the city today.

In the spirit of Richard Haas's Pabst mural, this chapter examines the creation of Milwaukee's civic German landscape of the late nineteenth century and how that landscape figures into the design and planning schemes of the late twentieth century. In particular, we will examine two distinct German landscape Renaissances. The first extends from roughly 1890 to 1918 and represents a period that witnessed a flowering of public, commercial, and residential German-influenced building. This "invention" of the German Renaissance style took place under the guidance of a German civic culture that was becoming entwined in the very fabric of the city. Such embeddedness also pointed

Figure 19.1. Pabst Building, ca. 1892, the year it was built (Reprinted from *Architectural Record* 1 [April–June 1892]: 472)

Figure 19.2. Study for Centre Theatre Arcade, Wisconsin Avenue, Milwaukee, 1981, gouache on board, for the mural by Richard Haas (Reproduced from R. Haas, *Richard Haas, an Architecture of Illusion* [New York, 1981], 109)

toward assimilation, a process intensified greatly by extreme tensions during World War I.

After more than a half-century of neglect, Milwaukee began its second German Renaissance, or more precisely its "recycling" of the heritage it once scorned. Although too late to save the Pabst Building, redevelopment and architectural schemes in the late 1970s began to acquire a taste for the city's German heritage. From the Old World Third Historic District to the Grand Avenue Mall, Milwaukee's developers have recycled the historic and cash value of the landscape that, as late as 1981, they demolished almost indiscriminately. Yet, as Haas's mural further demonstrates, landscape—in this case, the city's German landscape—is rarely accepted uncritically. At crucial moments, discordant voices have erupted over Milwaukee's German landscapes. Their pattern matches the ebb and flow of *Deutschtum*, that is, German ethnic identity or those possessing it.[2]

Inventing the German Renaissance: 1890–1918

Early German Spaces within America's Deutsch-Athen

Even before the Civil War, as one nineteenth-century visitor noted, Milwaukee "was known among German-Americans as 'Deutsch-Athen.' Comparatively speaking, it deserved the name."[3] As with other fast-growing Great Lakes cities, Milwaukee drew a considerable number of immigrant groups; the Germans, however, were unsurpassed in Wisconsin's metropolis. In 1850, when more than 60 percent of the city was born in Europe, two-thirds came from the German lands, and in 1870, when just under half the city's population was born abroad, fully one-third of Milwaukee's residents had been born in Germany[4] (Table 19.1). Remarkably, by 1890 nearly 90 percent of Milwaukee's population had either been born outside the United States or their parents had been, making it the most "foreign" of all American cities; Germans accounted for the vast majority of these ethnics. However, it was more than sheer numbers that earned Milwaukee its flattering reputation as America's "German Athens." The Germans' numerical strength combined with their unprecedented influence in the city's cultural, political, and economic life to play a decisive role in shaping the social and physical fabric of Milwaukee.

During the years surrounding the Civil War, travelers to the city frequently deemed it "the German city of America," commenting on the German schools and elaborate Gothic church spires that pierced the sky. Others were struck by the *Deutschtum*'s unique, public landscapes

Table 19.1. Germans in Milwaukee, 1870–1970

Year	Total Population	German-born (% of total)	German stock[a] (% of total)
1870	71,440	31.6	—
1880	115,587	27.2	—
1890	204,468	26.8	59.6
1900	285,315	18.9	70.3
1910	373,857	17.3	44.6
1920	457,147	8.7	41.1
1930	578,472	7.1	27.4
1940	587,472	4.8	20.1
1950	637,392	2.9	15.6
1960	741,324	3.0	13.6
1970	717,110	1.4	7.0

Source: *U.S. Censuses of Population, 1870–1970.*
[a] German stock refers to all German-born persons or American-born persons with at least one parent born in Germany.

that created spaces for the Germans' festive culture, the attribute most distinct from "native," or American, life.[5] One such traveler, upon hearing German spoken as frequently as English, observed in 1884: "No one who visits Milwaukee can fail to be struck with the semi-foreign appearance of the city. Breweries are multiplied throughout the streets, lager beer saloons abound, beer gardens, with their flowers and music and tree or arbor-shaded tables, attract the tired and thirsty in various quarters. German music halls, Gasthausen [*sic*], and restaurants are found everywhere, and German signs are manifest over many doors."[6] The lager beer saloon and, in particular, its summertime equivalent, the outdoor beer garden, impressed visitors to the city as unique departures from an otherwise typically American landscape.

More than simply places to obtain inexpensive beer, the tavern and beer garden helped create a sense of community by bringing together Germans from across the deep economic, regional, and political cleavages that normally divided the ethnic group. The best taverns, many of them owned by the brewers, offered good beer, food that was often free, stimulating conversation, music, and, occasionally, a singing host. By 1890, Schlitz alone owned more than 50 well-built, distinctive taverns in the Milwaukee area. The best known of them all and truly a city landmark, the Schlitz Palm Garden (built in 1888), made use of high, vaulted ceilings, stained-glass windows, a pipe organ, rich oil paintings, and lush palms to create a social and physical atmosphere found "nowhere else in the United States"[7] (Figure 19.3).

Figure 19.3. Schlitz Palm Garden, ca. 1890 (Courtesy of the State Historical Society of Wisconsin, neg. number WHi [A61] 6239)

Such spaces served the important function of taking the place of nonexistent public parks. Following the well-established tradition in German-speaking countries, Milwaukee's beer gardens devoted considerable attention to ornamental landscape design and amusements (Figure 19.4). Ernest Ingersoll, a writer for *Harper's New Monthly Magazine*, visited Milwaukee in 1881 and provides a nice description of the beer garden's unique ambiance:

There being so many Germans and so much good beer, those out-door pleasure parks so dear to the German mind are in great number and handsomely appointed. They are pretty places, being laid out with well-shaded walks. Parterres of flowers, rustic colonnades, rock-work, fountains and all the other accessories of landscape gardening [are found] on a miniature scale. There are also bowling alleys and billiard saloons attached, and in the middle stands a concert hall, where a band discourses music to the crowd.[8]

By 1890, dozens of German beer gardens, the largest of which is said to have accommodated 12,000 people, afforded badly needed open space. With a total seating capacity of 105,000, half the city's residents could find a place on any given Saturday afternoon. The beer gardens, along with the tavern, provided the most tangible evidence of the growing effort by Milwaukee's Germans to manipulate the built environment.[9]

Figure 19.4. Summer night in a Milwaukee beer garden, 1881 (From Ernest Ingersoll, "Milwaukee," *Harper's New Monthly Magazine*, April 1881, 702–718, illustration on p. 717)

This manipulation was not, of course, greeted with uniform enthusiasm by the city's non-German population. Negative stereotypes often depicted the Germans as beer-drinking oafs who preferred to mingle with their own kind rather than join American society. Such nativist views, combined with attempts to repeal the right to use the German language in public schools, frequently put Milwaukee's *Deutschtum* on the defensive and limited German influence on the landscape to places of worship or informal socializing. By the early 1890s, however, the city's Germans had reached a point where they could challenge such views and, during a key public demonstration of political and cultural unity, ushered in a new period of unparalleled influence that forever changed Milwaukee's built form.[10]

"We are not strangers in this land"

On October 6, 1890, more than 12,000 marchers and tens of thousands of onlookers took to Milwaukee's streets to celebrate the 200-year history of *Deutschtum* in the United States. Elaborate tableaus depicting scenes from that history perched atop a dozen floats. Preceded by such grand scenes as "The Founding of Germantown in Pennsylvania," "General Washington and His German Bodyguard," and "The Germans in the Civil War," the one that attracted the most attention was the one closest to home, "Deutsch-Athen." The parade attracted members from each of the city's nearly 100 German organizations and concluded at a beer garden on the city's south side. There, speeches by Milwaukee's mayor, the state's governor, and the German communities' leaders combined with music, dance, and beer to create the warm sociability the Germans called *Gemütlichkeit*.[11]

The celebration's leaders hoped to achieve more from the occasion than a good time. They also sought to communicate a serious threefold message to the city's non-Germans: that Germans had remained politically passive too long; that they, too, had played a major role in creating the United States; and that they had permitted native-born Americans to relegate Germans "to a secondary class, by ridiculing our customs and language." Summed up by the day's motto, "We are not strangers in this land," the celebration's program concisely spelled out the German leaders' message to the city:

The celebration of this so-called "German Day" may be regarded as an outward sign of the newly-awakening self-assertion on the part of our German-American brothers. . . . We hope to correct the false impression, that this country has only native Puritanism to thank for its greatness. . . . Let us rally all our forces, not to make a great show of our power through violence, but simply to prove that thousands of loyal [German-]American citizens think and feel from a far more enlightened standpoint than many ignorant politicians.[12]

A show of history "in a right light," rather than force, was to be the German way to "throw off the yoke of servility."

The points of contention with the "natives" that rankled German leaders the most centered on recent attempts to curtail the use of the German language in schools, as previously noted, and the ever-present threat to "time-honored Germanic usages" (i.e., beer consumption).[13] Indeed, the 1890 German-American Day signaled a departure from other major German festive celebrations in its direct confrontation with American politics. Earlier festivals such as the Schiller Centennial of 1859, the Humboldt Centennial in 1869, and the Peace Celebrations of 1871, commemorating German victory in the Franco-Prussian war, had also sought to communicate a sense of ethnic pride and cohesion, but it was a pride of cultural *potential*. Symbols—such as Friedrich Schiller and Gambrinus, the legendary Flemish king credited with inventing beer—were derived largely from the Old World. American politics, generally too divisive for the already divided Germans to broach, were kept on the periphery.[14] By 1890, however, Milwaukee's Germans were celebrating achieved *contributions* and their equal place in the city's social and political structure.

For all the defensive language of the 1890 German-American Day, the occasion marked the place of *Deutschtum* in the city's power structure rather than their exclusion from it. Bolstered by the newly discovered confidence wrought by increasing wealth, Milwaukee's Germans "virtually called the tune of local politics" in the decades surrounding the turn of the century. Germans such as the Kerlers, Brumders, Bests, and Uihleins were becoming as important as the Wards, the Allises, and the Plankintons. As they sought to demonstrate in the 1890 celebration, Germans by this time were usually considered separate from other immigrant groups and on near economic and political par with native-born Americans.[15] They could, indeed, not be taken as simply strangers in this land.

From Strangers in the Land to Master Builders: The German Renaissance

Prior to the 1890 celebration, public structures of all sorts, from office buildings to government edifices, reflected the dominant styles of the day and the generally non-German background of their financiers and architects. German landscape influences remained tucked away from the city's central stage of power. This was soon to change, however, and the demonstration of ethnic pride and unity so evident at the impressive 1890 celebration ushered in a period, previously unknown and since unmatched, of German-inspired building. Political and economic power achieves particularly strong resonance when inscribed upon the landscape, in view for everyone to see. The city's Germans capitalized on this principle and, from 1890 to 1915, built homes, industrial complexes, commercial blocks, and public buildings in such a way as to transform Milwaukee "from a typical American Victorian city to one with a decidedly Teutonic air." While other ethnic groups occasionally expressed their ethnic identity through unique religious structures and dwellings, only the Germans built a wide array of public and civic structures that reflected Old World patterns and motifs.[16]

Although spectacular German-inspired homes such as the Captain Frederick Pabst house, the Henry Harnischfeger house, and the Joseph Kalvelage house could be dismissed as quirks of the eccentric and wealthy, public and commercial buildings assumed a critical importance in the city's search for clarity and self-definition within a new, complex industrial order. During these years, Zona Gale observed that Milwaukee had "come to civic self-consciousness, [had] begun to find itself." While it remains "genuinely American," she hastened to add, the city's civic consciousness derived from its ability to "keep Nuremburg and Strassburg and Heidelberg in its veins." For her part, Edna Ferber saw the landscape of these years as providing reason to believe that "Milwaukee was as German as Germany."[17] Other cities, like New York, Chicago, and Boston, facing similar challenges, found the Italian Renaissance style to be most appropriate for their prominent new buildings. Only Milwaukee, with its unique ethnic power structure, employed its own interpretation of the dominant classical revival styles to create what architectural historians call the German Renaissance.[18] Eye-catching details, elaborate curved, stepped, or scrolled front gables, and the use of imaginative sculpture defined this Old World aesthetic.

A central characteristic of all these buildings and the many residential structures built during this "Golden Age" is the recurring German background of both the clients and architects. From capitalists Frederick Pabst and George Brumder to architects Otto Strack and Henry Koch—and the hundreds of lesser-known builders and clients—one is struck by the persistent connection to the Old World. Germany and Austria were undergoing a tremendous resurgence in nationalistic revivals at this time, which, in turn, inspired their New World counterparts in Milwaukee. Just as a German ethnicity was being "invented" in Germany in the late nineteenth century, so was it being invented in Milwaukee during the city's German Renaissance building boom. Nowhere else in the United States is such a deliberate and successful adaptation of transplanted urban design to be found.[19] Three buildings, in particular, facilitated the creation of a new civic spatial order infused with *Deutschtum:* the Pabst Theater, the Germania Building, and City Hall. Each possessed a look and sent a visual message very different from the structures that preceded them, as well as from those that were to follow.[20] Accordingly, we will consider each in detail.

By the early 1890s, Captain Frederick Pabst directed the world's largest brewery and began expanding his empire into the core of his adopted city. As one of the city's most prominent German leaders who had been instrumental in shaping Milwaukee's industrial base, Pabst was to become one of the key pacesetters in the move to remold the city's landscape into one that reflected its German character. The captain celebrated his extremely good fortunes with the building of both the Pabst Building and his Wisconsin Avenue mansion in 1892, two structures which inaugurated the German Renaissance style in Milwaukee. Simultaneously, Pabst enlarged his brewery with new buildings that made use of ornamental towers, massive arched portals, and Gothic arched windows to give it an Old World, castlelike character. And, perhaps most notably, he purchased the old Nunnemacher Grand Opera House in 1890, renamed it the Stadt (or city) Theater, and, when a fire destroyed it three years later, rebuilt it as the magnificent Pabst Theater. Milwaukee's German stage, long a great cultural and educational institution and not merely a commercial enterprise, received a breath of life from Pabst, who renewed the commitment to German-language productions. As the philanthropic culmination of the Pabst empire and the cultural showcase of "Deutsch-Athen," the Pabst Theater symbolized the central place of *Deutschtum* in Milwaukee's civic culture.[21]

Not surprisingly, Captain Pabst hired the German-born architect Otto Strack to rebuild his Stadt Theater in such a way as to rival in splendor any theater of its size in Germany or Austria. Strack's Berlin- and Vienna-based training made him popular among Milwaukee's German-American businessmen, who commissioned him for much of the new commercial and residential building taking place at this time. Strack's new theater took its inspiration from the baroque palaces of Germany's eighteenth century (Figure 19.5). The mansard roof above the stage area, the orange brick, terra-cotta ornament, and the decorative iron work, all reflect this aesthetic. Like Vienna's Burgtheater (1888), the Pabst was to house the traditional queen of Germanic arts in the baroque style, commemorating the era in which theater first joined together nobility, the clergy, and common people in a shared space. And, like the Burgtheater, the Pabst Theater provided a meeting ground in which *Deutschtum*'s differences of class, politics, and region, if not eradicated, were at least softened by the shared cultural-ethnic experience.[22]

Figure 19.5. Pabst Theater, ca. 1900, built in 1895 (Courtesy of the State Historical Society of Wisconsin, neg. number WHi [X3] 21796)

Meanwhile, the German publishing mogul George Brumder was paying close attention. From its modest beginnings in 1873, Brumder's German-language publishing empire rose to become nineteenth-century America's largest publisher of German books, newspapers, and magazines. The nation's largest German-language newspaper, *Germania,* joined with Milwaukee's other German papers to surpass the circulation of the city's English-language press two to one, and led the attack against the movement to remove the German language from the state's public schools. The tremendous success of Brumder's press outgrew its facilities, so in 1892 he hired local architect Eugene Liebert to begin plans on a building that would match the needs of his publishing empire.[23]

The first design closely resembled New York publisher Joseph Pulitzer's "colossal" office building, which housed the *New York World* and was, at the time, the tallest building in New York City. But as the financial panic of 1893 slowed construction, Brumder reconsidered the statement he wished to make, and chose to emphasize the paper's German character. The resulting Germania Building of 1896 made use of four large copper domes, placed at the corners of the roof, with spiked finials rising from their tops (Figure 19.6). The striking silhouette of these domes caused them to be known locally as kaiser's helmets for their

Figure 19.6. Germania Building, ca. 1900, built in 1896 (Courtesy of Milwaukee County Historical Society)

Figure 19.7. Milwaukee City Hall—Die Stadthalle, ca. 1900, built in 1893–1895. Postcard. (Courtesy of the State Historical Society of Wisconsin, neg. number WHi [X3] 50368; Ex coll. John Gregory)

similarity to the Prussian military headgear of the period. Majestic eagles, perched atop small orbs, surrounded the domes. The strongest Germanic symbol on the building took the form of a three-ton, 10-feet-tall statue of the goddess Germania, which stood above the entrance. It was not just any Germania, but an exact replica of the Germania over-looking the Rhine River near Rudesheim at the Niederwald Denkmal, or German National Monument, erected in 1871 to commemorate the creation of the new, unified German state.[24]

The supreme accomplishment of German Milwaukee, however, could be found several blocks away in the City Hall of 1895. Milwau-kee's explosive growth in the 1880s warranted the building of a new City Hall to house all the city's departments under one roof. Council members, citizens, and the press took considerable interest in the new building and delighted in a design competition that attracted 11 entries from architects as distant as New York and Boston. In the end, after much haggling and debate, the Milwaukee architect Henry Koch won the competition with a structure at once "modern . . . , substantial, fireproof, and ornamental and well adapted to the uses of city government." Of course, it was the unique structure and ornamentation that caught everyone's eye. The only entry in the design competition to display specific reference to German architecture, this public building prominently symbolized *Deutschtum's* attainment of dominance in the local culture.[25]

Patterned after features from actual sixteenth- and seventeenth-century German *Rathäuser,* or city halls, that Koch had seen in Germany, his City Hall casts a distinctively Old World look on Milwaukee's downtown streetscape (Figure 19.7). The tower and 12 dormers are ornamented with the distinctively curved gable, or roofline, that is the distinguishing feature of the German Renaissance style. Moreover, the many sculptural details, such as cherubs and the heads of foxes and wolves, were incorporated to indulge the German taste for richly carved ornament. Koch's City Hall made Milwaukee's claim as America's German Athens concrete and palpable. Indeed, nowhere else in the world, outside Germany, could such a structure have been built.

The Great War: Veiling Milwaukee's German Landscape

In retrospect, what to many appeared to be the "Golden Age" of *Deutschtum* in Milwaukee may be seen more accurately as its twilight. While "the old flavor still permeated the city" in 1910, Ernest Meyer reminisced two decades later, "things had changed a good bit."[26] George Brumder's German-language publishing empire reached its peak around 1900, and soon began expanding its interests into fields like cement and insurance to make up for sharp declines in subscription rates. The Pabst Theater, while still maintaining considerable influence and importance, witnessed a decline in the quality of German plays being offered. And, Christopher Bach, the stalwart of German music in Milwaukee for more than 50 years, gave his farewell concert in 1910. Perhaps most significant of all, the decade following the 1890 celebration was marked by a slow but steady decline of German spoken in streets, shops, and homes. While George Brumder, Frederick Pabst, and others were creating strong statements of German unity and identity in Milwaukee's downtown landscape, everyday life for the city's *Deutschtum* was becoming an amalgam of two previously separate worlds. Like a braided stream, elements of the city's diverse cultures intertwined in such a fashion as to make it increasingly difficult to separate the "German" from the "American."[27]

The commonly accepted view that World War I served as the "catalyst that jelled the Americanization of the German population of the U.S." essentially holds true for Milwaukee during these troubling years, though one could have hardly predicted such an outcome at the war's onset.[28] Before American involvement in 1917, Milwaukee's German community presented a nearly unified front in favor of the German-led Axis powers. Most lent tacit support to the fatherland and agreed with Brumder's *Germania-Herold* that the British, not the Germans, were to blame for the crisis. Support for Germany reached a climax in 1916, when more than 170,000 Milwaukeeans attended the seven-day Charity War Bazaar at the city's Auditorium, in which $150,000 was raised for war relief—for Germany.[29]

The heavy-handed Americanization and suppression of Milwaukee's *Deutschtum* was to come only with the entrance of the United States into the war. Actions such as the 1916 Charity War Bazaar infuriated Milwaukee's superpatriots. All notions of tolerance for dissent were quickly squelched, and Milwaukee's Germans, once a "model" ethnic group, suddenly attained pariah status. Perhaps it was because Wisconsin's loyalty was so suspect throughout the nation (it became nicknamed the Kaiser's State during these years) that its citizens reacted so violently to dissent. Everything remotely identified with *Deutschtum* became targets for the privately organized Loyalty Legion's charges of treason and sedition.[30]

Once the Loyalty Legion succeeded in removing references to the "German Athens of America" from city guidebooks in May 1918, it got down to the business of reworking the landscape itself. Its first target was inspired by an act committed in Green Bay one month earlier: there, a band of "patriotic" vigilantes hauled down the city's "symbol of German Kultur" from the second floor of a downtown building (ironically, they found out a few hours later that the statue they had smashed was the Goddess of Liberty). With equal high-mindedness, Milwaukee's superpatriots followed the advice of one *Milwaukee Journal* reader who courageously identified himself or herself simply as "an American" and suggested removing the Schiller-Goethe monument in Washington Park, a city landmark since 1902.[31] Although the German "poet heroes" never actually left their pedestal, they sat out the remainder of the war under a dark cloak.

Brumder's Germania Building suffered a worse fate. Lieutenant A. J. Crozier, of the British and Canadian Recruiting Office, led the charge against the publisher's building and, in particular, its namesake, the Germania statue. "Everytime I look out of my office the sight of that statue at a time like this makes me see red," Crozier wrote to W. C. Brumder at the time. The Canadian officer also fired off letters to the

Milwaukee Journal objecting to the anti-American message that Germania sent forth and, after weeks of bitter confrontation, finally forced the German publisher to remove the giant figure in the dead of night. The removal so thrilled one true-blooded patriot that he felt compelled to offer the following piece of advice: "It is surprising that the statue was tolerated so long. Since they do not know what to do with it, I suggest that it be made into bullets and 'delivered' to the Germans, where it belongs." Although Crozier did not succeed in getting the "kaiser's helmets" removed with the Germania statue, he did make certain that the name Germania was chiseled off the front of the building and oversaw its renaming to the less offensive Brumder Building.[32]

Across the river at the Pabst Theater tensions were equally high. At least 10 different war-related plays were performed at the Pabst between 1914 and 1917, plays that reflected the attachment and the loyalty which Milwaukee Germans felt toward the Old World. The plays such as *Vorwärts* (Ahead), *Der König ruft* (The King Calls), and *Es braust ein Ruf!* (There Is a Call!) stirred strong emotional feelings within the audience and endlessly annoyed the city's non-Germans. Annoyance turned to rage when an anti-German mob placed a machine gun in front of the Pabst Theater and demanded the cancellation of a performance of Schiller's *Wilhelm Tell*. When the company attempted to stage a German comedy in its place at the presumably safer Freie Gemeinde Hall, the city's high sheriff had to disband an unruly, uniformed crowd which announced its intention to "break up this Hun show." One year later in 1918, these sentiments took a more official, if less violent, turn when the *Milwaukee Journal* called on the Pabst Theater to discontinue German-language productions. The theater resisted at first, but ultimately the pride of America's German Athens refrained from all German-language productions during the final, bitter season of the war.[33] When the curtain closed on German theater at the Pabst, it was perhaps a symbolic finale to the celebration of Milwaukee's German landscape. Despite the social and political clout Germans held in the city, in the end, they too were still regarded as ethnics, as only part American, and therefore, the suppression of the very signature they had embossed on the city's landscape appeared justifiable.

Recycling the Renaissance: 1975–1990

"Beerless Milwaukee"

The years that followed the Great War witnessed tremendous social and geographic changes for both Milwaukee's downtown and the urban region as a whole. The absolute number and percentage of Germans in the city declined so dramatically that by 1970 only 7 percent of Milwaukee claimed to be of direct German descent (Table 19.1). The fallout from the war's severe antagonism toward anything ethnic led to the national immigration restrictions of 1924, and though many portions of the city continued to use German as its first language, its days were numbered. Add Prohibition to these demographic trends and the character of Milwaukee's *Deutschtum* can be seen to retreat to the safer environs of the home and workplace. Milwaukee all of a sudden seemed considerably less German, "no longer . . . a city transplanted from the Rhine to the banks of the Milwaukee River."[34] One visitor in the early 1920s saw the Deutsch-Athen in a light that would have been unrecognizable only 10 years earlier: "The down-town streets of Milwaukee have nothing distinctive in layout or architecture. There are no striking buildings or vistas. This might be the downtown of any one of a dozen unindividualized Middle-Western cities. Perhaps in the days before Milwaukee became a great industrial center, in the heyday of beer and freedom, it was a gayer place."[35] Perhaps not quite using both eyes, the visitor nonetheless captured a remarkable truth: the reality and the image of the city had changed significantly in the course of only a dozen years.

Although Prohibition was eventually repealed, other forces were at work to ensure that the epithet Deutsch-Athen was to become a quaint name from the past. Following the trends of other American cities, Milwaukee's Germans suburbanized as quickly as their recently purchased Fords and Chevys would carry them away from the neighborhoods they once called home. The downtown they left behind was hit especially hard after World War II. The devastating combination of rapid suburbanization, shifting economic currents, and ill-conceived modernist civic policy initiatives all but gutted the city's once prosperous retail core and transformed the downtown area into a sea of unsightly parking lots and deadened one-way streets.

In cities across the country the response to this decline, and more generally to an emerging postindustrial order, was to reinvent and rebuild downtown areas as sites of specialty consumption, tourism, and festival.[36] Recycling dereliction into heritage, commodifying local history into design schemes and marketing packages, and appropriating images and physical relics of the past into "cultural capital" have become the hallmarks of these redevelopment initiatives. At its cornerstone, this response has relied on a process of *commodification*, or the rendering of places into items that can be advertised and sold in much the same way that one would peddle cigarettes and automobiles.[37]

As a source of physical relics and images from which to launch a redevelopment campaign, Milwaukee's German Golden Age proved irresistible. Developers, architects, and local boosters have selectively refashioned historical *images* into architectural facades or pastiches of local history. This reintroduction of German images into Milwaukee's downtown landscape is rooted in three parallel trends that have dominated the rebuilding of cities in the last two decades: historic preservation, the building of festival marketplaces, and postmodern architecture.

Historic Preservation: A Slice of Munich Revisited

Although historic preservation had been long praised for "giving people a sense of time, place and meaning in terms of where they live," it was only in the 1970s that it also became understood as a potential catalyst for downtown development.[38] In Milwaukee, preserving icons of the city's German Renaissance has become a centerpiece of downtown revitalization. The Pabst Theater, for example, was meticulously restored in 1978 and is now marketed as the focal point of Milwaukee's burgeoning theater district. Even the once highly contentious Germania Building has been reanimated. In 1980, a local preservationist acquired the building and inaugurated a three-year restoration plan for the downtown landmark. The first order of business involved changing the name of the building back to Germania. And although the Germania statue has never been returned to its lofty perch, the raised stone letters spelling "Germania" (which were chiseled off in 1918) have been replaced, this time in an elaborate gold Old World–Gothic style font. Meanwhile, several of the city's now-defunct breweries have been refurbished as prestigious office and commercial space.

The most striking piece of historic preservation, however, was the rejuvenation of North Third Street, the center of gravity for the city's German community in the late nineteenth and early twentieth centuries. The seeds of Third Street's transformation were planted as early as the 1950s when two of the remaining German merchants on the street, Mader's Restaurant and Usinger's Sausage, renovated the exteriors of their storefronts to incorporate numerous German motifs. These facelifts, added onto nondescript buildings and usually scorned by historic preservationists, inspired developers 20 years later when they sought to rework a declining and geographically peripheral retail thoroughfare into a highly visible shopping district and tourist attraction specializing in the sale of "Old World," particularly German, goods (Figure 19.8).

Figure 19.8. Old World Third Street, ca. 1994 (Photo: S. Hoelscher)

Third Street's transformation from forgotten corner of the downtown into specialized retail enclave took place through the industrious efforts of a local architect and the Third Street Merchants' Association. Their vision, spelled out in a 1975 report, followed the lead of Mader's and Usinger's and promised "redevelopment in character with Milwaukee's cultural and ethnic heritage."[39] In particular, the report advised streetscape alterations which recapture a small piece of the lost German Athens by returning the turn-of-the-century urban texture to Third Street: concrete street intersections were to be replaced with cobblestone; modern vapor street lights were to be replaced with the original luminaries; and any nonhistorical building adornments, such as neon signs, were to be removed from the facades.[40] At least one of the merchants, who showed slides of Munich at the initial planning meetings, supported the remodeling of Third Street's storefronts to mimic those of Germany, including street adornments such as window flower boxes planted with geraniums and new business signs with a German Old World look.[41]

The plan's potential was realized in 1976. Although no one would mistake the new Third Street for "Old Milwaukee," much less Munich, the renovation did successfully replant a colorful piece of Milwaukee's Germanic heritage in the contemporary downtown landscape. The commemorative party celebrating Third Street's new life was classic Milwaukee kitsch, with Mayor Meier, the last of the city's German singing mayors, cutting a length of Usinger's wieners that had been hung across the width of Third Street. Although no public subsidies were used for the street's physical transformation, the Common Council became involved in the mid-1980s. Under the council's direction, the six-block downtown area of Third Street became designated as Old World Third Street, a semantic move that strengthened the street's uniqueness as a downtown specialty district and also fashioned a highly visible and cohesive German-heritage narrative along one of the downtown's major thoroughfares. Finally, two years later in 1987, this once-forgotten zone became designated as a National Historic District.[42]

Festival Marketplaces: The Return of the Schlitz Palm Garden

Unfortunately, many of the hallmarks of the first German Renaissance were plucked from the landscape long before this second Renaissance, forcing developers to build upon the ghosts of the once-popular icons. The most prominent example of such a place is the four-block Grand Avenue Mall: a festival-marketplace codeveloped by the city of Milwaukee, the Milwaukee Redevelopment Corporation, and the Rouse Company.[43] Like similar large-scale retail projects constructed in other declining downtowns—most notably Boston's Faneuil Hall and Baltimore's Harborplace—the developers of the Grand Avenue incorporated images of a romanticized past into both their marketing packages and their design schemes. The "sparkling new jewel in the crown of the downtown Renaissance," as the developers dubbed the mall, was sold as a unique public space that would delicately combine the best of modern convenience (including roaming security guards and ample parking) with the cheery atmosphere of a bygone era. "[The Grand Avenue] takes the best of the city's rich history and magnificent architecture and adds to it a festive modern marketplace," wrote the Rouse-Milwaukee Company in 1982 in a promotional brochure detailing the opening of the mall. Even the mall's logo—a symbol known in architectural terms as a quatrefoil—was described as representing "the joining of Milwaukee's splendid past to a bright future by symbolizing the new unity of the four block marketplace."[44]

Most important, however, the Grand Avenue Mall was the first modern real estate project in the city which specifically appropriated the images of Milwaukee's Germanic legacy. Developers promised that the congeniality of America's German Athens would be returned to the now recession-weary Milwaukee populace in the form of the new mall's Speisegarten, an interior dining court which symbolically recalls the city's once-numerous German beer gardens. "Like the Palm Gardens of 'Turn of the Century' Milwaukee and the summer Biergarten tradition popular in the Germanic regions of Europe, the Speisegarten or 'dining garden' will be a congenial, airy place for dining and relaxed good cheer."[45] The Grand Avenue's developers further strengthened their celebration of Milwaukee's German Golden Age by selectively appropriating the popular images of the city's late-nineteenth-century urban landscape within the design of the mall itself. The walls of the Speisegarten, for example, are decorated with a tableau of memorabilia that reference the city's historical association with the brewing industry, including hops, beer steins, and the like. Outside the Speisegarten, enlarged photos depicting glimpses of celebratory moments in Milwaukee's history are used consistently throughout the interior of the mall, providing a visual rhythm to otherwise sterile interior corridors of the new arcade. One collection in particular depicts the cultural significance of the Schlitz Company to the city. Entitled "A Tour through 1896," the visual imagery includes a huge beer stein overflowing with suds and an enlarged historical photo picturing festive Milwaukeeans gathering at the Schlitz Palm Garden. For the consumer and tourist the message is clear: Although Milwaukee may no longer be America's Deutsch-Athen, the spirit and optimism of that time can be recaptured downtown once again, or at least salvaged in the form of selectively recalled nostalgic imagery. And further, in these image-conscious times, beer and brats—the stereotypical hallmarks of Milwaukee's German population, and once the source of German defensiveness—are now something to celebrate.

Postmodern Architecture: Salvaging the German Renaissance

Postmodern architecture is a final frontier in which developers and architects have reintroduced historical German images to contemporary downtown Milwaukee. Broadly speaking, the spirit of postmodern architecture (1978 to the present) rejects the functionalism of the "glass-box" modern aesthetic, which dominated architectural design in the three decades following World War II, and incorporates aesthetic cues taken from regionally or historically specific architectural styles into contemporary design.[46] Detractors argue that postmodern architecture is nothing more than a resurgent romanticism in which the prominence of dreamy iconography signifies that we have truly entered the age of the image. Proponents, however, see the promise of a new architectural temper that respects local sensibilities, tradition, and scale. Whichever view one takes, recently constructed buildings in downtown Milwaukee serve as excellent examples of a postmodernism that recaptures regionally specific architectural styles that were crushed in the onslaught of postwar modernism. Here, the hallmark steep curvilinear gables of the German Renaissance style, like other icons of Milwaukee's late-nineteenth-century landscape, have emerged as an irresistible aesthetic for new postmodern building designs in the city.

Although the 1988 Milwaukee Center, with its high degree of sensitivity to both the neighboring Pabst Theater and City Hall, may be considered the first example of postmodernism in the city, the most dramatic statement is the $74 million, 34-story 100 East Wisconsin Building (Figure 19.9). The building's site, vacant for many years following the unfortunate demolition of the Pabst Building in 1980, seemed to offer the grounds for a creative statement, but the building's first design by an out-of-town architectural firm destined it to be yet another uninspired, late-modern, flat-topped high-rise. However, at the request of a Milwaukee architect, the designers reworked their blueprints after investigating the architecture of the original Pabst Building.[47] The result is striking. The 100 East Wisconsin Building explicitly mimics the hallmark motifs of this unique architectural heritage, uniting the steep,

Figure 19.9. Close up of the 100 East Wisconsin Building, 1993, built in 1987–1989 (Courtesy of the City of Milwaukee Department of City Development)

curved gables of City Hall and the demolished Pabst Building with the spiked finials and orbs of the Germania Building. Overall, the 100 East Wisconsin Building closely resembles City Hall, with its mass broken into three parts, directly paralleling the composition of its Victorian neighbor.

Contesting the German Narrative in the Downtown Renaissance

The renaissance which engulfed downtown Milwaukee in the 1980s, however, also produced a series of jarring contradictions. For one, Milwaukee continued to lose its German population. In fact, by 1980 the combination of suburbanization and changing immigration patterns had reduced the proportion of the city's population who claimed a German ancestor, no matter how distant or remote, to less than 37 percent (Table 19.2).[48] Even more jarring, the funneling of capital investment into the downtown resulted in numerous ironic juxtapositions of an economically healthy central business district—increasingly geared

Table 19.2. German ancestry in Milwaukee, 1980 and 1990

Year	Total population	German-born (% of total)	German ancestry[a] (% of total)
1980			
City	636,212	0.8	36.7
Metro area	1,397,143	0.8	47.6
1990			
City	628,088	0.5	33.7
Metro area	1,432,149	0.6	48.4

Source: *U.S. Census of Population, 1980 and 1990.*

[a] The 1980 census marked the first time that a general question on ancestry (ethnic identity) was asked in a decennial census. The question was based on self-identification and was open-ended: What is this person's ancestry or ethnic origin? In cases where respondents listed multiple ethnicities (e.g., German-Italian-Irish), the first two were recorded. Therefore, the aggregate totals for each ethnic group are not mutually exclusive.

toward the needs of regional tourists, conventioneers, and corporate activities—with adjacent, predominantly black, central-city neighborhoods. While local boosters were praising the hundreds of millions of dollars that had been invested in downtown projects during the 1980s, factors such as deindustrialization, stifling residential segregation, and political disenfranchisement had produced staggering rates of poverty in the city's black community.[49]

These discrepancies did not go unnoticed locally, and numerous black leaders consistently criticized the redevelopment of the downtown at the expense of central-city residential neighborhoods. The most zealous activist was an outspoken alderman, Michael McGee, who represented the city's poorest neighborhood. In the spring of 1990, pressed for a symbolic issue around which to rally his supporters, the alderman directly attacked the German images of the downtown renaissance. Milwaukee's black community took to the streets, like the Germans of 1890, to assert their place within a diverse American civic culture.

In particular, McGee and his many supporters unleashed a campaign to designate the entire length of Third Street as Dr. Martin Luther King Drive. Most of the street was already named for Dr. King, but where it enters the newly revitalized downtown, the street name changes abruptly to Old World Third Street. McGee and his supporters believed that this semantic segregation vividly parallels the geographic polarization of postindustrial Milwaukee. A protest march chanted, "King Drive all the way."[50]

The renaming campaign came to an abrupt end three days after the march with the infamous "Sausage Scare." Alderman McGee informed the media that a black underground group, with whom he was not affiliated, had injected rat poison into the products made by Usinger's Sausage Company, a century-old cornerstone of the Old World Third Street project. In response, the company pulled 80,000 pounds of bratwurst and sausage from local stores, but neither they nor the U.S. Department of Agriculture could find anything more life threatening than the usual doses of fat and cholesterol.[51] Whether the Sausage Scare was a hoax cooked up by McGee, as the mayor and some of his supporters suggested, has never been determined. Nevertheless, by questioning the racist underpinnings of the semantic cordoning off of King from Old World Third, the German narrative had been challenged. The people who marched in the spring of 1990 to reclaim their street and their place, voiced a message ever so reminiscent of that of the Germans who marched the very same streets in the fall of 1890: We are not strangers in this land.

Conclusion

The landscapes of Milwaukee's two German Renaissances tell remarkably contrasting stories of *Deutschtum*'s place in the city (Figure 19.10). By 1890, the German community had expanded its claim to power by constructing a public and civic landscape that reflected its growing economic and political influence. No longer content with merely the landscape of informal socializing, key German businessmen hired German-trained architects to fashion a city based more on models derived from Hamburg and Nuremburg than on those from Boston or New York. This first German Renaissance has, in its turn, proved irresistible for current redevelopment efforts. While the historical connection to German culture was once severed and architects then took their cues from other American cities, it is the local, German history that once again provides the inspiration for this second Renaissance.

However, as landscape so ably highlights identity, so too can it become a rallying point for protest. The story of the invention and recycling of Milwaukee's German Renaissance is not merely a twice-told tale, well-known and familiar from repeated tellings. To return to the Pabst mural that began this chapter, landscape is always in danger of being forgotten, neglected, or erased and, with it, the memories of the accomplishments and struggles of its creators. Perhaps one of the most important contributions of Milwaukee's second German Renaissance lies in its ability to keep alive the memory of a time so decisive to the

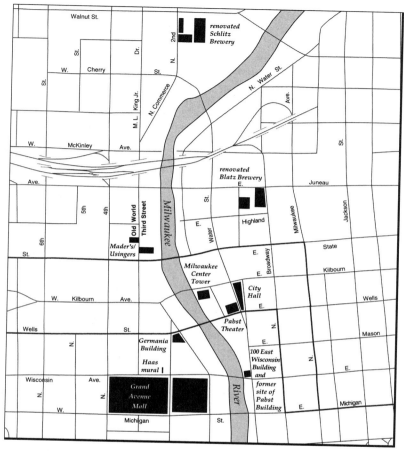

Figure 19.10. Map of downtown Milwaukee, 1996: the Germanic legacy in the built environment

growth of Wisconsin's largest city and, at the same time, to provide a visible point of contention for those whose struggles continue.

Notes

We would like to extend thanks to the library staff at the Max Kade Institute for German-American Studies at the University of Wisconsin–Madison for their expert assistance in tracking down arcane materials and to Les Vollmert of the City of Milwaukee Department of City Development for sharing his considerable knowledge of Milwaukee's nineteenth-century German architecture.

1. R. Haas, *Richard Haas, an Architecture of Illusion* (New York, 1981), 110. P. Goldberger's essay therein provides a useful introduction to Haas's work, "Introduction," 9–26. On the modernist enthusiasm to give Milwaukee's historic structures "a good shave," see L. Vollmert, C. Hatala, and P. Jakubovich, "Preservation Strategies," in their *Milwaukee Historic Ethnic Architecture Resources Study* (Milwaukee, 1994), 7.

2. Throughout this chapter, we make no distinction between *German* and *German-American*. For our purposes, the terms are used to mean the same thing: a community of German descent, tied together by heritage, language, and a sense of being German. The term *Deutschtum* summarizes this sense of German ethnic identity and was used frequently by both German Americans and others who described conditions in Milwaukee and other American cities. The best treatments of the Germans as an American ethnic group are the writings of Kathleen Neils Conzen; see her following works: *Immigrant Milwaukee 1836–1860: Accommodation and Community in a Frontier City* (Cambridge, Mass., 1976); "Germans," in S. Thernstrom, A. Orlov, and O. Handlin, eds., *Harvard Encyclopedia of American Ethnic Groups*, vol. 1 (Cambridge, Mass., 1980), 405–425; and "German-Americans and the Invention of Ethnicity," in F. Trommler and J. McVeigh, eds., *America and the Germans: As Assessment of a Three-Hundred Year History*, vol. 1 (Philadelphia, 1985), 131–147. Historian Dieter Berninger's recent article on Milwaukee's Germans during World War II makes use of a similar definition: "Milwaukee's German-American Community and the Nazi Challenge of the 1930s," *Wisconsin Magazine of History* 71 (Winter 1987–1988): 118–142. For a general treatment of the strategic uses of the past in urban development, see the outstanding work of M. C. Boyer, *The City of Collective Memory: Its Historical Imagery and Architectural Entertainments* (Cambridge, Mass., 1994).

3. Henry Villard (1856), quoted in B. Still, *Milwaukee: The History of a City* (Madison, 1948), 70.

4. Still, *Milwaukee*, 111–112; Conzen, *Immigrant Milwaukee*, 44–84.

5. E. Goes, "Milwaukee: The German City of America" (1898), quoted in B. Still, "The Growth of Milwaukee as Recorded by Contemporaries," *Wisconsin Magazine of History* 21 (March 1938): 262–292. For an excellent description of Milwaukee's German- and Polish-styled churches, see *Milwaukee Ethnic Church Tour: The Rich Heritage of Immigrant Architecture* (Milwaukee, 1994). On the central importance of festive culture for Milwaukee's Germans, Ernest Bruncken wrote: "The forms of social intercourse to which the Germans are accustomed, differ so much from those prevailing among the people of English, and particularly New England, extraction, that this would for a long time remain a barrier to the amalgamation of the two elements, even if it were not for the difference of language." E. Bruncken, "How Germans Become Americans," *Proceedings of the State Historical Society of Wisconsin* 45 (1898): 101–122, quotation on pp. 112–113.

6. W. Glazer, *Peculiarities of American Cities* (1884), quoted in H. R. Zimmermann, *Magnificent Milwaukee: Architectural Treasures, 1850–1920* (Milwaukee, 1987), ix.

7. Pleiss and Heck, *Schlitz Palm Garden,* brochure for the Grand Opening (Milwaukee, n.d. [probably 1888]); T. Mueller, "Milwaukee's German Cultural Heritage," *Milwaukee History* 10 (Autumn 1987): 95–108; R. Oldenburg, "The German-American Lager Beer Gardens," in his *Great Good Place* (New York, 1989), 89–104; and *Milwaukee Ethnic Commercial and Public Buildings Tour: The Rich Heritage of Immigrant Architecture* (Milwaukee, 1994), 28–29. Fred Holmes writes wistfully of the German lager beer saloon: "Social life today offers few meeting places like the old German saloon. Compared with it, the modern tavern is an arrogant pretender." *Side Roads: Excursions into Wisconsin's Past* (Madison, 1949), 67.

8. E. Ingersoll, "Milwaukee," *Harper's New Monthly Magazine,* April 1881, 702–718, quotation on pp. 717–718.

9. Landscape Research, *Built in Milwaukee: An Architectural View of the City* (Milwaukee, 1981), 112–114; Mueller, "Milwaukee's German Cultural Heritage," 101–104; and Conzen, *Immigrant Milwaukee,* 157–158. The figure on total seating capacity comes from the *1890 U.S. Census, Social Statistics of Cities.* This immense capacity becomes all the more remarkable when compared with other cities in which Germans composed a substantial portion of the population: beer gardens in Cincinnati had room for 5,000; Cleveland could seat 9,000; and St. Paul had room for 1,000.

10. Upon seeing souvenir postcards that presented a stereotypical portrayal of the Germans, beer baron Frederick Pabst responded, "The suggestions in some of the cards are positively indecent and many of them are actually insulting. Milwaukee residents, and especially the Germans, are not drunkards, as some of the cards would suggest. People who know the Germans understand that the caricatures are of the meanest and most contemptible nature." *Milwaukee Free Press,* February 17, 1906. The basic work on nativism in the nineteenth and early twentieth centuries remains J. Higham, *Strangers in the Land: Patterns of American Nativism, 1860–1925,* 2d ed. (New York, 1978).

11. W. Hense-Jensen and E. Bruncken, *Wisconsin's Deutsch-Amerikaner bis zum Schluss des neunzehnten Jahrhunderts,* vol. 2 (Milwaukee, 1902), 214; G. Meyer, *The German-American (Die Deutschamerikaner): Dedicated to the Celebration of the German-American Day* (Milwaukee, 1890), no pagination; Still, *Milwaukee,* 261–262.

12. Meyer, *The German-American.*

13. On the German opposition to the legislated threats to the Bennett Law, which threatened the use of the German language in schools, see: L. J. Rippley, *The Immigrant Experience in Wisconsin* (Boston, 1985), 44–59; and R. E. Wyman, "Wisconsin Ethnic Groups and the Election of 1890," *Wisconsin Magazine of History* 51 (Summer 1968): 269–293. For an introduction to the temperance movement in Wisconsin and the position of the state's Germans, see J. Apps, *Breweries of Wisconsin* (Madison, 1992), 59–65.

14. K. N. Conzen, "Ethnicity as Festive Culture: Nineteenth Century German America on Parade," in W. Sollors, ed., *The Invention of Ethnicity* (New York, 1989), 44–76.

15. Still, *Milwaukee,* 259–262; G. Korman, *Industrialization, Immigrants, and*

Americanizers: The View from Milwaukee, 1866–1921 (Madison, 1967), 42–43; and P. Woehrmann, "Milwaukee German Immigrant Values: An Essay," *Milwaukee History* 10 (Autumn 1987): 78–94.

16.   L. Vollmert, C. Hatala, and P. Jakubovich, "Ethnic Architecture in Milwaukee," in *Milwaukee Historic Ethnic Architecture Resources Study*, 15.

17.   Z. Gale, "Milwaukee" (1910), quoted in Still, "The Growth of Milwaukee," 291–292; E. Ferber, *A Peculiar Treasure* (New York, 1939), 131–132. For an excellent description of the spectacular German homes during this period, see: *Milwaukee Ethnic Houses Tour: The Rich Heritage of Immigrant Architecture* (Milwaukee, 1994).

18.   Landscape Research, *Built in Milwaukee*, 86. Historians have employed the terms *German Renaissance* and *Flemish Renaissance* interchangeably to describe this period and style. For consistency, we will use the former term.

19.   C. E. Schorske, *Fin-de-Siecle Vienna* (New York, 1981), 32–46. On the general trend toward "inventing traditions" in late-nineteenth-century Europe, see E. Hobsbawm, "Mass-Producing Traditions: Europe, 1870–1914," in E. Hobsbawm and T. Ranger, eds., *The Invention of Tradition* (Cambridge, England, 1983), 263–308.

20.   Each of the important public buildings that preceded the German Renaissance of 1890 followed some variation of the Italianate. *Central Business District Historic Resources Survey*, vol. 3 (Milwaukee, 1986), 40.

21.   Apps, *Breweries of Wisconsin*, 122–132; Zimmermann, *Magnificent Milwaukee*, 133; Mueller, "Milwaukee's German Cultural Heritage," 104–108; Vollmert, Hatala, and Jakubovich, "Ethnic Architecture," 15; and Conzen, *Immigrant Milwaukee*, 176.

22.   *Milwaukee Ethnic Commercial and Public Buildings*, 10–11; Schorske, *Fin-de-Siecle Vienna*, 37–38.

23.   S. Mallman, "The Brumders of Milwaukee," *Milwaukee History* 3 (Autumn 1980): 70; Still, *Milwaukee*, 264–265; H. R. Zimmermann, *Germania Building: A Milwaukee Landmark Restored* (Milwaukee, 1982), 6–10.

24.   Zimmermann, *Germania Building*, 10–23; *Milwaukee Ethnic Commercial and Public Buildings*, 12–13.

25.   J. Krause, "Brick by Brick, Council Battled over Stately Hall," *Milwaukee Journal*, January 23, 1995; *Milwaukee Ethnic Commercial and Public Buildings*, 8–9; Vollmert, Hatala, and Jakubovich, *Milwaukee Historic Ethnic Architecture*, 15.

26.   E. Meyer, "Twilight of a Golden Age," *American Mercury* 29 (August 1933): 456–464, quotation on p. 456.

27.   Mallman, "The Brumders of Milwaukee," 66–80; G. Becker, "German Theater in Milwaukee, 1914 to 1918," *Milwaukee History* 13 (Spring 1990): 2–10; J. Eichoff, "Wisconsin's German-Americans: From Ethnic Identity to Assimilation," *German American Studies* 2 (1970): 50; and Still, *Milwaukee*, 266–267. The metaphor of ethnicity as a braided river valley, so much richer than the customary metallurgic metaphor of the melting pot, is derived from K. N. Conzen, "Mainstreams and Side Channels: The Localization of Immigrant Cultures," *Journal of American Ethnic History* 11 (Fall 1991): 5–19.

28. L. J. Rippley, "Ameliorized Americanization: The Effects of World War I on German-Americans in the 1920s," in Trommler and McVeigh, eds., *America and the Germans*, vol. 2, 217–231, quotation on p. 217. The standard work on the effect of the war on German Americans remains F. C. Luebke, *Bonds of Loyalty: German-Americans and World War I* (De Kalb, 1974).

29. Of Milwaukee's German position in 1915, Francis Hackett wrote: "When German Kultur is disparaged or when they hear that Germany is to be 'crushed,' no one can doubt their origin. As between Germany and the Allies, they are for Germany, but this does not mean that the bulk of them desire at practically any cost, even to America, the triumph of German arms." F. Hackett, "How Milwaukee Takes the War," *New Republic*, July 17, 1915, pp. 272–273. On the 1916 bazaar, see Still, *Milwaukee*, 455–456; and Rippley, *Immigrant Experience*, 97–98.

30. L. L. Carey, "The Wisconsin Loyalty Legion, 1919–1919," *Wisconsin Magazine of History* 53 (Autumn 1969): 33–50; and J. D. Stevens, "Suppression of Expression in Wisconsin during World War I," unpublished Ph.D. diss., University of Wisconsin–Madison, 1967.

31. Stevens, "Suppression of Expression," 158; *Milwaukee Journal*, May 24, 1918; H. Mezke, *Gedenkblatt herausgeben von der Schiller-Goethe Denkmal-Gesellschaft* (Milwaukee, 1905); and Rippley, *Immigrant Experience*, 113.

32. Zimmermann, *Germania Building*, 24; *Milwaukee Journal*, May 13, 16, 17, 20, 1918.

33. Becker, "German Theater in Milwaukee," 2–10; O. Ameringer, *If You Don't Weaken: The Autobiography of Oscar Ameringer* (1940; Norman, 1983), 336; *Milwaukee Journal*, September 29, 1917, April 21, 24, June 5, 1918.

34. Wisconsin's WPA Guide wrote of Milwaukee on the eve of World War II: "Milwaukee no longer seems a city transplanted from the Rhine to the banks of the Milwaukee River; the German theater has disappeared with the beer gardens; 'Milwaukee German' is heard less and less frequently. . . . Although the city has more than 2,000 taverns, only a few have inherited the look of the *Bierstuben* where whole families gathered when *Gemütlichkeit* reigned and Milwaukee was the *Deutsch Athen* of America." Writers' Program of the Works Project Administration of the State of Wisconsin, *Wisconsin: A Guide to the Badger State* (New York, 1941), 242.

35. W. Waldron, "Beerless Milwaukee," in his *We Explore the Great Lakes* (New York, 1923), 181, 188.

36. B. Frieden and L. Sagalyn, *Downtown Inc.: How America Rebuilds Cities* (Cambridge, Mass., 1990).

37. R. D. Sack, *Place, Modernity, and the Consumer's World: A Relational Framework for Geographical Analysis* (Baltimore, 1992), 2–4; G. Kearns and C. Philo, eds., *Selling Place: The City as Cultural Capital, Past and Present* (Oxford, 1993); S. Ward and J. Gold, eds., *Place Promotion* (New York, 1994); D. Harvey, *The Condition of Postmodernity* (Oxford, 1989); and Boyer, *City of Collective Memory*.

38. T. D. Bever, *The Economic Benefits of Historic Preservation* (Washington, D.C., 1990), 1; and M. Wallace, "Preserving the Past: A History of Historic

Preservation in the United States," in his *Mickey Mouse History and Other Essays on American Memory* (Philadelphia, 1996), 177–222.

39.   W. Wenzler and Associates, *Downtown Third Street: A Proposed Development Program* (Milwaukee, 1975).

40.   Wenzler and Associates, *Downtown Third Street*, Appendix, pp. 9 and 11. Ironically, by these criteria, the part of the streetscape that first attracted redevelopment—the ersatz German fronts of Mader's and Usinger's—should have been removed. As "additions" to the original buildings, the fronts of the two businesses are considered "noncontributing" structures to the historic district. *Old World Third Street Historic District—National Register of Historic Places Nomination Form* (Washington, D.C., 1987).

41.   Wenzler and Associates, *Downtown Third Street*, 9–11 and Appendix.

42.   *Old World Third Street Historic District.*

43.   Construction of the $70 million complex was completed in late 1982 and was financed through an elaborate network of both private and public money, including $12.6 million in Urban Development Action grants and $23 million in tax increment financing. The construction of the Grand Avenue Mall was significant for numerous reasons. For one, it dramatically reversed the decline of downtown as a retail center. Prior to the mall's construction, downtown Milwaukee rated a lowly 14th in sales volume among shopping centers in the metropolitan area. After its opening, downtown's retail sales volume skyrocketed to first place, eclipsing even the larger regional malls in the suburbs. Also, the Grand Avenue Mall was the first major real estate project in the city to make substantial use of recycled buildings and adaptive reuse, an urban redevelopment trend which closely paralleled the rising tide of historic preservation. The mall's design combined the elegant Plankinton Arcade, a 1915 enclosed retail galleria that was modeled after the famous arcades of Europe, with a modern trilevel sky-lit arcade. See: A. Fleischmann and J. Feagin, "The Politics of Growth-oriented Urban Alliances," *Urban Affairs Quarterly* (December 1987): 207–232; and H. Gillette, "The Evolution of the Planned Shopping Center in Suburb and City," *Journal of the American Planning Association* (Autumn 1985): 449–460.

44.   Rouse-Milwaukee Inc., *The Grand Avenue: Milwaukee*, vols. 1–2 (Milwaukee, 1982).

45.   Rouse, *The Grand Avenue*, vols. 1–2.

46.   For a detailed account of postmodern architecture, see C. Jencks, *The Language of Post-Modern Architecture* (London, 1987); and P. Portoghesi, *Postmodern: The Architecture of the Postindustrial Society* (New York, 1982). Unlike most examples of postmodern architecture, which involve an indiscriminate mixing and borrowing of unrelated architectural styles, Milwaukee's postmodern buildings are clearly rooted in one vernacular style, mainly the German Renaissance.

47.   The architect was the same designer who was responsible for the transformation of Old World Third Street in the mid-1970s: William Wenzler. Wenzler excepted, the return of German images in Milwaukee's postmodern

architecture is the work of out-of-town architectural firms. William Wenzler, telephone interview, March 1995.

48.   See also, T. Norris, "Tracing the German Flavor: Sure Milwaukee Has Strong German Roots, but What Does That Heritage Mean Today?" *Milwaukee Journal*, June 24, 1990.

49.   G. Spires, "There Are Remedies for Racial Red-lining in Milwaukee," *Milwaukee Journal*, June 7, 1990.

50.   Ironically, the part of Third Street already named for Dr. King constituted the most vibrant core of the former German neighborhood and, in many ways, maintains a higher degree of historical integrity than the "Old World" section of the street. B. Miner, "Move to Rename 3rd Street Debated," *Milwaukee Journal*, June 7, 1990.

51.   "Usinger's Completes Recall after Threat," *Milwaukee Journal*, June 23, 1990.

20

The Cultural Landscape of Wisconsin's Dairy Farming

Ingolf Vogeler

The lasting aspects of many European settlement processes are visible in fields and farmsteads across the state of Wisconsin. Although many different crops are grown and many different kinds of livestock are raised, the ubiquitous farmsteads stand as landmarks in the open countryside.[1] In the past, farmsteads reflected the self-sufficiency and the multifaceted aspects of family farming. And even though agriculture has become more specialized today, the various buildings still found on farmsteads speak to our collective cultural heritage. Throughout Wisconsin history, farmhouses and horse and livestock barns were invariably surrounded by small outbuildings: chicken coops, pigpens, granaries, grain bins, corncribs, potato storehouses (in a few specialized agricultural areas), furrowing houses, tool and machine sheds, windmills, silos, root cellars, smokehouses, summer kitchens, and privies. Not every farm, of course, had all these buildings. Today, many of these buildings have lost their function and have been abandoned or used for alternative purposes, but observant travelers can still probably see most of these buildings within a day's travel. This chapter focuses on the sometimes relic and sometimes modern, but always visible, cultural landscape of Wisconsin's most widespread and prevalent type of farming activity—dairying.[2] The aim is to raise our consciousness of the distinctive functions and appearances of the many features that constitute Wisconsin's dairy landscape.

Dairying in Wisconsin: The Historical Context

Wisconsin's agricultural economy began its historical shift from wheat to dairying in the 1850s. Wisconsin was the third major frontier area in the United States to make that transition. By 1850, upstate New York and the "backbone counties" of northern Ohio had already become dairy farm areas.[3] Intensive dairying in Wisconsin developed rapidly and spread across the state from the 1860s to the 1890s. Although most early settlers kept just a few cows for their own use, some farmers were already practicing commercial dairying by 1860. These early dairy farmers were widely scattered in the southeastern counties of the state. Kenosha, Racine, Milwaukee, Walworth, and Green counties were the largest per capita producers of butter.[4] From this southeastern core, dairying diffused to other parts of the state. With the aid of the new refrigerator railroad cars, introduced in 1871, Wisconsin cheese began to be shipped long distances to East Coast markets. At the 1876 Centennial Exposition in Philadelphia, Wisconsin cheese and butter rated second after New York's. By 1920, Wisconsin had the largest percentage of pure-bred dairy cattle in the Midwest.[5]

Several major influences account for this rapid spread and development: (1) the settlement of New York dairy farmers; (2) the research and extension undertaken by the College of Agriculture at the University of Wisconsin in Madison to encourage dairying; and (3) the readiness of foreign-born immigrants to participate in the practical application of dairy knowledge. New Yorkers, who came to Wisconsin in large numbers during the 1830s, 1840s, and 1850s, brought with them the knowledge and skills of scientific and commercial dairying. Prior to Wisconsin's prominence in dairying, New York was known for good breeding and well-run dairy operations. In Wisconsin, New Yorkers frequently headed local movements to build cheese factories and to organize breeder associations and other kinds of dairy-improvement societies. They demonstrated that the old practice of breeding "dual purpose" cows, for both beef and milk, could be replaced with breeding for milk only.[6]

The College of Agriculture at the University of Wisconsin actively encouraged dairying through research and teaching. Research at the college resulted in the Babcock milk tester, bacteriological tests for detecting diseases, and practical methods of pasteurizing milk. To disseminate new knowledge and techniques, the college provided short courses and winter courses (the first in 1887) to dairy farmers. It also held farmers' institutes throughout the state, the earliest in 1886, at which scientists and farmers shared experience and knowledge.

Finally, Wisconsin dairying benefited from the enthusiasm and hard work of numerous immigrant farmers. By the early 1870s and 1880s, many old Yankee pioneer settlers were eager for the easy profits to be made in the new wheat areas of western Minnesota and the Dakotas. They sold their farms to the newly arriving Germans, Scandinavians, Bohemians, and other immigrants, who had no choice but to work "hard and persistently, the long year through" to make a living. To them, milking cows twice a day, feeding and tending cows, delivering milk to the factory, and working in the fields was all in a day's work. In many parts of the state, these European immigrant farmers became the backbone of Wisconsin's dairying economy.

Cultural Landscape of Dairy Farming

The relative importance of midwestern dairy farms in the United States is best illustrated by a cartogram that depicts the size of each state by the number of its dairy farms (Figure 20.1). By using relative rather than absolute space, the cartogram demonstrates the national importance of Wisconsin and adjacent states. Dairying in the northern Midwest, particularly Wisconsin, is a major source of farm income. In every Wisconsin county except Vilas and Milwaukee, dairy farms, defined by having half or more of their income derived from dairy products, ranked first (or in the case of Kenosha and Oneida, second) among the 13 farm types identified in the 1987 U.S. Census of Agriculture. The widespread prevalence of dairy farms in the state creates a distinctive cultural landscape, defined by the various kinds of land uses, buildings, and cattle associated with dairying.

Large barns, with attached milk houses and tall silos, are the salient features of this cultural landscape. Dairy barns are ubiquitous along the interstate and state highways in the northern Midwest. Their presence helps make the rural landscape interesting and attractive. To aesthetically inclined travelers, a red barn in a wintry scene or a set of white barns against the greens of summer is beautiful. The large dairy barns are also the subject of paintings and postcards. For some, barns may have only aesthetic value, but barns are actually practical structures, reflecting traditional rural values of hard work, cooperation, and love of the land. Especially on dairy farms, barns are the daily focus of economic activities. In many ways, the great barns of the dairy region

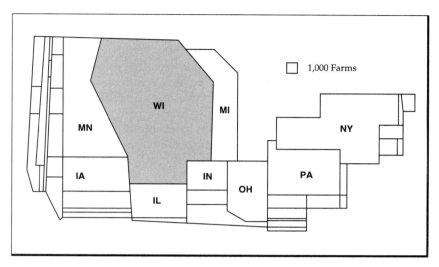

Figure 20.1. Cartogram of the United States based on the relative number of dairy farms per state

are testimonials to a way of life distinctive to a labor-intensive, livestock farm economy. The crop patterns associated with dairying also add a distinctive motif to the region's countryside. In the Wisconsin dairy landscape, barns, cornfields, hay crops, pastures, and woodlots intermingle to provide an ever-changing array of colors through all the seasons.

Dairy Barns

Dairy barns, being the largest of the many outbuildings that make up the Wisconsin dairy landscape and having unique features, catch our attention in the landscape and demand description and explanation.[7] Barns are machines. Their purpose is to increase usefulness, that is, to increase productivity and safety and to simplify tasks.[8] When U.S. agriculture began industrializing, farm magazines, farm building companies, and government agencies began to promote more efficient ways of doing things, including, of course, dairying.[9] As the Midwest, especially Wisconsin, became the center for dairy farming, this region produced a host of publications dealing with dairy barn design, construction, equipment, and practices. The Department of Agricultural Engineering at the University of Wisconsin in Madison, in particular, published detailed drawings and lists of materials necessary for various kinds of dairy barns.[10] Consequently, barn designs rapidly became standardized. As farmers in Wisconsin specialized in dairying, they developed distinctive structures. Historically, dairy barns required (1) lots of space in the loft for loose hay to feed dairy cows during the long, cold, and snowy winters, and (2) rows of windows in the stone or concrete basements to provide light for feeding and milking. Although the number of dairy farms has declined dramatically, especially since the 1950s, and newer dairy technologies have emerged, earlier patterns are still very much present in the cultural landscape.

The shape of roofs on dairy barns reflects the changing need for more storage space and, therefore, also the age of the buildings. Barn roofs come in three types: gable, gambrel (often incorrectly called hip), and round. Gable-roofed barns are the oldest and simplest barns to build. By the beginning of the twentieth century, when dairying became a midwestern speciality, gambrel-roofed barns became numerous. The advantage of gambrel over gable roofs is increased storage area for hay without increasing the height of the building. When farmers enlarged their dairy herds, they built new gambrel-roofed barns as new separate structures or added them to existing gable-roofed barns.

Perfected in the 1920s, round-roofed barns had even more loft space and therefore were very suitable for dairy farming.[11] Farmers could store more hay in round-roofed and pointed Gothic-roofed barns than in gambrel-roofed barns. The Gothic roof with its pointed top and the rainbow roof with its flatter top were also promoted by agricultural experiment stations. Local lumberyards provided preassembled rafters for these roof shapes, which were most popular in the late 1910s and 1920s. As free-standing structures, round-roofed barns are heavily concentrated in western Wisconsin and in the Upper Peninsula of Michigan. This is because frontier-stage (often log) structures in this last-settled part of the upper Great Lakes states were being replaced at about the time round-roofed barns were gaining popularity.[12]

In his study of the distribution and diffusion of barns in northeastern North America, Noble identified only five barn types in the northern midwestern dairy region, including Wisconsin. These five types—English, German, three-bay, Wisconsin, and round—accounted for 25–100 percent of all the barns in each county of the region. The English and German barn types are found only in a few places. Three-bay and Wisconsin barn types are distributed widely throughout the northern Midwest. Round barns are scattered thinly.[13] A brief description of the three most common barn types in Wisconsin follows.

Although three-bay barns are similar to German bank barns in both appearance and function, both having three internal parts or bays, three-bay barns were perceived as American buildings and hence had wider acceptance than the German barns, which had forebays that hung over the basements.[14] Three-bay barns, often built over basements and with ramps leading to the upper stories, are common throughout Wisconsin, especially in the southern parts of the state.[15]

In the last quarter of the nineteenth century, the Agricultural Experiment Station at the University of Wisconsin in Madison promoted an improved barn design, one better suited for the growing dairy industry of the state. "Wisconsin" dairy barns are normally only about 35 feet wide and 100 feet long.[16] Longer versions of this barn type result in noticeably rectangular shapes. These barns characteristically have rows of windows along the basement walls. Noble claims: "More county agents reported more occurrences of this barn than of any other except the raised three-bay, which may well confirm the effectiveness of this barn for dairy farming. In many counties of Wisconsin, between half and three-quarters of all barns are of this type."[17]

Round barns are a distinctive Wisconsin contribution to U.S. rural architecture. This barn type reflects the private and public technological innovations that occurred at the turn of the century to make farming more efficient and profitable. In 1889, F. H. King at the University of Wisconsin at Madison designed a distinctively practical barn which had a greater storage capacity, requiring fewer building materials, and afforded easier feeding (because of a central silo) than rectangular barns. Innovative farmers adopted this new style of barn throughout Wisconsin in the decades between 1890 and 1930. Round barns are not particularly clustered in any part of the state. As an innovative barn type, they were adopted wherever farmers needed new barns to replace old ones and were willing to experiment with the "latest" farm technology. A 1983 survey demonstrated that round barns were essentially restricted to the midwestern Corn-Soybean and Dairy belts: Wisconsin had 180 round barns; Michigan, 25; Indiana, 154; Iowa, 160; and Nebraska, 36.[18]

Today, few farmers know how to build traditional wooden barns, and the teams of neighbors needed for erecting these barns are no longer available. Instead, today's farmers build low-slung metal buildings usually constructed of sheet metal and steel girders because they are cheaper to construct and maintain and easier to use than the multistory barns built only 50 years ago. Whereas in the nineteenth and early twentieth centuries, barns were built to last for several generations, pole barns are not expected to last more than 40 years. As a recent innovation, pole barns are associated with prosperous agricultural areas where farmers can borrow enough capital to build new buildings. Dairy farming is often not very common in such prosperous areas. As a consequence, pole barns are proportionately less important in the heart of the northern midwestern dairy region, especially in north-central Wisconsin.

Silos and Milk Houses

Wisconsin has been a pioneer in the design of silos. The first above-ground silo was built in 1880 in Oconomowoc, and the world's largest silo was built in 1898 at Lake Mills. It was 64 feet in diameter and 60 feet high and held as much as a 200-acre yield of corn. By 1888, the United States had 91 silos; 60 of them were in the township of Lake Mills.[19] The introduction of silos and silage was an innovation that allowed farmers to keep larger herds of dairy cows during the winter at a lower cost, because silage was less expensive than dry feed grains. Nevertheless, many arguments were made against this new storage technology and silage: cows would lose their teeth, silage would burn out their stomachs, calving would be difficult, and silage would affect the quality of milk. Some even claimed silage made their cows drunk. As late as 1908, a few creameries still refused to accept milk from farmers who fed their cows silage.[20]

Almost none of the old, square, wooden silos are left, but many round silos, which were perfected in the 1890s and come in many heights, widths, and materials, may still be seen. During the 1910s, round wooden-stave silos were replaced by masonry silos, then by poured-concrete silos, and then cement-stave silos. Cement companies promoted these silos throughout the Midwest. During the 1920s, tile blocks were used for silos and barn basements alike. Farm-building design handbooks recommended that silos be placed at the end of the "feeding alley," or the end of barns, and this is where most of the dairy silos are indeed found. Alternatively, some silos were placed on the long side of barns.[21] By 1924, Wisconsin had more silos (100,060) than any other state, including such dairy states as New York (53,300) and Michigan (49,000).[22]

By the late 1940s, a new kind of silo was developed: blue-and-white thermoslike Harvestores became the latest innovation in storing silage. Perfected in Wisconsin, Harvestores were constructed of fiberglass bonded to sheets of metal. The first 100 Harvestores were manufactured in 1949. Today these blue silos are found throughout the northern midwestern dairy region, especially in the southern half of Wisconsin, where the high cost of these silos could be more readily financed.

Figure 20.2. An original gambrel barn and its two small silos, with Gothic and shed barns added. The variety of roof styles and silos reflects the changing nature of dairying. (Photograph by the author)

Indeed, Harvestores are good indicators of high-value farming areas and high farm indebtedness, because they cost about twice as much as similar-sized cement silos. Dairy farmsteads in the more prosperous parts of southern Wisconsin are frequently characterized by two or more Harvestores. In the northern parts of the state, dairy farmsteads reflect the harder physical and economic conditions of farming; hence, Harvestores are generally absent. Today, almost all dairy farms have at least one silo; many have three or four, of different ages, heights, and building materials (Figure 20.2).

The most recent way of storing silage has again changed the appearance of the dairy landscape. In the last five years, "trench silos" have appeared in large numbers on larger dairy farms. Concrete walls, or a trench in the ground, hold silage, which is covered by plastic sheets (often black in color) that are held down by old car tires to prevent the plastic from blowing off. An even simpler way of storing silage is to blow it into a plastic tube (frequently white) that lies directly on the ground, usually somewhere near the farmstead. These long, white plastic "silage bags" are quite striking in the green landscapes of summer and the brown of fall.

Milk houses attached to dairy barns are another distinctive element of the dairy landscape. When milk was stored and transported in milk cans, milk rooms were usually located inside barns. But after World War II, separate milk houses were built. Bulk storage containers and health codes that required separate buildings were important in promoting the change. Since the 1960s, large-scale dairy farmers have shifted to labor-saving, automated, pole barn milking parlors, with pipelines and bulk storage tanks.

Dairy Barns: Topographic Map Symbols and Field Evidence

The architectural characteristics of dairy barns are readily visible to anyone driving along Wisconsin's rural roads, because the distinctively narrow Wisconsin-type barns have a definitive rectangular shape. Also, as farmers increased their dairy herds, they enlarged their existing barns, thereby making the narrow structures even longer. These dairy barns can also be identified on U.S. Geological Survey topographic maps, which are available at most University of Wisconsin campuses. To demonstrate the varied distribution of rectangular-shaped farm buildings in Wisconsin and their strong association with dairying areas, a survey was made of 24 topographic maps (at the scale of 1:24,000) selected from counties with the highest concentrations of dairy cows (7,000 or more per county). For each map, the number of rectangular and square barns (shown as white shapes) associated with houses (shown as black squares) was counted along a one-mile-wide traverse which extended for six miles across each map (Figure 20.3). By dividing the number of buildings in each strip by six, the number of buildings per square mile was calculated. The results show that Wisconsin has the highest density of rectangular barns per square mile, with 8.3; other midwestern dairy states have lower densities: Michigan has 6.5, and Minnesota has only 3.8.

The distribution and relationship of rectangular and square barns were also examined. Rectangular buildings seem to be found mostly in counties with large numbers of dairy cows. Overall, the frequency of square barns varies independently of rectangular barns. The number of rectangular barns, for example, decreases markedly as the amount of land in forests, bogs, ponds, and lake increases, yet square barns continue to be common. Thus, on the sands plains of Juneau and Adams counties,[23] with their small number of dairy cows, rectangular barns almost completely disappear, yet square barns continue to be common, albeit less so than in the more productive areas to the north and south.

The relationship between dairy cows and rectangular-shaped build-

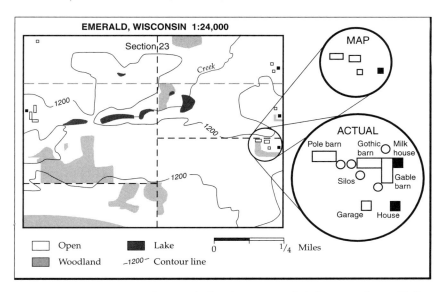

Figure 20.3. Map of a dairy farmstead, showing map symbols in comparison with the actual layout of the farm

ings on maps was tested by correlating the number of dairy cows in each county through which the survey traverse passed with the number of rectangular and square buildings in the traverse sample. The correlation between rectangular barns and dairy cows in Wisconsin is strong and statistically significant, whereas the relationship between square barns and dairy cows is very weak and not statistically significant.[24] Associating rectangular buildings with dairy farms in rural areas is, then, a very efficient yet conservative technique for delimiting dairy farm areas.

To test the relationship between the shape of rural buildings as depicted on maps and those actually found in an area, a random sample of 42 farmsteads was examined from the Emerald and Jewett topographic sheets (1:24,000) in St. Croix County, Wisconsin. Despite their depiction on topographic maps, essentially no barns were exactly square, that is, having dimensions of 1:1. Fieldwork indicates that barns shown as squares on the maps varied in ratios from 1:1.5 to 1:3.5 and that barns shown as rectangles had ratios of 1:4.0 or greater (Figure 20.3). Rectangular barns were either originally larger structures or, more commonly, smaller barns that had several additions added to them. Yet the rectangular and square shapes of barns on maps correlated significantly with similarly shaped actual dairy barns.[25] Rectangular-shaped barns generally indicate (1) larger and more prosperous individual farm operations and (2) more productive farming areas, reflected in a higher density of farm buildings.

In the midwestern dairy region, several other types of farm buildings are rectangular and therefore might appear as rectangular shapes on topographic maps. But each is slightly different from dairy barns. Tobacco sheds, as they are called in Wisconsin, are narrower and often shorter than dairy barns,[26] and poultry sheds are longer. Using county census data, a sample of broiler houses and tobacco sheds in Wisconsin was located on topographic maps. The owners of these farmlands were identified in plat books, and they were asked over the telephone to verify the actual uses of these buildings. Although tobacco sheds are easily identified in the field, they are not distinctive on topographic maps because their dimensions are not sufficiently different from those of the other buildings depicted on these maps. Broiler houses, on the other hand, are clearly differentiated on maps; they are noticeably longer than rectangular dairy barns.

Because the topographic maps that were used for the traverses were published from the mid-1970s through the 1980s, the number of structures still standing does not match the number shown on the maps. Some old barns are gone; some new pole barns have been erected. Certainly fewer barns are actively used for dairying. The map evidence and the barn count on the ground show that, in the sample area and throughout the midwestern dairy region, dairy farming was once much more important than it is today. Government agricultural statistics, of course, confirm this as well. Despite the decline in the number of dairy farms, dairy barns are memorials to past agricultural practices, and they continue to indicate the spatial extent of dairy farming in Wisconsin, with larger acreages under operation by fewer farms than in the past.

Other Dairy Landscape Elements

A high density of fences is characteristic of the dairy region, where small-sized farms and fields have traditionally required a great deal of fencing. Whereas cows were once sent out to pasture after each milking, today many dairy farmers feed their cows in or near the barns. Occasionally, a dairy herd can still be seen in the fields, but this is rapidly becoming a "photo opportunity." Yet, wire fences largely remain, especially around pastures and wood lots, because it takes more effort than it is worth to remove them. Actively used pastures are now enclosed by nearly invisible two- or three-strand electric wires attached to metal posts. Although they are effective barriers for cows and are "cost effective," they do not add to the visual landscape.

Because milk had to be picked up from dairy farms once a day in the past and even today is picked up at least every other day, regardless of weather conditions, rural roads in Wisconsin have "hard" or "improved" surfaces and are very well maintained. Almost all section roads, in fact, are paved. Indeed, rural gravel roads are most uncommon in Wisconsin, in sharp contrast with the Grain Belt of the Dakotas, where gravel roads are the norm. These paved "back" roads, which are today so ideal for recreational bikers, are an infrastructural benefit derived from Wisconsin's dairy industry.

The smaller size of dairy farms, compared with grain farms, generally results in a high density of rural population, even today, when fewer dairy farms operate larger acreages. The nonfarm rural population has increased as families occupy farmhouses and use barns for hobby or part-time farm purposes while they work in nearby towns and villages. All or most of the farm land is being cultivated and supports the larger-scale farm operations. The high density of rural settlement in dairy regions is still very noticeable on topographic maps.

Conclusion

Now that the large wooden dairy barns of an earlier agricultural era are falling down and being replaced by steel buildings, we are beginning to appreciate the old ones by visiting "working" farm and ethnic museums, such as Old Wisconsin; biking on hard-surface roads in the countryside where barns remain; buying a painting, photo, or book on barns;[27] collecting old barn boards for paneling living rooms; or eating at barnlike "Country Kitchens." In addition, barn-related themes such as black-and-white cows and "cheese-head" postcards, T-shirts, hats, and coffee mugs are available throughout Wisconsin in truck stops and boutiques alike. Both locals and travelers can enjoy the "Down Home Dairyland" radio show on Wisconsin PBS stations. The cultural landscape of dairy farming, much modified as a strictly functional agricultural activity over the last century, remains an aesthetic expression of Wisconsin's land and life.

Notes

1. For a comparative analysis of farmsteads, see G. T. Trewartha, "Some Regional Characteristics of American Farmsteads," *Annals of the Association of*

American Geographers 38, 3 (September 1948): 166–225. Also see F. F. Kahn, "The Use of Aerial Photographs in the Geographic Analysis of Rural Settlements," *Photogrammetric Engineering* 17 (1951): 759–771.

2.   L. Durand, Jr., "The American Dairy Region," *Journal of Geography* 4B (1949): 1–20.

3.   J. F. Hart, *The Land That Feeds Us* (New York, 1991), 167–168, 177.

4.   J. I. Clark, *Wisconsin Agriculture* (Madison, 1956).

5.   T. R. Pirtle, *History of the Dairy Industry* (Chicago, 1926), 32.

6.   J. Schafer, *A History of Agriculture of Wisconsin* (Madison, 1922).

7.   L. Durand, Jr., "Dairy Barns of Southeastern Wisconsin," *Economic Geography* 19 (January 1949): 37–44.

8.   F. M. White and C. I. Griffith, *Barns for Wisconsin Dairy Farms*, Agricultural Experiment Station of the University of Wisconsin, Bulletin 266 (Madison: 1916), 4.

9.   E. Burnett, *Modern Farm Buildings* (New York, 1913); T. E. French and F. W. Ives, *Agricultural Drawing and the Design of Farm Structures* (New York, 1915); and J. C. Wooley, *Farm Buildings* (New York, 1946).

10.   White and Griffith, *Barns for Wisconsin Dairy Farms*, 1.

11.   A. G. Noble, *Wood, Brick, and Stone: The North American Settlement Landscape*, vol. 2: *Barns and Farm Structures* (Amherst, Mass., 1984), 68.

12.   Noble, *Wood, Brick, and Stone*, 46, 67–68; and *Beginning Again: Immigrant Architectural Heritage in the Western Lake Superior Region*, Ethnic and Folk Arts Institute, University of Wisconsin–Superior, 1978.

13.   Noble, *Wood, Brick, and Stone*, chap. 4.

14.   R. Bastian, "Southeastern Pennsylvania and Central Wisconsin Barns: Examples of Independent Parallel Development," *Professional Geographer* 27, 2 (May 1975): 200–204; and R. F. Ensminger, *The Pennsylvania Barn* (Baltimore, 1992), 98.

15.   S. Schuler, *American Barns* (Exton, Pa., 1984), 20.

16.   White and Griffith, *Barns for Wisconsin Dairy Farms*; and Noble, *Wood, Brick, and Stone*, 46.

17.   Noble, *Wood, Brick, and Stone*, 61. White and Griffith (*Barns for Wisconsin Dairy Farms*, 56) indicate that "the rectangular plank-frame construction is the most standard and satisfactory."

18.   L. J. Soike, *Without Right Angles: The Round Barns of Iowa* (Des Moines, 1983). Also see L. T. Jost, *The Round and Five-or-More Equal Sided Barns of Wisconsin* (n.p., 1980); and *Beginning Again*, which describes several round barns in northern Wisconsin.

19.   See Noble, *Wood, Brick, and Stone*; and also N. S. Fish, "The History of the Silo in Wisconsin," *Wisconsin Magazine of History*, vol. B (Winter 1925): 160–170.

20.   See I. Vogeler, *Wisconsin* (Boulder, 1978).

21.   B. D. Halsted, ed., *Barns, Sheds, and Outbuildings: Placement, Design, and Construction* (1881; Brattleboro, Vt., 1983).

22.   Noble, *Wood, Brick, and Stone*, 72.

23.   P. E., McNall, H. O. Anderson, A. R. Albert, and R. W. Abbott, *Farming in the Central Sandy Area of Wisconsin*, University of Wisconsin, Agricultural Experiment Station, Bulletin 497 (Madison, 1952).

24.   The Pearson correlation between the number of dairy cows and rectangular barns is .622, with a significance level of .04; with square barns, the correlation is .008, with a significance level of .97.

25.   Having a chi square of .86, with 1 degree of freedom, and a significance level of .000.

26.   K. Raitz and C. Mather, "Norwegians and Tobacco in Western Wisconsin," *Annals of the Association of American Geographers* 61, 4 (December 1971): 684–696.

27.   Examples are E. Sloane, *An Age of Barns* (New York, 1967); E. Arthur and D. Witney, *The Barn: A Vanishing Landmark in North America* (Toronto, 1972); W. L. Wells, *Barns in the U.S.A.* (San Diego, 1976); and Schuler, *American Barns*.

21

A Pretty Strange Place
Nineteenth-Century Scenic Tourism in the Dells

Steven Hoelscher

How were all those wondrous objects formed among the pond'rous rocks?
Some primeval grand upheaval shook the land with frequent shocks;
Caverns yawned and fissures widened; tempests strident filled the air,
Madly urging foaming surges through the gorges opened there;
With free motion toward the ocean rolling in impetuous course,
Rushing, tumbling—crushing, crumbling rocks with their resistless force;
And the roaring waters, pouring on in ever broad'ning swells,
Eddying, twisting, seething, whirling, formed the wild Wisconsin Dells!
—John Clerke, *Through the Dells* (1875)

During the 40 years after the Civil War, the Dells of the Wisconsin River grew to become the most important sightseeing destination in the burgeoning Old Northwest. The water-carved canyons, gulches, and sandstone rock formations attracted nature-loving tourists by the tens of thousands, arriving by train from cities and villages across the country. Leaving behind their jobs in Chicago's banks, Philadelphia's department stores, and St. Louis's schools, nineteenth-century tourists sought inspiration and a cessation of the rigors of daily life in what one writer in 1879, without a trace of irony, called "a treasure-house of the wildest scenery [equal] to the famed beauties of the Canons of the Yellowstone or the picturesque Watkin's Glen in New York." For an anonymous journalist the following year, the Dells constituted "one of the weirdest and strangest freaks of Nature" such that "every feature of the trip [up the river] has something strangely fascinating. Everything is so entirely different from every-day life that we can almost imagine ourselves in a different world." Strange, weird, wild, and picturesque, the Wisconsin River Dells achieved for a short period the status of regional cultural shrine.[1]

Today such a description seems as curious as the landscape that the two nineteenth-century writers sought to chronicle. Ever greater numbers of tourists continue to travel to the Dells, but today they encounter a very different place. Water slides, a greyhound race track, wax museums, sprawling miniature golf courses, and T-shirt shops define the cultural landscape of the Wisconsin Dells today. Almost as an afterthought, boat companies plead with parents to "See the *Real* Dells," reminding them that by taking a boat ride they are doing "First Things First." Much like Coney Island around the turn of the century, the Wisconsin Dells today are a monument to the enormous appeal of mass culture with its prodigious emphasis on consumption and controlled leisure. From Robot World to the Shipwreck Lagoon, and from Storybook Gardens to "Ripleys Believe It or Not," the Wisconsin Dells today are, as one Chicago resident put it recently, "a pretty strange place."[2]

What has not changed is the perennial attraction of the weird, the strange, and the inexplicable. But whereas today "strangeness" is constituted in an overwhelming and unabashed mass commercialism that highlights haunted mansions and chasm-leaping German shepherds, in the nineteenth century it was the strange "freaks of Nature's handiwork" that lured tens of thousands of visitors each year. More directly, nineteenth-century tourists who visited the Dells went there specifically to see, to gaze at, landscapes worthy of a three-day train ride. This distinctive type of tourism, which I will call scenic tourism, arose to prominence in the United States beginning roughly in the 1820s and paved the way for the emergence of the mass tourism we find today. But, in its overwhelming emphasis on visual experience and the association of the tourist with the pilgrim seeking redemption and inspiration in landscape, nineteenth-century "scenic tourism" remains distinct from the more recreation-oriented tourism of this century. While a "search for natural beauty and other environmental amenities [remains] a prime impulse for travelling in North America," it now competes with such diversions as water slides and gambling and, as a consequence, is often marginalized.[3]

The nineteenth-century promoters of tourism in the Wisconsin Dells devoted considerable energy to fashion a strange and alluring river landscape out of what they perceived to be raw wilderness. Through wood engravings, written descriptions, and, most important, photographs, both local and national writers and artists developed a landscape iconography that accentuated the Dells' distinctiveness or, as it was put at the time, "its weird and grewsome [sic] . . . formations."[4] These landscape images at times were contradictory and occasionally complementary, but were always put forth with a Babbittesque zeal for place promotion. And in the case of photographer Henry Hamilton Bennett, they achieved an artistic brilliance that has yet to be matched anywhere else in the state.

This chapter offers a historical slice of the early years of tourism in the Wisconsin Dells by examining the literary and visual sources that tourists themselves pored over before embarking upon, and after returning from, a scenic excursion to the Dells. These photographs, engravings, and travel descriptions offer a window into the world of the nineteenth-century tourist and help us to understand the historic attraction of the Dells as tourist space.

From Travail to Early Scenic Tourism

The earliest recorded scenic tour of the Dells by European Americans occurred nearly 150 years ago when the natural scientist Increase A. Lapham led an expedition there in the fall of 1849. Native Americans had traveled the river long before that, of course, making its watershed their home and utilizing its abundant natural resources. Likewise, fur traders, missionaries, and, later, lumbermen navigated the treacherous stretch as part of a key leg in the journey from the river's far northern reaches to its confluence with the Mississippi.[5] But it was not until Lapham's excursion that we read of a trip to the Dells for scenic, rather than utilitarian, purposes.

As early as the mid-1840s, the Dells of the Wisconsin River had a reputation as the region's greatest natural wonder. A geological survey described the Dells in 1847 as possessing "singular and beautiful effects. Architraves, sculptured cornices, moulded capitals, scrolls, and fluted columns are seen on every hand; presenting, altogether, a mixture of the grand, the beautiful, and the fantastic."[6] Lapham, firmly believing in the importance of "knowing the beautiful scenery of Wisconsin . . . our own State," traveled from his comfortable Milwaukee home to take in such wonders.[7]

Increase Lapham's tour to the Wisconsin Dells, like that of so many

Figure 21.1. Drawing by Increase A. Lapham, *The Dells of the Wisconsin River, 1849* (Courtesy of the State Historical Society of Wisconsin, neg. number WHi [X3] 35617, from the Lapham Papers, "Geological Notes of a Tour to the Dells . . . ," Box 2, 1844–1849)

other educated elites of his day, mixed the gathering of scientific data with the pleasure of seeing new and interesting landscapes. His diary of the trip blends a disinterested description of local vegetation, hypotheses regarding the origins of the many layers of exposed sandstone, and disappointment over the lack of fossils with an excitement over encountering a sublime landscape and an awe at the depth and raw force of the river at its narrowest point. The trip was an arduous one, as his diary reveals: "Next morning our eagerness to see 'the Dreadful Dells' induced us to leave our beds at 5 o'clock and drive 4 miles to the 'Dell House.'" There the tour group felt "fortunate to find a man who was willing to paddle us up through the gorge in a small boat . . . for navigation of the Dells is very difficult, requiring much skill and experience to guide the rafts"[8] (Figure 21.1).

Lapham's concern for safety was clearly justified. Raftsmen feared the treacherous stretch through the Dells. Indeed, as Robert Fries points out in his classic history of lumbering in Wisconsin, "both material and man-made obstacles made the rafting of lumber hazardous to property and life."[9] This difficulty of running the river combined with the absence of transportation facilities to delay the onset of scenic tourism for a decade after Lapham's tour. To follow the literary critic Paul Fussell, Lapham's arduous scenic tour of the Wisconsin Dells involved considerable travail and should be seen as a distinct precursor to the development of scenic tourism. *Travail,* the root word of *travel,* implies "trouble," "effort," and "work"—conditions that defined the travel experience for Lapham and others in the days before scenic tourism took hold.[10]

The utter dependence of the nascent tourism industry on transportation technology is demonstrated by the fact that only with the news of the imminent arrival of the railroad do we find the first indication of organized efforts to promote scenic tourism. Anticipating the arrival of the railroad by seven months, the local paper of Kilbourn City (renamed Wisconsin Dells in the 1930s) trumpeted in the spring of 1856 that "the wild, romantic scenery of the 'Dells' will always make them a place of resort for seekers of pleasure."[11] Following this charge, a riverboat pilot issued an advertisement some months later: "For Recreation Resort to the Dells! Where depressed spirits can be alleviated, gloom and melancholy soon be dispelled and the mind become greatly invigorated. Leroy Gates has purchased a pleasure boat for the purpose of penetrating the numerous occult caves of the Dells."[12] As the region's standing river pilot, whose main responsibility focused on guiding lumber rafts through the Dells during periods of dangerously high water, Gates knew the river well. For at least a dozen years, he guided small numbers of visitors to some of the more easily accessible sites and inlets. Also important, Leroy Gates was the village photographer as well, operating a studio in Kilbourn City from at least 1860.[13] Although he worked in portraiture and is not known to have photographed the Dells landscape, the early connection between the river, sightseeing, and photography begins with Gates, in the years just before the Civil War.

Despite the good efforts of Leroy Gates and a handful of hotel operators, visitors trickled into the Dells in only modest numbers. Little coordination existed between businesses in Kilbourn City and the railroad, nor had word of the region reached many beyond the narrow readership of the local newspaper. That newspaper pinpointed a further hindrance to any sustained tourist effort when it reported in 1869 that "most visitors do not stay long enough to go to all the points because carriages cannot go among the ravines and none but the healthy have the courage to penetrate the depths or climb their height."[14] Only the most agile could climb over the fallen trees, wade through the swelling brooks, and stomp through the marshes that encumbered the way to the choicest scenic views. Travail, rather than scenic pleasure, continued to define the tourist experience.

Equally inhibiting was the small scale of the scenic boat operation. The largest local hotel in 1869, the Tanner House, boasted a two-seat buggy to take hotel guests to the head of the Dells and a large rowboat to bring them back. By April 1873, the *Wisconsin Mirror* half-heartedly reported that there will be "at least fifteen good craft [rowboats] ready for use in a few days" for the upcoming season.[15] With such severe limitations on the number of sightseers that could be catered to, it is little wonder that the arrival of the first passenger steamship to Kilbourn City in June of 1873 prompted the *Wisconsin Mirror* to exclaim: Hurrah for the steamer and jolly excursions through the Dells!" That summer witnessed the launching of two steamers, the *Modocawanda* and the *Dell Queen,* each capable of carrying 70 passengers. The latter was replaced three years later by a steamboat able to carry 300 passengers. Thus began the tradition, persisting to the present, of excursion boats plying the Dells. Scenic tourism, stripped of a degree of its earlier hardships and annoyances, can be said to have begun.[16]

Regional Structure and the Rationalization of Scenic Tourism Space

The decades following the arrival of the railroad and excursion steamboats witnessed the inauguration of mass scenic tourism in the Dells. In 1874, more than 6,000 sightseers—mainly from the Midwest and the South—toured the Dells (a number that was to increase 10-fold over the next 30 years). By the mid-1880s Kilbourn City was within an easy day's train ride of all the major cities of the Midwest.[17] Celebrating the direct link between tourism in Wisconsin and railroads as he marveled at the speed with which "fashionable tourism" had displaced the "world of the savage," one writer exclaimed in 1884:

Our vast and daily growing railroad system has of late years been opening up ten thousand new and inviting fields for fashionable exploration. Regions, where but a few years since savagery and solitude reigned unbroken, now annually swarm with gaily dressed seekers of health and pleasure. . . . Gay yachting and rowing parties now skim the mirror-like smoothness of lakes and lakelets, which not many moons agone [sic] were only stirred by the prow of the Sioux or Winnebago birch-bark canoe.[18]

Replacing tourists for Native Americans, the railroads helped pave the way for regional development and profits. The Chicago, Milwaukee, and St. Paul Railroad, in addition to bringing passengers to the Dells, began to promote the budding resort area through maps and guidebooks, making it *the* key site along its route. Like the railroads of the Great West, the Milwaukee Road hoped to "add an enchanting scenic beauty to its standard efficiency," thereby clearing a path to increased ticket sales.[19]

The Wisconsin Dells fulfilled another function for the Chicago-based railroads and their financial backers as well. Performing a role similar to the cultural institutions that legitimized Chicago's claim to metropolitan status, the scenic wonders demonstrated that the city's hinterland possessed grandeur on a par with Watkin's Glen, Niagara, and Yosemite.[20] Scenic tourist space became an essential component of regional and metropolitan identity. Chicago-based boosters touted the virtues of a tour of the Dells as a logical response to the nationalist need for grand and romantic scenery:

The more one rambles over our own magnificent continent, and the more one sees of its never-ending glories, sublimities and beauties, the greater must be his contempt for the average American tourist who turns his back on scenes as transcendently grand, varied, and enchanting as ever the sun in all his celestial rounds looked down upon. . . . Nature never constructed a bigger combined idiot and cheap humbug than the American who goes into bogus raptures over the lakes and crags of Switzerland or Italy, while he has never seen or cared to see Niagara, the wild Yo Semite realm, Yellowstone Park, or the Canyons of Colorado. . . . But none of these more famous regions of resort surpass, if they even rival, in charms and attractions the comparatively unknown DELLS OF THE WISCONSIN.[21]

Likewise, steamboats offered much more than merely a way to move passengers up and down the river with greater efficiency. This means of travel enabled tourists to appreciate the landscape while it passed before them as a moving panorama. Charles Lapham's map *Dells of the Wisconsin River* (ca. 1875) exemplified and added to this new way of experiencing the river's landscape (Figure 21.2).[22] This long, narrow map highlights the various scenic features encountered on a steamboat ride. The river's dense concentration of grottoes, caves, and canyons—all named and ready to be identified—appeared before the tourists on the deck of the steamboat and could be pinpointed on the map. Whereas the elder Lapham's tour (and those of other proto-tourists like him) was hampered by long treks on foot or by a carriage on a bad road, the steamboat enabled tourists now to turn their attention to each passing scene. Indeed, by 1884, one tourist, a schoolteacher from Rhode Island, could describe the Dells landscape from the safe distance of the steamship *Eolah* as a "weird, strange-rock *panorama* . . .

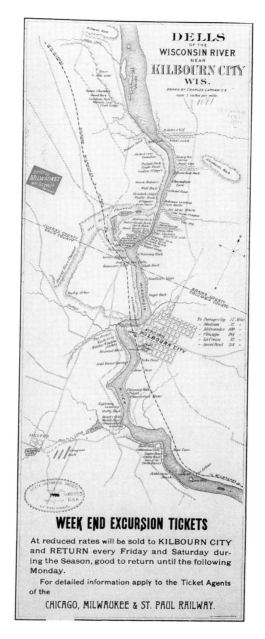

Figure 21.2. Map by Charles Lapham, *Dells of the Wisconsin River near Kilbourn City Wis.*, ca. 1875 (Originally issued by the Chicago, Milwaukee, and St. Paul Railway; courtesy of the State Historical Society of Wisconsin, neg. number WHi [X3] 50290)

431

Table 21.1. Tourist accommodation in selected regional centers, 1886 and 1903

	Number of hotels[a]	Guest capacity	Range of daily rates, $	Average daily rate, $[b]	Range of weekly rates, $
			1886		
Kilbourn City	10	241	1.00–2.50	1.90	5.00–8.00
Oconomowoc	23	1,121	1.50–3.00	2.60	5.00–17.50
Waukesha	30	2,180	1.50–3.50	2.70	5.00–14.00
Lake Minnetonka, Minn.	12	2,300	2.00–3.50	3.00	12.00–21.00
			1903		
Kilbourn City	13	700	1.00–2.50	1.70	5.00–12.00
Oconomowoc	13	813	1.00–3.50	2.60	5.00–21.00
Waukesha	29	2,378	1.50–6.00	2.70	5.00–40.00
Lake Minnetonka, Minn.	8	900	1.50–4.00	2.80	7.00–20.00
Minoqua[c]	18	750	1.50–2.00	1.90	7.00–12.00
Sturgeon Bay[c]	5	219	1.00–2.00	1.70	5.00–10.50

Sources: *Tourists' Guide to Summer Homes in Wisconsin, Iowa, Minnesota, and Dakota* (Chicago, 1886); *Summer Homes on the Chicago, Milwaukee and St. Paul Railway* (Chicago, 1903). Both guidebooks were published by the General Passenger Department of the Chicago, Milwaukee, and St. Paul Railroad and may be considered comparable.

[a] Includes hotels and private residences.

[b] The average daily rate is calculated as the weighted arithmetic mean, weighted by the number of guests at each location.

[c] 1886 data are unavailable.

which the swift-running water has worn into various grotesque and fantastic shapes."[23]

Along with the transformation of the Wisconsin River landscape into a panorama came initiatives to standardize and rationalize the tourist industry within Kilbourn City itself. Not only did the guest capacity triple between 1886 and 1903 (Table 21.1), but numerous "improvements" were made throughout the Dells during these years. The most important included: trout ponds, billiard tables, and croquet grounds in Cold Water Canyon; regularly scheduled steamer excursions; orchestra music aboard the boats for moonlight excursions; and a walk of logs and boards up Witches' Gulch.[24] As different entrepreneurs entered the scene, new amenities were added, so that by the turn of the century, boosters could point to a new 400-acre city park; stone walks and terraces around precipices; wire suspension bridges over chasms; rustic seats throughout the canyons; vistas opened through forest groves; an eating house and dance pavilion; a large, landscaped park by the train station; a new pavilion at the boat landing with "large and commodious waiting rooms"; and an intercounty carnival. Perhaps the most outrageous "improvement upon nature" was the addition of sea lions held captive in underwater cages for the amusement of sightseers.[25] These so-called improvements carried an importance beyond merely making the stay in Kilbourn City and a tour of the Dells more entertaining. The walkways, especially, helped transform the area into a genteel space to be experienced in much the same way that one would experience a stroll through a quiet city park (Figure 21.3).

That genteel space, moreover, attracted tourists from a wider range of society than some of the more elite resorts elsewhere in the state and throughout the country. From early on, the Wisconsin Dells drew visitors from the emerging middle class, but also from the "mechanics and laboring classes," who were drawn to the location's cheaper rates. Guidebooks made the distinction, time and time again, that unlike "fashionable crowded resorts that come at too high a price," the Dells could be enjoyed more satisfactorily by saving money.[26]

That a stay in the Dells was indeed a *relative* bargain is demonstrated by comparing the cost of lodging in Kilbourn with accommodations in other regional tourist destinations (Table 21.1). In both 1886 and 1903, the average daily rate at a Kilbourn hotel or boarding house was substantially lower than at accommodations in Oconomowoc and Waukesha, Wisconsin, or Lake Minnetonka, Minnesota. Also apparent is the explosion in Kilbourn City's guest capacity during this period compared with the relative stability of Waukesha or dramatic decline in Oconomowoc and Lake Minnetonka. These numbers, along with the growing popularity of both the northwoods region and Door County around the turn of the century, are evidence of a geographic and social shift in midwestern tourism beginning at this time: a move toward outdoor recreation and away from genteel relaxation associated with spas. While elite resorts, such as Gifford's in Oconomowoc, retained some of their glamour as gathering spots for well-to-do urbanites, the trend moved toward middle-class retreats in ever more peripheral locations. The Wisconsin Dells sat squarely in the middle of this dynamic.[27]

H. H. Bennett and the Creation of the Dells

Inseparable from the emergence of Kilbourn as an important tourist destination is the photography of Henry Hamilton Bennett. While H. H. Bennett may himself have credited "God or Natural forces" for the physical construction of the rock formations that he so lovingly photographed, in a very real sense his own work created the Dells,[28]

Figure 21.3. Engraving by Alfred A. Waud, *In Rood's Glen* (Courtesy of the State Historical Society of Wisconsin, neg. number WHi [X3] 50287, from W. C. Byrant, ed., *Picturesque America or, the Land We Live In*, vol. 2 [New York, 1872], 515)

for visual images, like language, can play an active role in the creation of place. And, while photography alone cannot mold nature, it can, like language, "direct attention, organize insignificant entitites into significant composite wholes, and in so doing, make things formerly overlooked—and hence invisible and nonexistent—visible and real."[29] Bennett's genius lay precisely in his ability to create "significant composite wholes" out of the disparate geological formations of the Dells and translate them into nodes along a linear tourist route.

From the late 1860s until just before his death in 1908, Bennett tirelessly photographed and rephotographed scenic views in the Dells, which, in many cases, he named himself.[30] Witches' Gulch, Boat Cave, Fat Man's Misery, Cold Water Canyon, and Artist's Glen are just a few of the many names that Bennett bestowed on the landscape that he was beginning to photograph. Naming photographic views served the function of helping make a familiar view into the alien and unfamiliar and thus worthy of making a three-day trip and purchasing a photograph.[31]

Bennett's busiest and most productive years occurred during the Dells' rise to regional and national prominence and contributed substantially to that ascension. The decade of the 1870s, in particular, launched his career. Bennett's enormous productivity is demonstrated by the fact that within a seven-year period the number of views that his studio could offer the tourist expanded 10-fold, from a little over 200 in 1872 to more than 2,500 by 1879. The next decade proved even busier, with greater demand for ever more distinctive views and Bennett's increasing technical capacity for photomechanical reproduction.[32] Indeed, by the early 1880s, Bennett had expanded his business well beyond serving visitors who stopped by his studio. He was shipping his photographs throughout Wisconsin and to Chicago and Milwaukee, and even filling occasional orders from places such as Toledo and Boston.[33]

Unbeknownst to him, Bennett's timing was impeccable, for the period of his greatest activity also coincides with the "golden age of landscape photography" in America. Extending from roughly the end of the Civil War until slightly before the turn of the century, the "view tradition" in American photography achieved a status since unmatched. The landscape pictures of Timothy O'Sullivan, William Henry Jackson, Carleton Watkins, and others in the West largely defined that tradition as they sought to depict the wild and seemingly barren spaces of the Far West. O'Sullivan's West, in particular, is presented as a "boundless place of isolation," as a vast space that is unmarked, unmeasured, and only in the process of being mapped. Even

Figure 21.4. Albumen print by H. H. Bennett, *Lone Rock with Canoe, Wisconsin Dells,* ca. 1880s–1890s (From the Collection of the H. H. Bennett Studio Foundation)

the photographs of Watkins and Jackson, fitting more easily into a picturesque-sublime tradition, emphasize the potential of the West for future exploration, definition, and exploitation.[34]

By contrast, in Bennett's skilled hands, the Wisconsin Dells became "a fairy-story landscape, rugged and wild in half-scale, with enchanted miniature mountains and cool dark caves"[35] (Figure 21.4). Losing their occasional threatening demeanor and stripped of any hint of social or environmental tension, Bennett's Dells assumed an air of tranquility and calm. The sculpted sandstone formations along the river became subjects for an outdoor portraiture, posed and presented in their best light in order to make a good and lasting impression. The sublimity with which Bennett was first struck upon his earliest encounter with the Dells melted into a picturesque and semiwild "fairy-story" landscape.[36] More a garden than an awesome and imposing wonder, the Dells that Bennett created offered something pretty and strange, yet nonthreatening and accessible. Exploration came to mean scouting with picnic baskets and Sunday bonnets.

Meticulous in execution and returning to the same scenes over and over again, Bennett composed his views according to the principles of the picturesque aesthetic, framing the landscape with foregrounded elements.[37] Often this was achieved by rearranging a scene by trimming out underbrush of limbs or trees and by adding logs or a canoe to a placid river foreground. As seen through the stereoscope, Bennett's chosen photographic medium, the foregrounded elements leaped out at the viewer, creating an illusion of depth and movement.

Most often, the picturesque was achieved by peopling the landscape with well-dressed Victorian men, women, and children (Figure 21.5). Perched on overhanging rocks high above the river, picnicking on a sandbar, contemplating the grandeur of nature from a point of solitude, or floating lazily downstream, the figures in Bennett's photographs did much more than merely add a foregrounding element, however. Their well-heeled presence conveyed the message that this leisure landscape could be experienced safely and, therefore, inspirationally. Replacing the hostility of the rapids that intimidated Increase Lapham only a generation earlier with an innate hospitality, Bennett created what one art historian called a "gentle playfulness" in the Dells of the Wisconsin River. Even the scale of the landscape seemed shrunk to the size of a playground, providing the ideal setting for a leisurely escape from the city.[38] The genteel charm of this landscape invoked notions of a pretend-land, of a place where nature could be encountered on the emerging middle-class terms of control, safety, and inspired recreation.

In this way, Bennett's photographs go hand in hand with the creation of a nineteenth-century tourist aesthetic based on the belief that nature could be exploited not only for its extractive resources but, for the first time, its recreational and restorative potential as well. Recall that Bennett's Dells, while contemporary with the stunning views of the western photographers, depict a landscape that had already witnessed a half-century of intensive use through lumber rafting and more than 100 years of fur trade and route exploration. Moreover, the surrounding countryside was becoming well integrated into an urban network of railroads, markets of all sorts, and investment capital. The frontier, as Bennett's Portage historian neighbor, Frederick Jackson Turner, was soon to point out, had passed through this region a generation earlier. Bennett's photographs helped create and transform a space in which people were becoming ever more confident into a place of leisure retreat.

Figure 21.5. Albumen print by H. H. Bennett, *Steamer in the Narrows, Wisconsin Dells,* ca. 1880s (From the Collection of the H. H. Bennett Studio Foundation)

Another View of the Dells:
Guidebooks and Scenic Albums

Between 1870 and 1900, another medium appeared to supplement Bennett's stereoscopic views of the Dells. Aimed at roughly the same special audience of middle-class scenic tourists, a new guidebook genre made explicit use of Bennett's well-known views to create a way of seeing and experiencing the newly discovered pleasures of, as one author characterized it, "the marvelous and picturesque in nature." The writers of the guidebooks and scenic albums—local boosters, newspaper men, freelancers hired by the railroads, as well as more well-heeled aesthetes and poets—could hardly allow elements of the "marvelous and picturesque" be left to chance. Rather than allowing the tourist to wander aimlessly through the region, the guidebooks enlarged Bennett's fairy-tale landscape by directing and guiding visitors to specific points of visual interest.[39]

In most instances, Bennett's photographs served as models for the dozens of wood engravings that appeared in guidebooks and scenery albums, whether published by the Chicago, Milwaukee, and St. Paul Railroad, local boosters, or in magazines and books of a more national scope (Figures 21.3 and 21.6). His familiar techniques of foregrounding and accentuating depth appear even in this format, as does his depiction of the Dells as a semiwild playground. In Alfred Waud's engraving *The Jaws,* appearing in William Cullen Bryant's popular *Picturesque America* scenic album (1872), Victorian tourists pass underneath a spectacular overhanging cliff as waterfowl skim across the placid stream. Beyond these jolly rowboaters, a rainbow arches; and beneath it, raftsmen make their way through the long shadows of the setting sun for one last time that day (Figure 21.7).

These wonderfully illustrative wood engravings usually accompanied lengthy text descriptions of the prominent sites along a railroad or, in the case of *Picturesque America*, across the continent. Like the engravings, the text descriptions of the Dells depicted a landscape peppered with distinctive, otherworldly features. With encyclopedic predictability and comprehensiveness, the guidebook writers' journey took the shape of a steamboat excursion first through the upper and then the lower Dells. Often they were written as testimonials by "impartial and honest" fellow travelers, called upon to relay their delightful experiences. In every case, however, the narrative structure depended on highlighting, point by point, the many scenic features that Bennett had named only a few years earlier. Frequently, landscape became a metaphor from the skilled pens of the guidebook writers. The

Figure 21.6. Engraving by Joseph Fleming, *Dells of the Wisconsin River* (Courtesy of the State Historical Society of Wisconsin, neg. number WHi [X3] 50288, from *Gems of the Northwest: A Brief Description of Prominent Places of Interest Along the Lines of the Chicago, Milwaukee and St. Paul Railway* [Chicago, 1886], 33)

Figure 21.7. Engraving by Alfred R. Waud, *Dalles of the Wisconsin, "The Jaws"* (Courtesy of the State Historical Society of Wisconsin, neg. number WHi [X3] 50289, from Byrant, *Picturesque America or, the Land We Live In,* vol. 2, 519)

Jaws of the Dells was not merely a narrow portion of the river, but was "guarded by two immense, standing-like sentinels on duty and sternly looking down in their stately grandeur." Nor in Witches' Gulch would a sightseer simply stroll through a pleasant canyon; she would "see the face of a hideous old hag, which has given the ravine its name." In this way, the landscape became anthropomorphized, imbued with a human quality designed to capture the tourist's interest and to make explicit the connection between naming and viewing.[40]

Descriptions alternated between the morose and the picturesque to an extent that sets them apart from iconographic treatments of the

Dells. In the former case, as in this description of Witches' Gulch, strangeness and death are accentuated: ". . . one of the weirdest, wildest, strangest freaks of Nature's handiwork in sculpture, architecture and grave-digging, all combined, that any human eye ever beheld. [The Gulch has been] fitly named for it seems a place where all the witches . . . might have concocted hell-broth, planned 'deeds without a name,' and danced with genii and demons in elfin revelry."[41] Other picturesque descriptions focused on the sights visible to the tourist by boat, but here they are presented in grand terms of sweeping vistas and breathtaking views. One writer, after telling of the Gulch's terrifying entrance, makes pains to point out that just beyond "we have an elegant view of the river, its bluffs and many islands, which is often compared to a similar view on Lake George."[42]

Moreover, the panorama that spurred these literary descriptions in the guidebooks took the visitor through a storied landscape. John Sears, in his history of nineteenth-century tourism, notes that Americans were hungry for locales saturated with religious and folkloric symbolism, for sacred places bathed in history and depth.[43] Thus, the tourist gaze became all the more attractive when it acquired "associations" with a romantic past; it was the medium of text that allowed authors to expand into the realm of "romance and lore." As James Maitland put it, the Dells offered more than "weird, impressive surroundings," for "connected with [those surroundings] are reminiscences of Indian days, and tales and traditions of the wild raftsmen's life of early white settlement. Almost every spot along the banks of the river for miles hereabouts is identified with some legend of tragic intent."[44] For Maitland and countless other guidebook authors, tragedy and suffering became starry-eyed romance and yet another way to differentiate the Dells from ordinary life.

The imagined presence of Native Americans, in particular, fulfilled a perceived need for antiquity. Idealizations of both prehistoric groups and the neighboring Ho-Chunk peoples provided the writers and their audiences with legends and stories, thereby "investing the locality with the charm most necessary to all resorts—romantic tradition."[45] And yet, these legends or "primitive associations" were mere artifacts of a vanished race, no different from the artifacts that the smart traveler might find in Pompeii. Not surprisingly, no mention is made of the forcible removal of the local Ho-Chunk or of their return to the region. When Native Americans are described in the present tense, the writers frequently apologized for their "current absence of sentiment or romance" and begged readers to remember that "they are in their most degenerate state."[46]

Only slightly less important, but equally misleading, were descriptions of the river's lumber-rafting past. The Wisconsin River became the state's most important conduit for the burgeoning nineteenth-century lumber trade, with as many as 100 rafts passing through the Dells in a single day as late as 1867. In these early years, before the real growth of scenic tourism, people living in the vicinity would "frequently resort there during the rafting season to see the rafts pass through" the narrow section of the river. A major attraction was the frequent and often violent destruction of the rafts as they passed through the Narrows and over the Kilbourn Dam.[47]

With the coming of the railroad and attendant passing of the rafting era, such sights became ever less common. In their place, however, arose myths, legends, and songs that eclipsed anything that the river and its working history had to offer. J. J. Brown's early guidebook set the trend in mythicizing the raftsmen. In *The Tourist's Guide to and through the Dells,* Brown takes considerable poetic license with his renaming and description of Notch Rock: "Notch Rock, or Raftsmen's Dread, where at the high stage of water the current sucks and whirls with whirlpool force, and often draws the raft to destruction, unless manned by a crew: Who are firm in the hand, and cool in the head, For often while passing, they strike the Raftsmen's Dread."[48] That Brown was nearly alone in renaming this landmark was of little consequence, for this playing with mythic, and distant, danger ran through nearly every guidebook printed in the nineteenth century.

Brown's ennoblement of the raftsmen through a heightened sense of danger is important, and representative of its genre, for two reasons. First, his heroic depiction of the raftsmen neglects the omnipresent social strife that accompanied the raftsmen's journeys through the Dells. As Wisconsin River historian Rick Durbin has shown, the history of rafting through the Dells is one of severe social tensions that resulted in the anarchic destruction of two dams and the loss of dozens of lives.[49] Second, "danger" is used seductively, as a way of heightening a sublime experience, yet kept now at arm's length. Guidebook writers such as James Maitland reveled in the risks to life and limb a trip through "these treacherous waters" used to bring, but now, thanks to the "staunch and handsome boats" that regularly ply the river, the traveler can soak in the misty delights of danger from the safety of the *Dell Queen.*[50]

Conclusion: "The Dells Are Closed"

By the turn of the century, scenic tourism in the Wisconsin Dells began to wane. Thousands of tourists from around the region continued to travel to see its remarkable natural wonders, of course, and this continues to some extent today. We are reminded by immense billboards, after all, to see "first things first," and indeed many travelers do. And yet, profound changes to the tourism industry, the natural landscape, and the social scene have combined to make today's Dells a place that H. H. Bennett would hardly recognize. The end of excursion trains in 1908, the rise of automobile travel and the attendant individualism engendered by it, the heightened need for a greater variety of attractions to "supplement nature," and, most important, the final damming of the river at Kilbourn City in 1907, all gave rise to a new, recreation-based form of tourism.[51] When the local manager of the soon-to-be-disbanded Wisconsin Dells Company declared apocalyptically (if also cynically) in 1908 that "the Dells are closed," he was reflecting on and anticipating such changes.[52] As the local newspaper put it at the time, with these changes, "the class of visitors will [no longer] be confined to mere sight-seers, but will include people who are looking for diversion and recreation."[53]

And yet, for 40 years after the Civil War, scenic tourism in the Dells created a unique space within the region. That space was a tourist space anchored on the great inland metropolis of Chicago. While the Dells never achieved a national status that their boosters endlessly dreamed about, for a period before the turn of the century, the water-carved sandstone features of the Wisconsin River attained a status of regional, cultural shrine. At once sublime and picturesque, threatening and safe, genteel and wild, the Dells were successfully divested of these inherent contradictions and reinvented as a "pretty strange place."

Notes

I would like to thank Ollie and Jean Reese of the H. H. Bennett Studio, Wisconsin Dells, for permission to reproduce the photographs in this chapter and for sharing the rich world of the nineteenth-century Dells with me. A slightly different version of this chapter was presented at the 1996 annual AAG meeting in Charlotte, North Carolina; I thank Deryck Holdworth for his insightful comments there.

1. J. Maitland, *The Golden Northwest* (Chicago, 1879), 29; *The Great West* (Chicago, 1880), 86. Two recent works which most forcefully argue that nineteenth-century tourist sites were imbued with the sacred and may be con-

sidered cultural shrines are P. McGreevy, *Imagining Niagara: The Meaning and Making of Niagara Falls* (Amherst, 1994); and J. F. Sears, *Sacred Places: American Tourist Attractions in the Nineteenth Century* (New York, 1989).

2. On the rise of Coney Island and its crucial place in the historical geography of mass leisure, see J. Kasson, *Amusing the Million: Coney Island at the Turn of the Century* (New York, 1978). T. Christensen, "Chicago Comics Give the Dells a Tweak," *Milwaukee Journal*, December 11, 1994, E1, E5. For a penetrating account of the contemporary Dells landscape, see D. E. Ross, "Placing Authenticity in a Leisure Landscape: Mind and Environment in Wisconsin Dells, Wisconsin," M.A. thesis, University of Wisconsin–Madison, 1989.

3. J. Jakle, *The Tourist: Travel in Twentieth-Century North America* (Lincoln, 1985), 53; and A. F. Hyde, *An American Vision: Far Western Landscape and National Culture, 1820–1920* (New York, 1990). On the rise of tourism in nineteenth-century America generally and the importance of certain key sites upon which to gaze, see Sears, *Sacred Places.*

4. "Sketches in Wisconsin," *Harper's Weekly*, September 5, 1885, 588–589. For a useful introduction to iconographic landscape analysis, see S. Daniels and D. Cosgrove, "Introduction: Iconography and Landscape," in D. Cosgrove and S. Daniels, eds., *The Iconography of Landscape: Essays in the Symbolic Representation, Design, and Use of Past Environments* (Cambridge, England, 1988), 1–10.

5. M. B. Bogue, "Exploring Wisconsin's Waterways," in *Wisconsin Blue Book* (Madison, 1989–1990), 101–298.

6. J. G. Norwood, *Owen's Geological Reconnaissance of the Chippewa Land District,* U.S. Senate, 30th Cong., 1st sess., 1848, Doc. 57, p. 108, quoted in L. Martin, *The Physical Geography of Wisconsin,* 3d ed. (Madison, 1965), 345.

7. I. A. Lapham, lecture to the Unitarian Church of Milwaukee, February 1, 1848 Increase A. Lapham Papers, Box 2, 1844–1849, State Historical Society of Wisconsin, Madison, 1849. For a man of science, Lapham's travels to the Dells were typical. See B. M. Stafford, *Voyage into Substance: Art, Science, and the Illustrated Travel Account* (Cambridge, Mass., 1984).

8. I. A. Lapham, "Geological Note of a Tour to the Dells," Increase A. Lapham Papers, Box 2, 1844–1849, State Historical Society of Wisconsin, Madison, 1849.

9. R. R. Fries, *Empire in Pine: The Story of Lumbering in Wisconsin, 1830–1900* (Madison, 1951), 71.

10. P. Fussell, "From Exploration to Travel to Tourism," in his *Abroad: British Literary Traveling between the Wars* (Oxford, 1980), 37–49.

11. *Wisconsin Mirror* (Kilbourn City), March 25, 1856.

12. *Wisconsin Mirror,* September 23, 1856.

13. J. Reese, "Wisconsin's Most Popular Vacation Spot: Boat Trips through the Dells," in M. Goc, ed., *Others before You: The History of Wisconsin Dell's Country* (Wisconsin Dells, 1995), 135; S. Rath, *Pioneer Photographer: Wisconsin's H. H. Bennett* (Madison, 1979), 19.

14. *Wisconsin Mirror,* August 11, 1869.

15. Reese, "Wisconsin's Most Popular Vacation Spot," 136; *Wisconsin Mirror,* April 5, 1873.

16. *Wisconsin Mirror,* June 28, 1873; A. C. Bennett, "A Wisconsin Pioneer in Photography," *Wisconsin Magazine of History* 22 (1938–1939): 268–279.

17. J. J. Brown, *The Tourist's Guide to and through the Dells of the Wisconsin River and Vicinity, Kilbourn City* (Kilbourn City, 1875), 7; and *Gems of the Northwest: A Brief Description of Prominent Places of Interest along the Lines of the Chicago, Milwaukee, and St. Paul Railway* (Chicago, 1886). Madge Patterson Van Dyke notes that in 1916 the "annual army of tourists [numbered] 50 to 60 thousand strong." M. P. Van Dyke, "The Story of Kilbourn and Its Vicinity," B. A. thesis, University of Wisconsin, 1916, 85.

18. P. Donan, *The Tourists' Wonderland: Containing a Brief Description of the Chicago, Milwaukee, and St. Paul Railway Together with Interesting General Descriptive Matter Pertaining to the Country Traversed by This Line and Its Connections* (Chicago, 1884), 9.

19. *Kilbourn and the Dells of the Wisconsin, with Views en Route from Chicago to St. Paul and Minneapolis* (Chicago, 1901), 9.

20. H. L. Horowitz, *Culture and the City: Cultural Philanthropy in Chicago from the 1880s to 1917* (Chicago, 1976). William Cronon hints at this point in his *Nature's Metropolis: Chicago and the Great West* (New York, 1991), 381–385.

21. P. Donan, *The Dells of the Wisconsin, Fully Illustrated* (Chicago, 1879), 3–4. Anne Farrar Hyde makes a powerful case for the utility of landscape in defining nationalist sentiments. Hyde, *An American Vision.*

22. The son of Increase Lapham, the map's cartographer worked for the Chicago, Milwaukee, and St. Paul Railroad for 58 years as a surveyor. The map of the Dells was his first project with the Milwaukee Road and quite possibly launched his career. *Milwaukee Journal,* January 8, 1935.

23. *Gems of the Northwest: A Brief Description of Prominent Places of Interest along the Chicago, Milwaukee, and St. Paul Railway* (Chicago, 1885), 33; emphasis added.

24. Brown, *The Tourist's Guide to and through the Dells,* 3–5; Bennett, "Wisconsin Pioneer in Photography," 274.

25. N. Wetzel, *The Dells of the Wisconsin: America's Most Beautiful Summer Home* (Kilbourn City, 1903); and Reese, "Wisconsin's Most Popular Vacation Spot," 148.

26. J. E. Jones, *A Description of a Noted Western Summer Resort: A Trip through the Dells of the Wisconsin River* (Kilbourn City, 1887). James Gilbert notes that the decades under discussion, from 1870 to 1900, witnessed a remarkable expansion of new middle-class occupations, and he raises the interesting point that tourism was one chief way to become *culturally* middle class. J. Gilbert, *Perfect Cities: Chicago's Utopias of 1893* (Chicago, 1991), 1–22. In his 1879 guidebook, Donan writes of the appeal to middle classes: ". . . what is especially worthy of mention in these times of pocket-pinching financial stringency, the charges for entertainment, transportation and everything the tourist needs, are absolutely primitive in their cheapness." He brags that a trip to the Dells cost "scarcely a third of the rates at many other places of resort," and these other places "do not furnish a tithe of the attractions and real enjoyment to be found here." Donan, *The Dells of the Wisconsin,* 5.

27. Only two resorts existed in the future Minoqua area when the Chicago, Milwaukee, and St. Paul Railroad reached it, and even when the railroad finally reached Sturgeon Bay in 1894, the 35-mile ride took over three hours. M. Dunn, *Easy Going: Wisconsin Northwoods, Vilas and Oneida Counties* (Madison, 1978), 18–20; and J. F. Hart, "Resort Areas in Wisconsin," *Geographical Review* 74 (1982): 192–217, reference on 206–7.

28. H. H. Bennett to H. C. Adams, February 2, 1906, in the H. H. Bennett Studio Archives, Wisconsin Dells (hereafter cited as HHBSA).

29. Y.-F. Tuan, "Language and the Making of Place: A Narrative-Descriptive Approach," *Annals of the Association of American Geographers* 81 (1991): 684–696, quotation on p. 685.

30. The catalogue that accompanied a 1970 exhibit on Bennett's work declared him to be "almost a legend in the history of American photography." *Bennett, Steichen, Metzker: The Wisconsin Heritage in Photography* (Milwaukee, 1970). On Bennett's life, see: Bennett, "Wisconsin Pioneer in Photography," 268; and, especially, Rath, *Pioneer Photographer*, for a good biographical treatment. The general background of Bennett's life for this section has been gleaned largely from these two sources as well as from conversations with his granddaughter, Jean Reese.

31. A. Trachtenberg, "Naming the View," in his *Reading American Photographs: Images as History, Mathew Brady to Walker Evans* (New York, 1989), 119–163. Guidebook author Col. P. Donan writes, "Every scene along the river is mirrored in [Bennett's] mind, and wears the name he gave it." Donan, *The Tourists' Wonderland*, 23. Donan is not alone in crediting Bennett with naming the sights in the Dells. See also F. H. Taylor, *Through to St. Paul and Minneapolis in 1881: Random Notes from the Diary of a Man in Search of the West* (Chicago, 1881), 7; Maitland, *The Golden Northwest*, 30; and Bennett, "Wisconsin Pioneer in Photography," 278.

32. On the numbers of views available, see *Wisconsin Mirror*, April 7, 1872; and Donan, *The Dells of the Wisconsin*, 7. Bennett's capacity to meet demand is revealed in a letter in which he informs a client that his studio enables him to produce close to 30,000 prints per month. H. H. Bennett to J. E. Porter, January 1, 1887, HHBSA.

33. H. H. Bennett, November 22, 1882, notebook entry, HHBSA; and H. H. Bennett, *Wanderings among the Wonders and Beauties of Western Scenery* (Kilbourn City, 1883).

34. W. Naef and J. N. Wood, *Era of Exploration: The Rise of Landscape Photography in the American West, 1860–1885* (New York, 1975), 12; P. B. Hales, "American Views and the Romance of Modernization," in M. A. Sandweiss, ed., *Photography in Nineteenth Century America* (Fort Worth, 1991), 205–257; and J. Snyder, "Territorial Photography," in W. J. T. Mitchell, ed., *Landscape and Power* (Chicago, 1994), 175–202.

35. J. Szarkowski, *The Photographer and the American Landscape* (New York, 1963), 4.

36. H. H. Bennett, April 23, 1866, notebook entry, HHBSA.

37. Sara Rath notes Bennett's indebtedness to Milwaukee businessman

William Metcalf for not only financial assistance but also technical advice on achieving a pictorial effect in his photographs. Rath, *Pioneer Photographer*, 43–44.

38. T. Bamberger, "A Sense of Place," in T. Bamberger and T. Marvel, eds., *H. H. Bennett: A Sense of Place* (Milwaukee, 1992), 5–6.

39. F. Wisner, *The Tourist's Guide to the Wisconsin Dells* (Kilbourn City, 1875), 3.

40. Wisner, *The Tourist's Guide to the Wisconsin Dells*, 26; and *Gems of the Northwest* (1885), 36. At times, guidebooks gently mocked the elaborate naming of features in the Dells as unrealistic and fanciful. One writer, for example, noted that "a huge wall in the midst of this canyon is known as the Devil's Jug, but what a devil's jug could be doing in Cold Water Canyon I'm sure I fail to know." *Gems of the Northwest* (1885), 36.

41. Donan, *The Dells of the Wisconsin*, 30.

42. *The Great West* (Chicago, 1880), 84.

43. Sears, *Sacred Places*, 72–86.

44. Maitland, *The Golden Northwest*, 29.

45. Jones, *A Description of a Noted Western Summer Resort*, 8.

46. Jones, *A Description of a Noted Western Summer Resort*, 7.

47. "Dells of the Wisconsin: Black Hawk's Cave," *Wisconsin Historical Collections*, vol. 5 (Madison, 1867), 298–299.

48. Brown, *The Tourist's Guide to and through the Dells*, 11.

49. R. Durbin, "The Kilbourn Dam: A Tale of Many Hopes, Dreams and Schemes," *Wisconsin Magazine of History* (forthcoming).

50. Maitland, *The Golden Northwest*, 32. Patrick McGreevy notes that for nineteenth-century travelers, terror could be considered sublime only if the actual danger was at a distance. McGreevy, *Imagining Niagara*, 10–11. Promoting the safety of the guided steamships fulfilled another function, as well: it discouraged independent travel and ensured that tourists would tour the Dells by the commercial boat companies.

51. The first two changes are nicely summarized in Reese, "Wisconsin's Most Popular Vacation Spot," 148–150. Of these changes, the most controversial is the creation of the 1907 Kilbourn Dam. The full impact of the dam on scenic tourism in Wisconsin remains to be written, but landscape architect John Nolen was one of the first to see its dangers. In his report on the future state park system of Wisconsin, Nolen emphasized that the Dells are "Wisconsin's most characteristic and precious possession in the form of natural scenery. They are unique. For picturesqueness, romantic scenery, for alternative suggestions of mystery and majesty, the features of interest are numerous and varied." But Nolen was quick to caution that "this is what the Dells are today. What they will be when the dam now under construction is completed, when the water is raised permanently eighteen to twenty feet above the present, is not easy to say." The Dells, of course, were not made a state park. J. Nolen, *State Parks for Wisconsin*, Report of the State Park Board of Wisconsin, 1909, 27–29. For his part, H. H. Bennett fully comprehended the impact the impending dam would have upon the landscape that he knew so well. As one

of the only people within Kilbourn City to challenge the dam and in a manner that anticipated John Muir's cry, seven years later, against the damming of Yosemite's Hetch Hetchy Valley, Bennett wondered, "Why should such beauty spots [as the Dells] be destroyed or even injured as has been by building dams and cutting trees here for the profit of perhaps a few men when the good God made it that all people for all times might enjoy the beauties?" Bennett died a little more than a year after the Southern Wisconsin Power Company broke ground on the dam. H. H. Bennett to Julia Lapham, January 1, 1906, HHBSA.

52. *Kilbourn Weekly Illustrated Events,* July 18, 1908. The newspaper was quoting Nat Wetzel.

53. *Kilbourn Weekly Illustrated Events,* September 7, 1906.

22

The Northwoods
Back to Nature?

Timothy Bawden

Wisconsin is a place of rituals guided by the changing seasons. Spring tickles the senses, a time for both reawakening and anticipation. By Memorial Day weekend, station wagons and minivans crowd the highways, crammed with the essentials to survive this and several more summer holidays to come. The parade begins in the cities below the state's southern border, gains momentum along the way, and ends in any one of countless retreats far to the north. U.S. Highway 51, stretching from border to border up the middle of the state, has long served as a favorite path for carrying urbanites infected with "cabin fever" to northern destinations (Figure 22.1).

Traveling north along this route from Portage, the rolling green fields of dairyland gradually give way to the flat expanse of Glacial Lake Wisconsin. Highway 51 hugs the eastern shore of the empty lake bottom, crossing numerous oddly linear streams, which are actually irrigation ditches carved into its sandy floor. The local relief is subtle, at best, for about an hour, until one reaches Rib Mountain near the city of Wausau. The "mountain" provides a welcomed relief and was once thought, albeit incorrectly, to be the highest point in the state. From here it is a steady climb, weaving back and forth over the Wisconsin River several times, into the Northern Highland—the portion of the Canadian Shield that extends into Wisconsin. Highway 51 narrows

450

from four lanes to two until the road is soon swallowed up by continuous pine and hardwood forest. The ubiquitous tavern, always in the
shadows of the trees, serves as a modern-day trading post, carrying
the gamut of provisions from brandy to bait. At each intersecting road,
stacks of arrow-shaped signs point in the direction of resorts, lodges,
and camps set back off the beaten path in this seeming wilderness. In
time, the only clearings become the many lakes, whose names prompt
a game of phonetic gymnastics: Kaubashine, Kawageusaga, Shishebo-
gama. . . .

This is the northwoods of Oneida and Vilas counties. Few other
areas of the Midwest are as much a part of the ritual retreat from the
city. To most, the allure of the northwoods is that it simply is what
the city is not: pristine, wild, unspoiled, simple. In short, the northwoods is close to nature, and that's the draw. But is it? What do we
mean by "natural"? Indeed, the environment of this region, and northern Wisconsin in general, is quite different from the cities, towns, or
dairyland countryside to the south. However, it is anything but untouched by human hand, today or in the past. It has been shaped and
reshaped by human action and, in most cases, for human purposes,
particularly during the past 100 years. The story of this place is as
much about people and society, or what we might call culture, as it is
about the physical environment, or what we might call nature. Therefore, perhaps the more poignant question is: What does the story of
the northwoods tell us about the relationship between the two—culture and nature?

The Assault on the Forest

No other element has altered the vegetation of northern Wisconsin as
quickly as humans or, perhaps more appropriately, as quickly as some
methods of forest clearing that humans have used. Various peoples
have occupied northern Wisconsin for over 10,000 years. In fact,
the earliest arrivals, referred to as Paleo-Indians by anthropologists,
hunted and gathered food on the heels of the receding glaciers.[1] Many
other nomadic groups followed over the centuries, and their greatest impact upon vegetation was undoubtedly the use of fire.[2] What
is sometimes referred to as the "original" vegetation of Wisconsin is
really only a snapshot condition of the forest at the time of white settlement in the nineteenth century. At the time of statehood in 1848, over
three-quarters of Wisconsin was covered in forest, and almost all of
northern Wisconsin. Seventy years later, virtually the entire forest of
northern Wisconsin had been decimated, either by the axe or by the

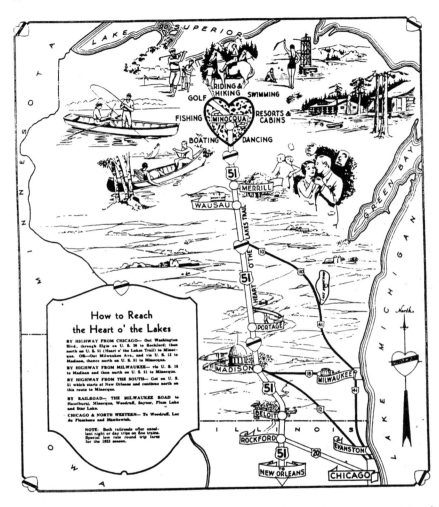

Figure 22.1. A promotional map to the "Heart of the Lakes" region, ca. 1930 (Reprinted with permission from the *The Lakeland Times: The First 100 Years: The Centennial Edition* (Minocqua, Wis., 1988)

fires that swept through the debris left behind. Less than 1 percent of that pre-European northern forest remains today.[3]

So how do we understand the vegetation of northern Wisconsin if it was almost completely removed a century ago? Land surveys from the mid-nineteenth century go a long way. At that time, surveyors were required to describe the general plant cover of every quarter-mile section.[4] From the field notes of these surveyors it has been possible to construct with fair accuracy a general picture of the northern forests just prior to white settlement. Much of the northern forest was a hardwood-conifer mix of hemlock, maple, and birch. The southern extreme of Canada's vast and dense boreal forest of spruce and fir extended down into the Lake Superior Lowland in the northwest corner of the state. The many swamps throughout the region were sometimes filled with large stands of conifers such as spruce, tamarack, and cedar. Elsewhere, pine forests thrived in abundance on the sandy glacial outwash plains in the northwest, north-central, and northeast areas.[5] It was here that the mightiest tree of the Great Lakes flourished: the white pine. These trees grew as high as 200 feet, as wide as 8 feet in diameter, and as old as 500 years.[6] They were the royalty among the trees of the northwoods, but by the nineteenth century they were also the lumberjack's prize.

In the early days of logging, before railroads, the only way to transport logs was via nature's highways: the waterways. Mill towns sprang up along the state's rivers and along the shores of Lake Michigan, transforming logs from the "Pineries" into lumber for the cities to the south and for homes on the prairies farther west. By 1857, as many as 107 mills lined the lower stretches of the Wisconsin River, but logging the rich and distant pineries of the far north was delayed until railroads could enter the upper valley.[7] The many rapids and constant bends in the northern stretches of the river were not suited to transporting logs. The key to opening this region was the completion of a line from Milwaukee to Ashland by the Wisconsin Central in 1877. It ran between the watersheds of the Chippewa and Wisconsin and quickly became the most important logging route of its time. During the 1880s, several other railroad lines penetrated the headwaters of the Wisconsin River, located in present-day Oneida and Vilas counties. The Milwaukee, Lake Shore, and Western Railway, which later became part of the Chicago and Northwestern, completed lines to Eagle River and to Rhinelander in 1882. Meanwhile, the Chicago, Milwaukee, and St. Paul Railroad acquired the Wisconsin Valley Line and linked Merrill and Minocqua in 1887.[8] These railroads ushered in a new era in the logging industry.[9] Not only did they provide an easier means for getting labor to remote camps; they also opened up new tracts of land where logging had been previously confined to stream banks. Log drives on rivers continued, but nature's highway, the river, had become just an auxiliary to culture's highway, the railroad.

The railroads extended deeper and deeper into the forests, until even the most remote lands were drawn into the machine. Back in

Figure 22.2. The Wisconsin Land and Lumber Company, owned by Wisconsin lumberman Charles Meyer, located its headquarters just over the state's northern border in Hermansville, Michigan, in 1883. Scores of towns were born of the logging era throughout the Pineries of Michigan, Wisconsin, and Minnesota. (Courtesy of the State Historical Society of Wisconsin, neg. number WHi [X3] 27580)

1852, Wisconsin congressman Ben Eastman bragged to his colleagues in the House of Representatives: "Upon the rivers which are tributary to the Mississippi, and also upon those which empty themselves into Lake Michigan, there are interminable forests of pine, sufficient to supply all the wants of the citizens for all time to come."[10] Just a few years later, Increase Lapham, perhaps Wisconsin's first conservationist, looked at the forests another way. In 1855 he lamented, "It is much to be regretted that the very superabundance of trees in our state should destroy, in some degree, our veneration for them. They are looked upon as cumberers of the land; and the question is not how they shall be preserved, but how they shall be destroyed."[11] The spirit of the times, or *zeitgeist* if you will, was driven by the euphoric growth of the nation. It was generally felt that the American character emerged from the ability of people to overcome the physical environment and utilize the abundant wealth of the land. Pleas for forest conservation were limited to intellectuals and "fanatics," to which Wisconsin senator Timothy Howe arrogantly replied that the "generation yet unborn" had done nothing for him and that he did not care to "sacrifice too much" for them[12] (Figure 22.2).

Wisconsin was among the leading timber producers in the country throughout the 1890s and actually held the top position for a number of years. But by 1900, all the profitable pineries of the state had been exhausted.[13] Timber production declined rapidly during the early 1900s. Many lumbermen began to worry as they watched the quantity and quality of white pine decline.[14] The production of hardwoods and hemlock brought a temporary fix, but this remedy simply reinforced what was already quite well known: the supply was indeed finite. For many lumbermen it was simply a matter of geography. To the west lay untouched evergreen forest and, to the south, abundant pines and hardwoods. But what about the northern lands they left behind in stumps? Here, history, not geography, dictated the answer. Farms follow forests.

The Farmer's Last Frontier

"With farms supplanting the forest, northern Wisconsin will not revert to a wilderness with the passing of the lumber industry, but will be occupied by a thrifty class of farmers whose well directed, intelligent efforts bring substantial, satisfactory returns from fields, flocks and herds."[15] In this statement, Dean William Henry of the University of Wisconsin's College of Agriculture echoed the sentiments felt throughout the state during the 1890s. It is perhaps no coincidence that the rhetoric fell on the heels of Wisconsin historian Frederick Jackson Turner's famous essay on the frontier, delivered to a Chicago group in 1893. The essay, entitled "The Significance of the Frontier in American History," posited that the westward-moving frontier is the source of the American character: it is what made America and Americans stand apart from the rest of the nations of the world.[16] For Turner, the frontier was the meeting point between "savagery" and "civilization." It was here that the most rapid Americanization took place as people from all backgrounds gathered and shed their cultural baggage to confront the wilderness as pioneers. He envisioned the frontier as being critical for furthering democracy and egalitarian values in America by providing the destitute with a chance to begin anew on free land. Turner saw a natural succession in American settlement: once the early pioneers cleared the land a new class of immigrants moved in to build schools and roads, followed by another, entrepreneurial class of people with the capital to build villages that would grow to towns and then to cities. The crux of the argument was that an agricultural foundation was necessary for subsequent development, and it was here that the root of American values were formed.

Although Turner was in fact looking west when he offered his thesis, the promoters of northern Wisconsin applied his rhetoric to the unsettled north. In 1899 Judge James O'Neill echoed Turner's progressive optimism: "Northern Wisconsin is great and prosperous, but her period of most substantial development is only now in sight. The next quarter century will bring her well up in productive wealth with the southern half of the State, resulting in a commonwealth of patriotic, progressive, and intelligent citizenship, rich in agriculture, manufactures, and commerce, and to which, as one of the great family of States, we may point with justifiable pride."[17]

Lumber companies, railroads, speculators, local newspapers and boosters, colonization companies, the state legislature, and even the University of Wisconsin advocated northern settlement.[18] Lumber companies had an obvious incentive to peddle their holdings, given that most of the timber had been removed. However, in some instances they found themselves in conflict with farming, because as they sold their land and settlement grew, the services required by the local population grew as well, meaning that the taxes increased on the land the companies continued to hold.[19] Railroads were also very eager to attract farmers to northern Wisconsin. The Wisconsin Central Railway, for instance, had nearly a million acres to sell and sent agents to cities on the East Coast and to Europe to scout for buyers.[20] They distributed brochures with titles such as *A Farm in Wisconsin Will Make Money for You from the Start: Crops Never Fail.* Enticing photographs of prosperous farms with captions touted the richness of the land.[21] Indeed, photographs were a persuasive way to advertise land in the 1890s. They gave credibility to the seller, and they did not require translation or even literacy. A salesman could be a swindler, but how could a picture lie? As one brochure proclaimed:

These pictures show, plainer than words, what an honest industrious man can do in a few years. The lands which are now being offered for sale to homeseekers and which are near to railroads, and within easy distance to good schools and churches, are being purchased by a thrifty industrious class. There are neither cyclones nor blizzards in this part of the country to make life miserable to the pioneer; but a healthy bracing climate, and rich productive soil, where abundant crops can be raised without fear of drought or other undesirable contingencies.[22]

Scores of land companies acquired logged-over lands with the hopes of unloading them on would-be settlers for a substantial profit. The Blue Grass Land Company of Minneapolis targeted Finnish immigrants for the settlement of a colony near Eagle River in eastern Vilas County. In an advertisement published in 1902, they described the area as "suitable for all kinds of farming . . . . the land is rich, clay-bottomed, making it the most productive hay land in America, just as good as those famed 'blue grass' lands of Kentucky." The advertisement went on to tout the affordability of the land for those who did not have the funds to farm wheat on the prairies, and promised that the company would donate free land for a church and a school and provide free railroad transportation. They even offered 40 acres of free land for the person who suggested the most appropriate and easy to pronounce Finnish name for the colony. The winning name was Toivola, meaning "vale of hope." Unfortunately, the name and the place had a very short life.[23] Similarly, the name of another Blue Grass Land Company town, Farmington, had a very short life, but, unlike Toivola, Farmington changed its name to St. Germain and went on to become one of the foremost resort spots of the northwoods.[24]

However, it was not only private interests that were caught up in the zealous promotion of cutover lands, and likewise, it was not only private interests that stretched the truth on the potential of the land for farming. In 1895, the Wisconsin legislature acted to establish a new Board of Immigration for the purpose of promoting the sale of northern Wisconsin land in Europe.[25] That same year it passed a mandate for the "dean of the College of Agriculture of the University of Wisconsin . . . to prepare a bulletin or hand-book describing the agricultural resources of Wisconsin, especially the newer and more thinly settled districts, with reference to giving practical helpful information to the home-seeker."[26] The result was the 200-page guide called *Northern Wisconsin: A Handbook for the Homeseeker*. The book came out just a year after the directive was made and was the product of on-the-spot interviews, photographs, and research undertaken by Henry and his associates. With detailed statistics and text, it covered every aspect of farming in northern Wisconsin from climate and crops to cattle raising. It encouraged agriculture but also noted some of the difficulties of land clearing and the problems associated with poor soil. But once again, the overall sales pitch was not made with words and numbers or maps; a good share of the targeted European audience probably could not read it anyway. As Henry put it, "This is an age of pictures . . . we have made this a picture book. We believe that when our readers have studied these views and read the legends accompanying them, they will have as correct an idea of the present agricultural conditions of Northern Wisconsin as is possible for one to obtain without actually visiting the country and studying the subject in person."[27]

The pictures for the *Handbook* were taken mainly during the summer and fall of 1895, which Henry admitted was an especially prosperous crop year. Nonetheless, it appears as though the bountiful harvests and

Figure 22.3. A sample photograph from *Northern Wisconsin: A Handbook for the Home-seeker*, ca. 1895 (Courtesy of the State Historical Society of Wisconsin, neg. number WHi [H44] 94)

crops in some of the photographs were more a product of the camera than they were of the sandy, stump-ridden soils from which they were purported to have been grown (Figure 22.3). No one worried about misrepresentation. By 1897, 50,000 copies of the *Handbook* had been distributed, and 60,000 pamphlets, illustrated with scenes from Henry's book, were printed in English, German, and Norwegian.[28]

Statistics show that the efforts of early promoters were at least partly successful. Between 1880 and 1900 the population of the 24 northernmost counties increased from 120,000 to 400,000, and about 20,000 new farms were established.[29] But the numbers fail to show that much of the increase in population was associated with the lumber industry and the growth of mill towns. Agricultural settlement was largely confined to the southern counties of the cutover. For many of the northern counties, agriculture was a much more difficult sell. Only 13 percent of Oneida County was under cultivation by 1910. In Vilas County, the share was only 3 percent.[30]

What dampened the optimism in this "vale of hope"? Clearing the land was among the biggest obstacles to farming in many parts, but certainly not the only one. The glaciers had deposited a mix of unsorted rocks and boulders across northern Wisconsin. Given enough time, the land could be sifted through sufficiently to make plowing possible. But in many places, this task could not be taken up until the clearcut forest debris was removed. White pine might have been the royalty of the forest and the prize of the lumberjack, but to the farmer it was just one big curse. Enormous stumps and roots littered the cutover lands. Moreover, and perhaps as nature's way of seeking a long, agonizing revenge, white pine took practically geological time to decompose. In the *Handbook*, Dean Henry informed prospective farmers: "Stump lands are usually seeded to grass and yield fine crops of hay or pasture for many years. The stumps standing among the grass of the meadows are removed without great difficulty after a few years."[31] It is certainly unclear what his concept of "difficulty" was, but suffice it to say, he was wrong, very wrong! Stump removal was a terribly arduous task (Figure 22.4).

H. L. Russell succeeded Henry as dean of the College of Agriculture in 1907 and continued the fervent promotion of northern farming. His devotion to development may have been driven by the fact that he had invested his own money in northern lands and was a leading stockholder in two colonization companies.[32] In 1915, the college opened a new land-clearing branch in the Department of Agricultural Engineering, with a primary focus on the techniques of stump removal, commonly perceived to be a major impediment to northern farming.[33] By the following year, members of the department took their operation on the road along the same tracks that had been used to carry the white pine south just decades earlier. They traveled the cutover by train, giving demonstrations of state-of-the-art stump pulling and the use of dynamite. The train traveled like a circus and was appropriately named the Land Clearing Special. Settlers came considerable distances to see the exhibition, some walking as far as 25 miles.[34]

Despite these endeavors, the first two decades of the twentieth century dramatically soured the delirious optimism heard during 1890s. Even the bright minds and cutting-edge "science" of the university could not overcome some basic facts of nature. In some places, even when the stumps were removed, the only crop that really thrived in the sandy soils was the potato. In other areas, where vast coniferous forests once stood, the soils were too acidic for growing just about anything but more coniferous forest. And even where the land permitted the growth of some crops, the climate always presented a risk. While the seasonal high and low temperatures do not vary dramatically across Wisconsin, the average length of the growing season, or days between the last and first killing frost, is considerably shorter in the northern part of the state. For example, in some parts of Vilas County it is less than 90 days, compared with over 160 days in southeastern Wisconsin.[35]

Figure 22.4. Stump pulling in northern Wisconsin ca. 1915 (Courtesy of the State Historical Society of Wisconsin, neg. number WHi [D482] 6801)

Sparked by a sharp decline in agricultural prices, land clearing and farm making in northern Wisconsin practically came to a halt during the 1920s. Despite all the energy that had been poured into agricultural settlement over the previous three decades, only 6 percent of these counties was in cultivated crops.[36] Colonization companies went bankrupt, farms were abandoned, and local governments, which relied on property taxes, were swept into dire financial straits.[37] When landholders could not pay their taxes, the land reverted to the county. A report, published in 1928, found that a quarter of the land in Wisconsin's 17 northernmost counties had been offered for sale as tax delinquent in 1927, and over four-fifths of it had gone unsold.[38] Moreover, local governments could not maintain services and infrastructure for the scattered settlers that were already there. An Oneida County town board member summed up the financial burden of isolated settlers well when he remarked:

A few years ago a man and his family came here from the southern part of Wisconsin. He bought a "forty" of land north of Highway No. 8 several miles from any neighbors and without any road leading to it. He asked the town board to build him a road. We of the town board refused. The next year, when there were two new members on our town board, a petition was circulated for a road and was presented to our board. The new members voted to build the road. It cost $12,000 to build that road. This man used the road just once—to move out of the town.[39]

Ultimately, the problem was not so much that things would not grow in the cutover environment, but rather that the activities being advocated at the time were not suitable to the realities of the environment, and, for the most part, no "science" could change that. While the promoters pressed on, albeit with more caution, a counter-current of opinion began to surface. The message was to consider planting the one thing that had proved successful in the north: trees.

Reforestation, Conservation, and Recreation

The idea of reforestation stirred up controversy across the state from the newsroom in Rhinelander to the courtroom in Madison. Already in 1897 there was enough legislative support to appoint the Commission on Forestry. The commission's first task was mainly a reconaissance to determine the condition of the land and the remaining forests in the northern counties. What they found was hardly profound: after 40 years of logging, nearly the entire area had been logged over. Moreover, they concurred that only 20 percent of the land in the northern counties was any good for farming in the first place. Their conclusion contradicted the hope of the farming frontier: "The land must be restocked and the young timber must be given a chance to grow on all lands which are essentially forest soil and not desirable for agriculture."[40] Forestry affairs increased briskly during the next 20 years, largely through the work of Edward M. Griffith. In 1903 the new State Forestry Commission (renamed the State Board of Forestry in 1905) was established, and Griffith, fresh from serving under Gifford Pinchot in the U.S. Bureau of Forestry, was appointed superintendent of state forests and state fire warden in the following year. He carried the conservation movement of the progressive era to Wisconsin.

Griffith acquired almost a quarter of a million acres of land in 1905, earmarked for state forests. This included a 40,000-acre reserve at the headwaters of the Wisconsin River, which was the first step in his endeavor to create an astounding 1.5 million–acre forest reserve out of all of the nonagricultural land in Forest, Vilas, Iron, Price, and Oneida counties.[41] A forest reserve in northern Wisconsin was not a completely novel idea. In 1878, the state had set aside 50,000 acres in Vilas and Iron counties with the provision that no authority be given to anyone to cut down any timber on these lands; the *reserve*, however, had fallen victim to the ax in 1897 without resistance.[42] This time, the program was more elaborate. The Board of Forestry created a system of fire patrols and began an educational program to reduce fire loss in timber and soil.[43] It established the first state tree nursery at Trout Lake in Vilas County in 1911. White and red pine seed from a local logging operation, as well as Scotch pine seed from Germany and Ponderosa pine from the Black Hills, were prepared there for future planting. Meanwhile, almost 200,000 seedlings and transplants from the Michigan Agricultural College were planted on tracts ravaged earlier by clear-cutting and subsequent fires.[44]

By 1912 the total state forest reserve had grown to 400,000 acres.[45] By then, however, vehement opposition had mounted against the forestry program in the counties most affected by the plans. The principal objections to a big forest reserve were that the lands taken by the state were removed from the tax rolls and that such reserves were retarding the development of northern Wisconsin. Two Rhinelander newspapers led the attack. They had two objectives: first, to promote the agricultural potential of the north country, and second, to undermine the credibility of Griffith and the Board of Forestry in general.[46] They, too, turned to the power of the camera to emphasize their case. One paper ran a picture of a man in a field of tall corn with a caption that read: "When corn grows like this picture shows it does in Oneida County, this is too good a County to be given over to reforestation schemes or for state manipulation."[47] Again, how could a picture lie? A prominent Oneida County lumberman declared that further reforestation would ruin the future of farm settlement and would only make the area a "home for wild animals and a playground for the idle rich of the cities."[48] Even Griffith's forest fire fighting program came under attack. One assemblyman contended that forest fires brought settlers to the burned-over areas and hence constituted a boon to the northern counties.[49] The overall message was clear: northern Wisconsinites were not interested in the "rejunglizing" of their counties (Figure 22.5).

By the fall of 1913, representatives of Oneida, Forest, Vilas, Iron, and Price counties formed an association to oppose further state forestry work. In that year, assembly representative Whiteside of Hurley introduced a bill that would halt state land purchases until the forestry program could be investigated.[50] Over 100 Oneida and Vilas county residents went to Madison to lobby for the Whiteside bill, and when the bill passed and a study commission was appointed, one of the Rhinelander newspapers heralded the victory, saying, "Surely this is a great honor to the County Board of Oneida County, also to the citizens who have done all in their power to defeat the Griffith *destruction* of Northern Wisconsin."[51] A legislative committee held hearings in several towns in Oneida and Vilas counties, surveyed the land on

Figure 22.5. A newly planted forest emerges amidst the charred tombstones of the old in northern Wisconsin ca. 1930s. (Courtesy of the State Historical Society of Wisconsin, neg. number WHi [X3] 49797)

horseback, and talked with local cutover farmers of their experience.[52] To the chagrin of forestry opponents, the committee gave the forestry program a clean bill, but recommended more be done to stimulate private forestry practices by revising the method of taxing standing timber. Meanwhile, a case protesting state land purchases came before the Wisconsin Supreme Court in 1913. In 1915 a decision was made in the case, and the court held that no further state funds could be spent for forestry on the grounds that such expenditures involved unconstitutional appropriations for internal improvements.[53] The anti-forestry coalition of northern Wisconsin had won a decisive victory. The decision ended Griffith's program. The northwoods visionary resigned shortly after.

In 1927, the legislature passed the Forest Crop Law, which made it possible for private landholders, whether companies or individuals, to raise timber without the risk of being taxed into insolvency or forced to cut prematurely.[54] The measure was not enough, however, since individuals did not have the resources to replant their cutover lands; furthermore, the bill did not address the problems associated with providing infrastructure for isolated settlers. But in 1929, the legislature passed two important, interrelated bills which set national precedence in rural land policy and paved the way for the region that exists today. First, the Forest Crop Law was extended to counties, which essentially allowed them to consolidate lands for county forests. Second, counties were empowered to zone for agriculture, forestry, and recreation.[55]

In 1933, Oneida County became the first county in the nation to adopt a comprehensive rural zoning ordinance. Under the ordinance, the county board had the authority to remove settlers from a forestry or recreation zone and abandon maintenance of a road to a settler in an area zoned for forestry. The "nonconforming" settlers received public assistance to relocate to more densely populated districts. The ordinance met with some local opposition, but was upheld by the attorney general, who wrote, "The cut-over area of northern Wisconsin speaks as eloquently against haphazard development as any city condition."[56] Vilas County adopted the plan a few months later, followed by 27 northern and central Wisconsin counties. Not only were "pioneers" not allowed to homestead and receive free land; they were removed from the land they owned for the sake of a collective good. Here was the antithesis of the nineteenth-century "spirit of the times." Here was the antithesis of Frederick Jackson Turner's model of succession. Here was precisely what Dean Henry's *Handbook* proclaimed would *never* happen. Farms would not follow forests. And why not? The goals of people and society, or what we might call culture, were not compatible with the realities of the physical environment, or what we call nature. Culture had to compromise, for its own sake. A resort owner from Three Lakes in Oneida County recognized this need in 1926: "Griffith at the head of the forestry department years ago told us what we would have to do with this and like areas, but we laughed at him and threw him out of office, now we are finding Griffith was right, he was far-sighted enough to see just what was going to happen to us. This land as a [forest] is going to be worth more to Wisconsin than the few struggling farms that may locate there in years to come." [57]

Over the following decades, Griffith's dream came closer to reality in tangible ways. During the 1920s, the forest reserves, acquired earlier by Griffith, became state forests. The first, Northern Highland State Forest, in Vilas County, was established in 1925. The second, the American Legion State Forest in Oneida County, was established in 1929. Federal purchases of 1.5 million acres during the 1930s established two national forests, Nicolet and Chequamegon. Meanwhile, counties throughout the state set aside a little over 2 million acres in forests of their own by the 1950s. [58] These forests developed quickly with help from federal New Deal programs during the 1930s. Millions of trees were planted by the Civilian Conservation Corps (Figure 22.6). They also built access roads, controlled fires, and even removed currant and gooseberry bushes to protect the white pine stands from blister rust. By the 1960s, over two-thirds of northern Wisconsin was forested. Forests covered more than 80 percent of Vilas and Oneida counties. [59] The white pine, the royalty of the forest, had once again become among Wisconsin's most prized resources, not for its value when cut, but for its value while standing. People today pay dearly for the smell of pine in the northwoods.

Conclusion

So the story has come full circle—somewhat. The forests today are not the ones the loggers saw a century and a half ago. True, the pines are there, as well as the deer, birds, and other forest varmints. But there is a significant addition to this community, and that is us. There is a tendency to think of the forests and the city as polar opposites: the city as a product of society and culture, and the forests as a product of nature. But the "nature" of the northwoods is in many ways a product of culture and society: from the species of trees that were transplanted there to those that were either forgotten or not "reintroduced"; from the fish that were stocked in the lakes to the predators that were eliminated from the forest; from the fires that have been cur-

Figure 22.6. Civilian Conservation Corps workers replanting trees in the American Legion State Forest in Oneida County, ca. 1930s (Courtesy of the State Historical Society of Wisconsin, neg. number WHi [X3] 31960)

tailed to save forests to the cutting practices used to "manage" them; or from the "forest zones" that keep settlers out to the "user friendly" infrastructure set in place to bring visitors in. In what category, then, does the northwoods fall—nature or culture? The question truly cannot be answered in these terms. People and the environment together have shaped this place: the lumberjack cutting the forests, the farmer pulling the stumps, the forester planting the trees, and the urbanite traveling along Highway 51. In each relationship there is an exchange between society and nature, each transforming the other, like the scene at the Minocqua depot in the late nineteenth century. The story, however, carries a message which goes beyond this historical account: the present is tomorrow's past. Our role in this ongoing communion will become a part of the larger story for future generations to tell. The capacity for people to transform the land will continually expand as new resource needs are created and new technologies are applied to harness them. In the previous century it was forests; in the next century it may be water resources, mining, or something we cannot yet envision. Herein lies the challenge: We must recognize this mutual relationship more readily than our historical counterparts did if we wish to create a future landscape that sustains Wisconsin's land and life.

Notes

1. R. C. Nesbit, *Wisconsin: A History* (Madison, 1973), 10.
2. J. T. Curtis, "The Modification of Mid-latitude Grasslands and Forest by Man," in W. L. Thomas, ed., *Man's Role in Changing the Face of the Earth* (Chicago, 1956), 721–736.
3. Curtis, "The Modification of Mid-latitude Grasslands and Forest."
4. For further explanation regarding this method of inventory and other early descriptions of the northern forests, see J. T. Curtis, *The Vegetation of Wisconsin* (Madison, 1959).
5. R. W. Finley, *Original Vegetation Cover of Wisconsin* (St. Paul, 1976).
6. Finley, *Original Vegetation Cover*, 205.
7. T. Kouba, *Wisconsin's Amazing Woods: Then and Now* (Madison, 1973), 49.
8. D. D. Scrobell, *Early Times: The Early History of the Minocqua Area as Seen through the Pages of the Minocqua Times Newspaper* (Minocqua, Wis., 1988), 4–7.
9. For a description of changes in transportation technology in the logging industry, particularly in Wisconsin, see the first six chapters of R. F. Fries, *Empire in Pine: The Story of Lumbering in Wisconsin, 1830–1900* (Madison, 1951), 3–100. William Cronon discusses the impact that the gradual shift from water to rail transportation in the lumber business ultimately had on Chicago and its hinterland in *Nature's Metropolis: Chicago and the Great West* (New York, 1991), 152–198.
10. As quoted in C. Twining, "The Lumbering Frontier," in S. Flader, ed., *The Great Lakes Forest: An Environmental and Social History* (Minneapolis, 1983), 124.
11. As quoted in Kouba, *Wisconsin's Amazing Woods*, 106.
12. Fries, *Empire in Pine*, 246.
13. M. Williams, *Americans and Their Forests: A Historical Geography* (Cambridge, England, 1989), 228.
14. Cronon, *Nature's Metropolis*, 200.
15. This was part of the introduction to W. A. Henry, *Northern Wisconsin: A Handbook for the Homeseeker* (Madison, 1896), 6. Henry was the dean of the College of Agriculture at the University of Wisconsin and was mandated by the state legislature to prepare this manual to inform potential settlers about the agricultural possibilities of northern Wisconsin. Incidentally, the wording of the mandate practically predetermined that the manual would be used to "promote" agriculture. Although Henry discussed some of the realities and limitations of the land with charts, maps, and text, the photographs that were included gave an overly rosy picture of the area. The *Handbook* stands as perhaps the most extreme contribution to settlement boosterism that the university has ever made.
16. A fine examination of Frederick Jackson Turner's life and his classic

essay "The Significance of the Frontier in American History" can be found in M. Ridge, ed., *Frederick Jackson Turner: Wisconsin's Historian of the Frontier* (Madison, 1986).

17. J. O'Neill, "The Future of Northern Wisconsin," *Proceedings of the State Historical Society of Wisconsin, 1899,* vol. 46 (Madison, 1899), 210.

18. There is a considerable volume of literature that has examined these promoters separately. The following offer useful overviews of cutover promotion for agriculture and settlement: V. Carstensen, *Farms or Forests: Evolution of a State Land Policy for Northern Wisconsin 1850–1932* (Madison, 1958); J. I. Clark, *Farming the Cutover: The Settlement of Northern Wisconsin* (Madison, 1956); A. Helgeson, *Farms in the Cutover: Agricultural Settlement in Northern Wisconsin* (Madison, 1962); and I. Vogeler, "The Northwoods Region" in his *Wisconsin: A Geography* (Boulder, 1986), 84–123.

19. Vogeler, *Wisconsin,* 106.

20. Clark, *Farming the Cutover,* 5.

21. Wisconsin Central Railroad, *A Farm in Wisconsin Will Make Money for You from the Start: Crops Never Fail* (Milwaukee, 1897), 3.

22. Wisconsin Central Railroad, *Scenes of Pioneer Life in Northern Wisconsin along the Lines of the Wisconsin Central Railroad* (Milwaukee, 1896), 1.

23. J. I. Kolehmainen and G. W. Hill, *Haven in the Woods: The Story of the Finns in Wisconsin* (Madison, 1951), 67–69.

24. M. J. Dunn III, *Easy Going: Wisconsin's Northwoods, Vilas and Oneida Counties* (Madison, 1978), 54–55. This "guidebook" is a particularly helpful source for short histories of communities and local lore, as well as descriptions of hospitality and recreation spots in Oneida and Vilas counties.

25. Helgeson, *Farms in the Cutover,* 29.

26. Henry, *Northern Wisconsin: A Handbook for the Homeseeker,* 1.

27. Henry, *Northern Wisconsin: A Handbook for the Homeseeker,* 2.

28. Helgeson, *Farms in the Cutover,* 31–32.

29. Helgeson, *Farms in the Cutover,* 40; Clark, *Farming the Cutover,* 10.

30. *The Lakeland Times: The First 100 Years: Centennial Edition* (Minocqua, Wis., 1988), 136.

31. Henry, *Northern Wisconsin: A Handbook for the Homeseeker,* 61.

32. Helgeson, *Farms in the Cutover,* 98–99.

33. Helgeson, *Farms in the Cutover,* 105.

34. J. I. Clark, *Cutover Problems: Colonization, Depression, Reforestation* (Madison, 1956), 3–5.

35. R. Finley, *Geography of Wisconsin* (Madison, 1976), 136–141.

36. Carstensen, *Farms or Forests,* 94.

37. Vogeler, "The Northwoods Region," 112.

38. Carstensen, *Farms or Forests,* 100.

39. Clark, *Cutover Problems,* 14–15.

40. E. D. Solberg, *New Laws for New Forests: Wisconsin's Forest-Fire, Tax, Zoning, and County-Forest Laws in Operation* (Madison, 1961), 28.

41. Solberg, *New Laws for New Forests,* 36.

42. Kouba, *Wisconsin's Amazing Woods,* 122.

43. Clark, *Farming the Cutover*, 15.

44. *The Lakeland Times: The First 100 Years*, 139.

45. Solberg, *New Laws for New Forests*, 36.

46. Carstensen, *Farms or Forests*, 65.

47. As quoted in Clark, *Farming the Cutover*, 17.

48. As quoted in Helgeson, *Farms in the Cutover*, 103.

49. Helgeson, *Farms in the Cutover*, 93.

50. Carstensen, *Farms or Forests*, 63.

51. Clark, *Farming the Cutover*, 18.

52. Clark, *Farming the Cutover*, 18.

53. Helgeson, *Farms in the Cutover*, 94.

54. Carstensen, *Farms or Forests*, 107. The Forest Crop Law allowed the owner of standing trees to pay an annual tax of 10 cents per acre until the trees were cut. Once they were cut, there was a tax of 10 percent on the crop's value.

55. For a description of Wisconsin's pioneer role in regard to rural zoning, see F. G. Wilson, "Zoning for Forestry and Recreation: Wisconsin's Pioneer Role," *Wisconsin Magazine of History* (Winter 1957–1958): 102–106. The most comprehensive examination of these policies is Solberg, *New Laws for New Forests*.

56. As quoted in *The Lakeland Times: The First 100 Years*, 140.

57. Carstensen, *Farms or Forests*, 195.

58. Kouba, *Wisconsin's Amazing Woods*, 175–209.

59. Statistics are from Kouba, *Wisconsin's Amazing Woods*, 239.

23

Northern Exposures
Bird's-Eye Views of Nature and Place
on Wisconsin's Lake Superior Coast
During the Summer of 1886

Eric Olmanson

Of or belonging to a bird's eye: as in *bird's-eye view:* a view of a
landscape from above, such as presented to the eye of a bird; a
perspective representation of such a view; also *fig.* a resume of
a subject.
—*Oxford English Dictionary*

In 1886, the ice broke up and moved out of Chequamegon Bay the first
week of May. Within days, boats began to arrive at and depart from the
competing harbors of Bayfield, Washburn, and Ashland. Just whose
harbor had the longest shipping season was subject to endless debate
in the local newspapers. The three Chequamegon Bay communities
did agree, however, that Chequamegon Bay contained the best harbors and that the Superior and Duluth harbors, on the western end of
the lake, always opened last and closed first. Despite this, the fiercest
rivalry was probably between Superior and its Minnesota neighbor,
Duluth. Even after railroads connected the region to the larger world,
having the best harbor on the lake was a matter of local pride and was
considered vital to the growth of a metropolis. The region remained
largely dependent on the great lake and subject to its whims. Open

water or ice, calm or rough, Lake Superior was a constant reminder that human activity was not entirely independent of nature's rhythms. When the big lake froze, parts of the local economy did too.

Spring thaw allowed iron ore from the just-opened Penokee-Gogebic mines near Hurley, as well as from mines in Michigan, to be shipped through Ashland Harbor to the steel mills of Pittsburgh, Youngstown, and Gary.[1] It also meant that coal and other commodities could come by water to be distributed inland. In Bayfield, the Boutin and Booth fisheries began their open-water season. Brownstone from mines on the Apostle Islands and the nearby mainland was shipped as far as Chicago and New York to build fashionable brownstone houses. And up and down the coast, logs arrived in unbelievable quantities from the seemingly interminable forests to supply the mills, which shipped their finished lumber to build railroads, fences, and buildings, not only on the Great Plains and in American cities, but also as far away as Latin America and Europe.[2] In 1886, Ashland was in the midst of an iron boom that was still a year from peaking. The entire region was buzzing with activity. Warmer weather also brought tourists by the thousands. Bayfield and Ashland newspapers complained that there were not enough accommodations for all the visitors.[3]

Bird's-Eye Views and Their Artists

On the 29th of May 1886, a regular reader of the *Ashland Press* would have read about a visitor with a special purpose. Henry Wellge, a Milwaukee lithographer and bird's-eye artist, was in town making sketches for a bird's-eye view of Ashland. The next week Mr. Wellge was busy canvassing for subscribers to the work. The *Press* declared the preliminary sketch "a splendid and *correct* view worthy of the artist. . . . It gives a fine view of the harbor and will make a beautiful picture to frame. It not only gives an outline but a detail view of all the public and private buildings, easily recognizable by their owners."[4] Wellge was one of the most prolific of the nineteenth-century viewmakers and, in 1886, was at the height of his career. A week after finishing at Ashland, he began sketching Bayfield. He sketched Washburn and Hurley the same summer. As he did with Superior, completed in 1883, Wellge created snapshot images of the major population centers of Wisconsin's Lake Superior coast during a critical period in their development. But how "correct" are these views? What sort of record do the bird's-eye views leave for the historical geographer to interpret over 100 years later? What can they tell us about the people, the place, and the time? Are they accurate portraits?

Bird's-eye views are generally reliable records of street patterns, buildings, and neighborhoods. They give an idea of the size of a town and how it was situated in the landscape. The main customers for views were the town residents. Therefore, the lithographs had to be realistic—but not too realistic. Historian John Reps recommends looking at the views as "flattering, carefully posed, and retouched portraits rather than as completely candid records of reality."[5] Just like other maps, bird's-eye views present only a certain vision of what a town is like; they *lie* in the sense that what they leave out is as important as what they leave in. They are both more and less than they appear at first glance. The viewmaker tried to reflect the local residents' idealized notion of their town: the town as viewed from its best vantage point, emphasizing its structures and its potential and omitting its social conflicts, weedy lawns, rundown buildings, garbage disposal problems, and stray dogs.

Because it is generally flat and had been systematically surveyed, the Midwest was an ideal place to practice the viewmaker's craft. Accurate street maps could be obtained and used by the viewmaker as base maps. The base map could then be easily adapted to two or three favorite perspectives, which the artist used over and over, boilerplate fashion, to save time. While Bayfield and Washburn were situated in terrains that defied regular grids, Ashland sat on relatively flat ground. The 1886 view of Ashland is a good example of the two-point linear perspective typical of most of Henry Wellge's bird's-eye views (Figure 23.1).[6] If lines are drawn down the middle of the streets and extended outward, they converge at two points beyond two borders of the print. The first vanishing point is well above the horizon to the right of the view. The second is to the left of the view near the horizon. This technique maximized the potential market for the final prints by pulling up the background so that even the most remote buildings would be discernible.

Lithographs, or chromos, have been called the democratic art.[7] This is especially true of bird's-eye views. The prints were affordable by average people. The *Bayfield County Press* (formerly the *Bayfield Press*) declared, after all, that "a copy [of the Bayfield bird's-eye] ought to be found in every home in town."[8] Bird's-eyes also depicted communities of all sizes. And finally, because of artistic conventions and shortcuts, bird's-eye views tended to make all communities look alike. Harbors are always busy; neat and well-planned streets seem to push the urbanized area outward; boats and trains and horse-drawn carriages are always in motion; smokestacks are churning smoke toward

Figure 23.1. Bird's Eye view of Ashland, 1886, by Henry Wellge (Courtesy of the State Historical Society of Wisconsin, neg. number WHi[X3]3161)

the sky. All towns seem to have been caught in the process of rapid expansion. Sometimes the impression was accurate.

At the same time that Wellge's bird's-eye view of Ashland was being delivered to subscribers, a Sanborn insurance map of Ashland was also being produced. Systematic comparison of the September 1886 Sanborn insurance map of Ashland with Wellge's 1886 bird's-eye view attests to the essential accuracy of the latter.[9] In almost every respect, the insurance map validates the bird's-eye, but there are revealing discrepancies. A building shown burned on the insurance map is intact in the bird's-eye view. Perhaps the viewmaker chose to show the building that was, or the building that would be rebuilt. Or maybe the building burned after Wellge left and before the insurance cartographer arrived. Most revealing, however, are the buildings and docks that had not yet been built. The September 25, 1886, issue of the *Ashland Press* reported that Wellge's

lithograph is now being delivered to subscribers by J. J. Stoner, of Madison, and is altogether a fair and accurate picture of what Ashland is. The public buildings that are *and those that will be* . . . are recognizable. . . . The new dock *to be built* by the Lake Shore Company is also upon the map, and *gives a good idea of how things will look in that locality in another summer.*[10]

Unlike a photograph, the bird's-eye view recorded not just what *was* there but what *would* be there (or at least what some people hoped would be there).

City views and their makers may be classified into two types: outsider and insider.[11] Outsider artists used traditional landscape techniques. They required no special knowledge of their subjects beyond what they could see. They simply chose the best vantage point and drew what they saw. Insider artists, on the other hand, required an intimate knowledge of their subjects because their views were drawn from an imaginary perspective somewhere above the horizon. Bird's-eye views were essentially maps. Potential customers were more likely to purchase a copy if they could locate familiar buildings, especially their own homes. Paradoxically, insider artists, like Wellge, were usually outsiders to the communities they drew. Therefore, insider viewmakers had to spend several days walking the streets, getting to know the town. They made sketches of individual buildings and took detailed notes about terrain as well as architecture. The best viewmakers attempted to produce very accurate and useful documents.

Wisconsin, in the latter part of the nineteenth century, seems to have been a leader in the production of bird's-eye views and, therefore, an ideal place to explore what these views can tell us about how

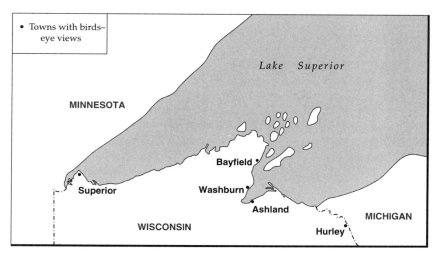

Figure 23.2. Map of Wisconsin's Lake Superior coast

contemporaries perceived their surroundings. According to John Reps, "Nearly all of the artists and publishers who dominated American viewmaking from the end of the Civil War until the turn of the century got their start in southern Wisconsin or found themselves doing business there or in nearby Chicago."[12] Milwaukee, in particular, had several advantages as a printing center, including a large population of skilled German immigrants; proximity to cheap and high-quality Fox River valley paper; comparatively low rents; a central location between the urbanized East and the rapidly growing West; and finally, the magnetic effect of already established immigrants, artists, and publishers.[13] Henry Wellge was a typical viewmaker. He was a German immigrant to Wisconsin and, though he traveled widely to produce his views, he was based in Milwaukee.

This chapter uses bird's-eye views to explore what Wisconsin's Lake Superior coast communities were like in the mid-1880s (Figure 23.2). Five general questions guide our analysis of these views: What sort of place is portrayed? How is the natural environment portrayed? What is the apparent relationship between the place and the natural environment? Does the view tend to reinforce or deny the regional distinctiveness of the place? And finally, what was it like to live in this place during the short summer of 1886? What I present here is a collage—a view of a region as expressed by the self-conscious views of five of its most prominent communities: Superior, Bayfield, Washburn, Ashland, and Hurley.

Figure 23.3. Bird's Eye view of Superior, 1883, by Henry Wellge (Courtesy of the State Historical Society of Wisconsin, neg. number WHi[X3]33272)

The View of Superior

At the western extreme of Wisconsin's Lake Superior coast is the city of Superior. Between 1890 and 1900 Superior grew so rapidly that it dwarfed its closest northern Wisconsin rivals. But in 1886 Ashland and Hurley were in the midst of the Gogebic iron boom, and Superior did not yet seem to pose much competition. When the view of Superior was completed in 1883, the community was a mere village. Like most nineteenth-century communities, Superior's dreams were much larger than its reality. Superior's main rival was Duluth, Minnesota. When the Lake Superior and Mississippi Railroad—chartered to run from St. Paul to Superior—bypassed Superior in favor of Duluth around 1870, the major advantage went to Duluth. Superior did not have a rail connection until 1881, when the Northern Pacific passed through on its way from Duluth to Ashland.

In Wellge's bird's-eye view of Superior (Figure 23.3), the street grid dominates the picture and leads the viewer's eye quickly out of town and across the lake toward the horizon, as if trying to draw attention away from the sparsely settled streets. There is no land in view except that occupied by the town of Superior itself. Superior's hinterland is behind the viewer, left to the imagination. Of the views we will consider, this is the only one that looks out toward the lake. The perspective really gives no idea of how the community sits in relation to the land, but it does give a sense of activity. Not counting canoes and small sailboats, about 30 steamboats and larger sailing rigs are pictured out in the harbor or heading toward its entrance. Even though the harbor is far from full, the bird's-eye greatly exaggerates the amount of commerce: the total number of recorded ship arrivals to Superior for the entire 1883 season was only 21. In addition to the boats, a total of six trains appear in the view. One of the trains occupies the center of the view. Trailing smoke gives a sense of motion to both boats and trains.

By emphasizing the street grid and by filling the harbor with imaginary boats, Wellge was able to portray an orderly, prosperous, and growing community. This is the very image that town boosters would have portrayed. The town directory is impressive for a village of fewer than 2,500 people. It shows that all the necessities and many luxuries were available to residents and visitors. Names like the Kuykendall House, the European Hotel, and the Scandinavian House suggest perhaps an element of class or an influx of immigrants. Two lumber mills, a shingle mill, and a brickyard made up the industrial base. A physician, a druggist, grocery stores, meat markets, and hardware stores provided necessities, while saloons, a jeweler, a photographer, and a bookstore provided for other needs. Churches and schools indicate a growing community with a solid institutional foundation.

The View of Bayfield

The railroad, which passed through Superior on its way to Ashland, bypassed Bayfield. There was an overland road between the two communities, but during the summer most people traveled by boat. Although in 1886 Bayfield's population was only around 1,400, it is the only community which looks complete in its bird's-eye view (Figure 23.4). It appears to be naturally bounded by water, wooded hills, and sky. It has no gaping open spaces, begging to be filled up. Unlike the view of Superior, individual vessels and buildings in Bayfield loom large. In the center foreground a lone figure guides a small sailboat to shore. Bayfield seems balanced, tranquil, and orderly. Smokestack industry, in the form of Pike's sawmill, is neatly segregated—on the other side of the tracks—from the rest of Bayfield. Although the smoke surely symbolizes progress and prosperity, the lack of numerous smokestacks was probably not considered a shortcoming because of Bayfield's growing reputation as a haven for hay fever sufferers. In August of 1886, the *Bayfield County Press* published a report by Madison surgeon Joseph Hobbins, who discussed several remedies for hay fever. He considered the most promising treatment to be the "climatic influence as it is found in Bayfield or the shore of Lake Superior and its vicinity in Wisconsin."[14] Hobbins thought the most significant factors were ozone and pine aroma.

In addition to pure air, Bayfield was known for water. An elaborate

Figure 23.4. Bird's Eye view of Bayfield, 1886, by Henry Wellge (From the Collections of the Library of Congress)

wooden piping system brought running water to the village at a very early date. As early as 1858 Bayfield was advertised as the "city of fountains."[15] On Wellge's bird's-eye view it is possible to identify several of these fountains. Although some boosters still hoped that Bayfield would become the dominant metropolis on the Great Lakes, most visitors came for its quiet charms. By 1886 Bayfield was a well-established stop for tourists. Even Mark Twain and his wife made a brief stop in Bayfield during the last week of June on the propeller *India*.[16] In the bird's-eye view, touring boats rather than cargo boats dominate the harbor.

The iron ore boom which was beginning to stir Ashland and Hurley had little effect on Bayfield. The *Bayfield Press* put a positive spin on things: "Bayfield is getting there slow but sure having no boom but enjoying a steady, slow growth that means permanency every time."[17] Indeed, Bayfield had not changed much since Portage editor Andrew Jackson Turner had spent the better part of his vacation there in 1877. In his newspaper, Turner wrote: "There is no more charming little vil-

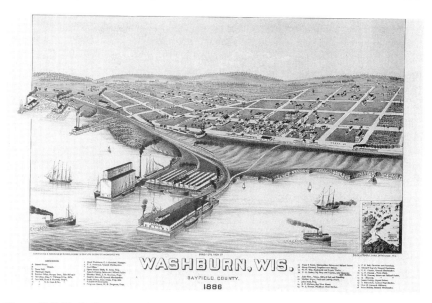

Figure 23.5. Bird's Eye view of Washburn, 1886, by Henry Wellge (Courtesy of the State Historical Society of Wisconsin, neg. number WHi[X3]34021)

lage in America. . . . Its beautiful yards, fountains and excellent hotel, all conspire to enchant our party. . . ."[18] While booming Ashland and Hurley were becoming famous for their slugging matches, saloons, and prostitution, Bayfield was known for its beauty and relative tranquility.

The View of Washburn

While northern Wisconsin towns like Bayfield and Ashland could claim a history going back beyond their establishment, in the 1850s, to the days of Radisson and Groseilliers and the Jesuits, the following was said of Washburn: "There are no romances of Indian lore in the history of its foundations — no poetical inspirations can be derived from its early history. When Washburn was platted, it was strictly a business consideration, and it has been a matter of business ever since."[19] Indeed, Washburn was platted as a terminal for the Chicago, St. Paul, Minneapolis, and Omaha Railroad, which arrived in 1883. The railroad docks dominate Wellge's bird's-eye view (Figure 23.5). No fewer than nine trains are depicted — coming, going, loading, and unloading.

In general shape, perspective, and proportions of sky, land, and water, the Washburn view is almost identical to the Bayfield view, but in every other way it could not be more different. While Bayfield has a "finished" look, Washburn has barely begun. The directory lists three sawmills, which are pictured to the left of the railroad docks, two newspapers, and a schoolhouse. A church is listed, with a large gap before the word *church*, as if the viewmaker was unsure of the denomination and planned to fill it in later. In 1886, Washburn had a population of around 900. By 1892, Washburn would grow enough to take the county seat from Bayfield (which had taken the seat from La Pointe back in 1858). Wisconsin's Lake Superior coast was a very dynamic place. Everything was in flux; anything seemed possible.

The View of Ashland

Wellge's view of Ashland tells a popular story (Figure 23.1). It is the story of the frontier—the conquest of culture over wilderness. Frontier progression can be seen moving from the shoreline, which is most developed, to the forest, which is least developed. Industry is transforming the land. The first step involved superimposing a grid on the landscape. Next, the cells of bounded land were cleared of trees. The stumps (for example, on the corner of Delaware and Fifth streets) were the last remains of the forest, which was cut and hauled to the sawmills on the shore, where it was transformed into lumber. After the stumps were removed, buildings were constructed. When a block went all the way from dense forest to dense buildings, the progression was complete. In Wellge's bird's-eye view, all the stages of the progression are visible.

In the case of Ashland, Wellge may not have exaggerated the busyness of the harbor. Excluding small excursion steamers and towing tugs, almost 1,000 ship arrivals to Ashland docks were recorded during the 1886 shipping season. Almost 50 million board feet of lumber were manufactured by the mills lining Ashland's shore. Forty-six cargoes of lumber were shipped to Chicago and other markets. More than 50 loads of coal, railroad iron, salt, oil, and other cargo were received. The dock on the left margin of the view shipped 721,981 gross tons of ore, and this was in the days of wooden ore boats, which could take a whole day to load while the trimmers distributed the cargo.[20]

The View of Hurley

Hurley was connected to Chequamegon Bay through Ashland by the recently opened Milwaukee, Lake Shore, and Western Railway. Most

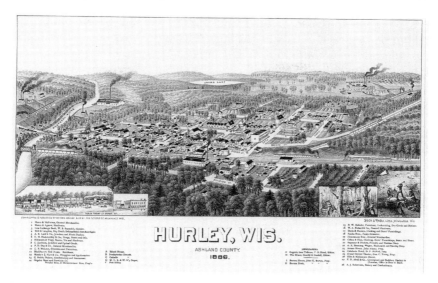

Figure 23.6. Bird's Eye view of Hurley, 1886, by Henry Wellge (Courtesy of the State Historical Society of Wisconsin, neg. number WHi[X3]34033)

of the Wisconsin iron ore that was shipped out of Ashland Harbor came from the mines near Hurley. Wellge's 1886 view summarizes Hurley (Figure 23.6). The story begins with recent history: Hurley in 1885. The left of two insets in the lower right-hand corner of the view shows men felling large trees in a dense forest. The other inset, to the right, shows miners working hard at the Colby Mine. Hurley was a mining boomtown, just recently hewn out of a dense forest. But already, according to the bird's-eye, there were a schoolhouse, two churches, two newspapers, as well as a couple dozen businesses providing everything from bakery goods to eyeglasses. Brown sandstone from the Bayfield Brownstone Company was brought in to build the more substantial structures and to create another link to Chequamegon Bay.[21]

One sawmill and three working mines are visible in the view. The Germania Mine is depicted in great detail in the right background. Most of the Gogebic mines were underground. Some were quite deep. The Montreal Mine eventually reached a depth of 4,335 feet.[22] In 1886 the miners would descend into the earth on long ladders. It was hard and dangerous work, arguably harder than being a lumberjack. Like the lumberjacks, the miners lived in isolated camps set up near their work. Both groups spent long periods cooped up in an exclusively male environment. When payday came, hoards of these men would head to town to drink, gamble, and pursue women. Because most of the miners worked across the Montreal River in Michigan, where law enforcement was more vigorous, Hurley became the "sin center" for the entire Gogebic Range.

Hurley's connection to mining is reflected in street names such as Iron, Copper, Granite, and Gold. The inset on the lower left side of Wellge's view shows the north front of Silver Street. This street, so modest and peaceful looking, would soon be notorious throughout the entire region. But that story was not part of the bird's-eye view. The viewmaker's pictures told a progressive, even heroic, story of future cities emerging from forests, of industry and hard work rearranging nature to create ordered, humanized places.

Bird's-Eyes, Boosters, and Northwoods Pastoralism

The visions of boosters and viewmakers were not necessarily identical. While boosters sold a vision of what their cities could become, viewmakers sold lithographs that depicted cities in the most positive light. The difference is subtle, but sometimes telling.

Newspaper editors were typically among the most prominent boosters of any given town. In part this was due to simple economics. The fortunes of a local newspaper were closely tied to local growth. More people meant more subscribers, and more businesses meant more paid advertisements. Both the *Bayfield Press* and the *Ashland Press* made use of bird's-eye views to promote their towns and illustrate growth. A striking example of the latter is the December 25, 1886, issue of the *Ashland Press*, which uses a pair of views to show Ashland's growth between 1876 and 1886.[23] The 1876 view depicts Ashland about a year before the first railroad to reach the city was completed. In the earlier view, buildings are barely discernible beneath the silhouette of trees.

One difference between the booster and viewmaker visions is found in the depiction of the natural environment. More than any other feature, trees define the region of which Wisconsin's Lake Superior coast is a part. Commercial bird's-eyes, like Wellge's views, tended to deemphasize the natural vegetation regime of northern Wisconsin. In his bird's-eye view of Superior, Wellge avoided depicting trees, almost entirely, by orienting the perspective toward the harbor. The forest was behind the viewer. In Wellge's Bayfield view, the coniferous trees do suggest the northwoods, but their regular, lined, and gentle parklike appearance is decidedly pastoral, not a landscape ideal commonly associated with the northwoods. The background forests of Washburn

ASHLAND, THE METROPOLIS OF THE NEW WISCONSIN.

Figure 23.7. Ashland, The Metropolis of the New Wisconsin, 1885, by Marr and Richards (Courtesy of the State Historical Society of Wisconsin, neg. number WHi[X3]2338)

and Ashland are depicted in a similarly stylized fashion. Even Hurley, just recently hacked out of the forest, seems to sit in a park of neatly planted trees. Commercial bird's-eye views often depicted an idealized natural setting.

Boosters and planners, on the other hand, thought of the northern forest as a commodity and an impediment, something that must be cleared and processed to give access to another commodity, the soil, which they thought would produce as abundantly as it did in more southerly regions of Wisconsin. In 1885 the *Ashland Press* commissioned a view of Ashland, The Metropolis of the New Wisconsin, which was done by Marr and Richards (Figure 23.7). Because this view was probably not sold separately, but was instead commissioned by a newspaper to illustrate their annual edition, its purpose was clearly to boost the image of Ashland. The artists did not need to show every building clearly in order to entice subscribers.[24] They had the leeway to take a different perspective. Ashland's shore faces the northwest. Streets run perpendicular to the shoreline (northwest-southeast) and parallel to the shoreline (northeast-southwest). Whereas Wellge chose to view Ashland from the west looking east, the Marr and Richards view is from the north looking south. The two views are almost mirror images of each other in terms of the general shape of land and water. Both views use a two-point linear perspective. But Wellge's vanishing points extend far beyond the borders of the print, and the one on the right is far above the horizon. Marr and Richards' vanishing points are

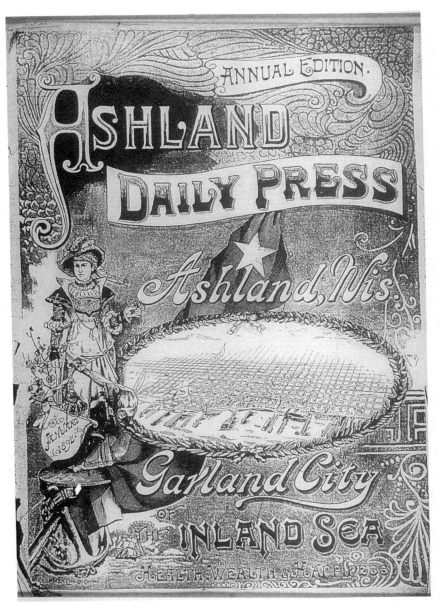

Figure 23.8. Cover of the 1892 annual edition of the *Ashland Daily Press*

just beyond the borders of the print and near the horizon. Marr and Richards' view is less maplike and more realistic.

The choice of viewpoint profoundly affected these views, and resulted in almost antipodal emphases. While Wellge's viewpoint emphasizes the heart of the built city, the Marr and Richards view emphasizes empty lots and the new Milwaukee, Lake Shore and Western ore dock, about a mile northeast of the center of the city plat. Described as the largest in the world, the ore dock was almost half a mile long and extends into the center foreground of the view. The dock was the main focus and explains the perspective, but the view also focuses on empty lots, in highlighted grid pattern. Rather than focusing on the already built landscape, the Marr and Richards view emphasizes undeveloped land, as if trying to lure investors.

Thus, the commercial bird's-eye and the locally commissioned boosters' view seem to depict two different places. In the former view the city seems to be naturally expanding inland, the gentle woods presenting little obstacle. In the latter view, just a year earlier, the forest appears to be almost impenetrable, and the city seems to be pinched between forest and lake and forced to expand along the shore rather than fight its way inland. Indeed it appears easier to expand into the bay than into the forests, as evidenced by the numerous docks. Even though Wellge's use of perspective pulls up the background so that distant buildings can be shown clearly, he still manages to subdue the forest as if the earth curves sharply back just beyond the city. On the original Wellge lithograph the Penokee-Gogebic range is faintly visible in the background. In contrast, the Marr and Richards view pulls up the range so that it looms mountainlike in the background. The forest is a thick wilderness, penetrated only by rivers and the thin trace of the Wisconsin Central Railroad.

Given the boosters' grandiose vision and their tendency to exaggerate natural advantages while denying natural disadvantages, such a realistic depiction of the natural environment is unexpected. But while I look at the Marr and Richards view and see wilderness engulfing a thin line of habitation calling itself the metropolis of the new Wisconsin, the boosters surely saw *resources*. There was iron in those hills, and immense forests meant an endless supply of timber for Ashland's mills. And although the boosters believed strongly that when the forests were cleared farms would eventually take their place, they also seem to have had a strong sense of the distinctiveness of their region. Their place was different from other places, and they knew it. Unlike Wellge's view, there is nothing pastoral about the Marr and Richards view of Ashland. But while the boosters might have had a clear-eyed vision of their region, they also had an idyllic notion of what their region could become, which outdid even Henry Wellge's pastoral representations. The cover of the 1892 annual edition of the *Ashland Daily Press* (no longer just a weekly publication) is suggestive of the ultimate relationship between the city and its natural environment (Figure 23.8). This unusual bird's-eye view could be interpreted as the final triumph of culture over wilderness. Nature is symbolically erased by civilization, personified as a woman.

A Final View

So what was it like to live on Wisconsin's Lake Superior coast? There is no easy answer. But in 1886 the place was not as isolated as its location might suggest. An advertisement in the *Ashland Press* announced, "Every train brings the latest styles and novelties in boots and shoes for the Boston Boot and Shoe House." In April and May, grocers advertised lettuce, radishes, onions, pie plant (rhubarb), bananas, oranges, pineapples, lemons, apples, strawberries, asparagus, spinach, and fresh oysters. Lake Superior whitefish was a staple. The first trial of the Ashland Lighting Company electric plant was a success. Meanwhile, Bayfield enjoyed "three trains each way daily," and although in July the *Bayfield Press* published a list of wants longer than its town directory, the village could already boast of a roller-skating rink and fresh roasted peanuts at the newsstand on Dalrymple's dock. Even when Chequamegon Bay froze shut on November 30, 1886, people could easily board a train and head elsewhere if they wanted a change of scene.

In the end I must conclude that Henry Wellge's bird's-eye views of Wisconsin's Lake Superior coast were not correct—but neither were they incorrect. It is hard to summarize so complex a place. When we think about the hard work and human deprivation associated with a regional economy based primarily on the extraction of wealth from northern Wisconsin's mines, forests, and stingy soils, Wellge's views seem unrealistic, even cartoonlike. This was a resource frontier, depicted by some as no more than a "hell-hole."[25] But I would argue that Wellge's viewpoint was an important one, because it focused on what was good in each town. Each resident could find his or her own home in the prints, and feel part of a progressive vision of what the communities could become. People tried to overcome the natural disadvantages of a place by the power of their imagination. Imagining was part of creating. In this way thought and beliefs are connected to the material world by action. The way people behave in a given place is limited as much by mental perceptions as it is by the physical environment. Geographer Yi-Fu Tuan has eloquently argued that words are vital to the creation of place.[26] For Wisconsin's Lake Superior coast in 1886, a picture was worth a thousand words.

Notes

Serious scholarship would be impossible without professional librarians and archivists. I must acknowledge the kind help I received from Harry Miller and Gerry Strey of the State Historical Society of Wisconsin Archives Reading Room.

1. According to Ingolf Vogeler, *Wisconsin: A Geography* (Boulder, 1986), 97, the Montreal and Cary mines near Hurley produced 94 percent of all the ore from Wisconsin's Gogebic range. Michigan mines produced about four times as much as Wisconsin did.

2. Vogeler, *Wisconsin*, 92.

3. See, for example, the *Bayfield County Press*, August 21, 1886.

4. *Ashland Press*, June 5, 1886; emphasis added.

5. J. W. Reps, *Views and Viewmakers of Urban America: Lithographs of Towns and Cities in the United States and Canada, Notes on the Artists and Publishers, and a Union Catalog of Their Work, 1825–1925* (Columbia, 1984), 70. For a similar statement, see G. Danzer, "Bird's-eye Views of Towns and Cities," in D. Buisseret, ed., *From Sea Charts to Satellite Images: Interpreting North American History through Maps* (Chicago, 1990).

6. Reps, *Views and Viewmakers*, 19.

7. P. C. Marzio, *The Democratic Art: Chromolithography 1840–1900: Pictures for a Nineteenth Century America* (Boston, 1979).

8. *Bayfield County Press*, October 2, 1886.

9. Sanborn insurance maps were also produced for Bayfield and Washburn during the summer of 1886, and these maps also attest to the essential accuracy of Wellge's bird's-eye views for those communities.

10. Emphasis added.

11. Reps, *Views and Viewmakers*, 17.

12. Reps, *Views and Viewmakers*, 209.

13. T. Beckman, *Milwaukee Illustrated: Panoramic and Bird's-eye Views of a Midwestern Metropolis, 1844–1908* (Milwaukee, 1978), 2–3. For another argument for the centrality of Wisconsin in the bird's-eye view business, see E. S. Maule, *Bird's Eye Views of Wisconsin Communities* (Madison, 1977).

14. *Bayfield County Press*, August 28, 1886.

15. *Bayfield, Lake Superior: Early History, Situation, Harbor &C.* (Philadelphia, 1858).

16. *Bayfield Press*, July 3, 1886.

17. *Bayfield Press*, July 10, 1886.

18. *State Register* (Portage), July 28, 1877.

19. *Ashland Daily Press*, Annual Edition, January 1892, p. 33.

20. *Ashland Press,* December 25, 1886. The information about the ore boats is from J. M. Dodd, "Ashland Then and Now," *Wisconsin Magazine of History* (December 1944): 189–196.

21. "Hurley, the 'new town on the range,' is long on iron but short on good building stone. . . ." *Bayfield Press,* May 22, 1886.

22. Vogeler, *Wisconsin,* 99.

23. Note that neither of these views is from an aerial perspective, so they are not really from a bird's-eye. Both are what Reps classified as outsider views, because the artist depicts only the nearest buildings.

24. The complete history of the Marr and Richards' view is so far a mystery. Both John Reps and Elizabeth Maule restricted their lists to bird's-eye views which were produced to be sold separately. Both Reps and Maule include this view in their lists, but there seems to be no evidence that it was produced separately. It is certain, though, that the view does appear in the June 6, 1885, edition of the *Ashland Press.* The copy cited by both Reps and Maule is held by the State Historical Society of Wisconsin. This copy lacks a town directory and other features typically found on commercial bird's-eyes. Furthermore, close examination of the print suggests that it was not a lithograph, but rather a wood block print, which was a technique used more commonly for views in newspapers and magazines than for views which were sold separately.

25. *Cincinnati Enquirer,* November 6, 1887, and December 17, 1887. See also *Encyclopedia of Wisconsin: A Volume of Encyclopedia of the United States* (New York, 1990), 234.

27. Y.-F. Tuan, "Language and the Making of Place: A Narrative Descriptive Approach," *Annals of the Association of American Geographers* 81 (1991): 684–696.

24

The Geography of Ojibwa Treaty Rights in Northern Wisconsin

Steven E. Silvern

In 1983 a federal appeals court ruled that the Wisconsin Ojibwa retained a treaty right to hunt, fish, and gather outside the boundaries of their northern Wisconsin reservations on the lands they had ceded to the United States in the nineteenth century. Following this ruling, northern Wisconsin lakes and boat landings became sites of intense, sometimes violent, conflict between Indian and non-Indian over the legitimacy and practice of these treaty rights. Courtrooms in Madison also became sites of intense debate between the state of Wisconsin and the Ojibwa over the nature of the treaty rights and the state's ability to regulate the exercise of these rights.

This chapter explores the contested geography of Ojibwa treaty rights in northern Wisconsin.[1] An important aspect of this cultural and political conflict was the determination of the geographic extent or definition of the treaty rights: Where could the treaty rights be exercised, and did they confer to the Ojibwa any territorial sovereignty over the ceded territory? Both in the courtroom and at the boat landings, non-Indians attempted to minimize, if not terminate, the exercise of the treaty rights by delimiting the geography of the rights. The state developed geographic-legal theories that restricted the space in which the treaty rights could be exercised. Protesters at northern Wisconsin boat landings also attempted to define the geography of treaty rights.

They hoped that their disruptive presence at the landings from which Ojibwa fishers launched their boats would deter Indian spearfishing of spawning game fish outside their reservations.

For many Ojibwa, treaty rights represent not only access to the fish and game of their former territory but also a means to share authority with the state over how the resources of that territory are protected and managed. The treaties, according to the Ojibwa, are a means to shape the development of northern Wisconsin. The Ojibwa argue that they should have a voice in determining whether or not any kind of potentially environmentally damaging activity, such as mining, can occur in the ceded territory. Their goal is more than access to lakes for fishing: they want a hand in shaping the geography of northern Wisconsin.

This chapter is divided into four parts. First I explore the history of the treaties and the state's efforts to restrict Ojibwa access to their former territory. Second, I explore the state's legal argument and its attempt to limit the geographic extent of the treaty rights. Third, I discuss the boat-landing protests and the associated drive to keep the Ojibwa on their reservations. Fourth, I look at the question of comanagement of land and resources and the tribes' attempt to use their treaty rights to shape economic development and natural resource use in northern Wisconsin.

Historical Geographies

The Wisconsin Ojibwa, according to their oral tradition, originated on the "shores of the Great Salt Water in the east," or the North Atlantic. In the late sixteenth and early seventeenth centuries the Ojibwa migrated westward up the St. Lawrence River to the Great Lakes, where they established a large fishing village at Sault Ste. Marie. In the late seventeenth century they migrated farther west to Madeline Island in Chequamegon Bay, where, by the end of the seventeenth century, they had built a large village. From this Madeline Island village, further migrations took place into Minnesota and south of Lake Superior into what is now the state of Wisconsin. They were attracted southward and westward by the abundance of furbearing animals, deer, wild rice, and other natural resources. Moving south from Madeline Island into present-day Wisconsin, the Ojibwa forcefully displaced the Dakota and Fox by the 1740s and established permanent villages at Lac Courte Oreilles, Lac du Flambeau, Rice Lake, Lake Chetek, and Long Lake.

By the mid-eighteenth century the Wisconsin Ojibwa, through their participation in the fur trade, were integrated into the European economy. The Ojibwa, at this time, were a seminomadic people whose hunting-and-gathering economy required them to travel seasonally as resources became available. During the winter, individual families moved south from villages along the shores of Lake Superior and inland lakes to occupy specific territories in central Wisconsin, where they hunted large game such as deer, moose, and elk. They also trapped beaver, martin, mink, rabbit, and muskrat, and they spear-fished through the ice. During the spring, families moved back north into the sugar bush to tap the maple trees and make maple sugar. They continued to hunt and ice fish. With the spring melting of lake and river ice, they speared fish at night with the aid of torches. In the summer, individual families came together and formed large villages adjacent to Lake Superior or smaller inland lakes. They would plant small gardens, hunt, and fish the lakes using gill nets. In the fall, they moved to the rice camps, where they would harvest wild rice, one of their most important food sources.

The Ojibwa quickly integrated European manufactured goods, obtained through the fur trade, into this complex seasonal round of subsistence activities. These goods did not radically alter Ojibwa culture but, rather, facilitated subsistence activities. In spite of their increasing integration into the European economy, the Wisconsin Ojibwa were able to maintain a degree of sociopolitical autonomy. In general, French and British government policies in the Great Lakes region were not aimed at physically displacing the Ojibwa from their territories. Both the French and British sought to exploit the Ojibwa and other native peoples as harvesters and sources of furbearing animals. These policies minimized the uprooting of Indian tribes. By contrast, the policies of the United States were aimed at expropriating Ojibwa lands and natural resources for permanent white settlement.[2]

In 1837, the United States negotiated a land cession treaty with the Wisconsin Ojibwa to allow for the exploitation of the pine forests of the Great Lakes region. In Wisconsin, this land cession extended from the St. Croix River east to the present cities of Crandon, Antigo, and Stevens Point, and from Stevens Point north to the city of Rhinelander, and from the cities of Osceola and Eau Claire north to Lake St. Croix (Figure 24.1). In 1842, another treaty was negotiated, and the Ojibwa ceded the lands north of the 1837 land cession. In Wisconsin, this land cession involved an area bordering the southern shore of Lake Superior. The United States sought these lands for their rich copper deposits. In both of these treaties, the Ojibwa reserved the right to harvest the natural resources in the ceded territory. They believed that access to the resources would enable them to maintain their traditional culture.

The Ojibwa remained in the ceded territory following the 1837 and

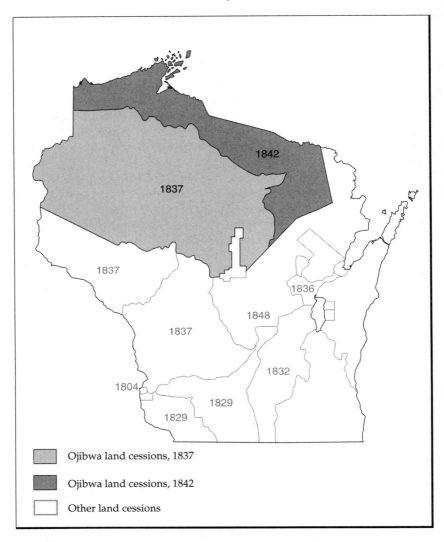

Figure 24.1. Land cessions by the Ojibwa cover much of the upper Midwest

1842 treaties. They participated as wage laborers in the mining and lumber industries, provided fruit, vegetables, and fish for local and national markets, and maintained their seminomadic, subsistence life-style. The Ojibwa continued to believe that the Americans wanted only the use of the timber and copper, not the land itself. But in 1850, President Zachary Taylor revoked the Ojibwa's treaty-reserved hunting, fishing, and gathering rights and ordered their removal to unceded lands in Minnesota. The Ojibwa refused to leave Wisconsin. Ojibwa leaders claimed that they had signed the 1842 treaty only after government negotiators promised they could remain on their lands as long as they did not make war or harass white settlers, neither of which they had done. The Ojibwa, seeking revocation of the removal order, sent a petition and delegation to Washington. In late June of 1852, the Ojibwa delegation met with President Fillmore and persuaded him to rescind the removal order and suspend all efforts at removing the Ojibwa from Wisconsin.

In 1854, the Wisconsin Ojibwa signed a third treaty in which they ceded lands in Minnesota along the north shore of Lake Superior. Before signing the treaty, the Wisconsin Ojibwa, afraid of future removal efforts, insisted that they be furnished with permanent homes in Wisconsin. The 1854 treaty, therefore, established permanent reservations in Wisconsin at Bad River, Red Cliff, Lac Courte Oreilles, and Lac du Flambeau. Two other Ojibwa bands, the St. Croix and Sokaogon (at Mole Lake), were not granted reservations and would remain landless until the 1930s, when they were granted small reservations (Figure 24.2). The four reservations created by the 1854 treaty would prove too small to provide for the subsistence needs of the Ojibwa. In the second half of the nineteenth century, many Ojibwa found wage labor as sawyers, log drivers, and graders for the railroads, and they sold surplus fish, deer, and wild rice for cash. The economy of northern Wisconsin, however, was unstable and unpredictable. Unable to rely on employment, the Ojibwa continued to engage in their traditional subsistence activities in order to meet their basic needs.[3]

At the end of the nineteenth century, Ojibwa hunters and fishers encountered increasing competition from commercial fishers, market hunters, and especially white sport fishers and hunters for the fish and game of the ceded territory. Tourists had been visiting Bayfield, near the Red Cliff Reservation, by passenger steamer lines since the 1860s.[4] By the 1880s, railroads connected northern Wisconsin to a large pool of potential visitors from Chicago, Milwaukee, and the Twin Cities. Tourists were attracted by the scenery, the relatively close location (300–400 miles from Chicago and Milwaukee and 100–300 miles from the Twin Cities), and the north's summer climate of warm days and cool nights. Deer and abundant gamefish rounded out the appeal of northern Wisconsin for this growing group seeking outdoor leisure.

At the same time that the Ojibwa were encountering increased competition from tourists for the fish and game of northern Wisconsin,

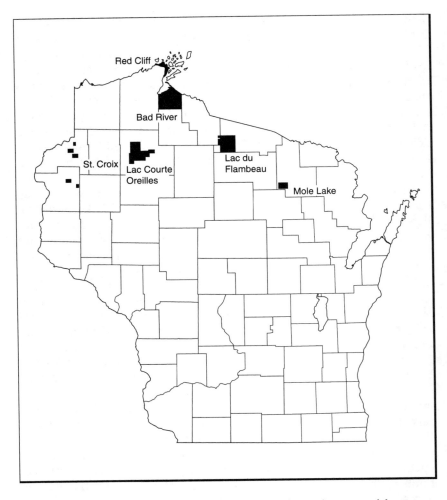

Figure 24.2. Ojibwa reservations are concentrated in the northern part of the state

their harvesting activities were also being restricted by the state of Wisconsin. The state first began to enforce fish and game laws against the Ojibwa and to restrict their access to fish and game in the ceded territory in 1889. In 1896, Attorney General W. H. Mylrea of Wisconsin issued a legal opinion which said that the state's fish and game laws applied to Indians in Wisconsin because the state had "unquestioned" police powers "to regulate and control the taking of fish and game" within its borders. Mylrea said that the treaty of 1842, which the Indians claimed exempted them from state laws, had been "abrogated" in 1848 by the congressional act creating the state. The act of creating the state "invested" the state with complete and "exclusive power over its territory." The opinion formed the basis for the state's policy of regulating Indian treaty hunting and fishing rights throughout the twentieth century.

Between 1889 and 1983, the state of Wisconsin, supported by decisions in the courts, enforced its fish and game laws on the off-reservation hunting and fishing activities of the Wisconsin Ojibwa. The state imposed fines and jail terms, impounded cars, and confiscated rifles and fishing equipment of those Ojibwa caught violating state fish and game laws off-reservation. Despite pressures from the state to assimilate and to abstain from subsistence activities, hunting, fishing, and the gathering of wild rice persisted as an important part of Wisconsin Ojibwa culture and economy.

Legal Geographies

In the late 1960s the Ojibwa and other Wisconsin tribes, influenced by the civil rights movement, looked to the federal courts for protection of their treaty rights to hunt, fish, and gather on and off their reservations. The Ojibwa, with the legal assistance of Wisconsin Judicare, an organization funded by the federal government and designed to serve the legal needs of low-income residents in northern Wisconsin, including the state's Indian population, initiated a number of test cases challenging state authority over their off-reservation usufructuary activities.

On March 8, 1974, two Lac Courte Oreilles (LCO) members, Mike and Fred Tribble, were arrested by Department of Natural Resources wardens on Chief Lake, outside the reservation, for violating state conservation laws. The Tribbles were found guilty in the Circuit Court of Sawyer County for possession of a spear for taking fish from inland, off-reservation waters and for occupying a fish shanty without name and address attached. On September 4, 1974, the LCO band, on behalf of the Tribbles and all its members, filed suit in the United States District Court for the Western District, in Madison, against the state of Wisconsin for interfering with the exercise of treaty-reserved rights to hunt, fish, trap, and gather in the area ceded in the treaty of 1837. On March 18, 1975, the original complaint was amended to include the area ceded by the treaty of 1842.

In their suit, the LCO maintained that hunting, fishing, and gathering rights, reserved in the treaties of 1837 and 1842, had never been abrogated and continued to have effect. The LCO sought a court ruling that they had the treaty right to hunt, fish, and gather wild rice free from state regulations on nonreservation lands and waters in northern Wisconsin. The state of Wisconsin argued that, while the Ojibwa had indeed reserved such rights in the treaties of 1837 and 1842, such rights had been extinguished by the president's 1850 removal order and by the treaty of 1854.

On September 20, 1978, federal district judge James Doyle ruled that the presidential removal order of 1850 was invalid and did not, as the state claimed, terminate the Ojibwa treaty rights. The Ojibwa, he said, had not misbehaved, and therefore the 1850 presidential removal order was "not authorized by the treaties of 1837 and 1842, was beyond the scope of the President's powers, and was without legal effect." Judge Doyle agreed, however, with the state's claim that the treaty of 1854 had extinguished the Ojibwa treaty rights. Doyle said that the parties to the treaty of 1854 intended to extinguish the Ojibwa right to hunt, fish, and gather on the ceded lands. He concluded that "the general right of the Lac Court Oreilles Band and its individual members to hunt, fish and gather wild rice and maple syrup in the area ceded by the Chippewa to the United States by the treaties of 1837 and 1842, free of regulation by state government, was extinguished."[5]

The LCO appealed Judge Doyle's decision to the United States Court of Appeals for the Seventh Circuit, in Chicago. On January 25, 1983, the Seventh Circuit, in what is referred to as the *Voigt* decision and *LCO* I, reversed Judge Doyle's 1978 decision. The appellate court ruled in favor of the tribes, declaring that the rights guaranteed them to hunt, fish, and gather off the reservation in the ceded territory had never been revoked or terminated either by the executive order of 1850 or the treaty of 1854.[6] The U.S. Supreme Court, on October 3, 1983, denied certiorari, upholding the Seventh Circuit's decision.[7] The case was returned to Judge Doyle for further determination of the scope of the treaty rights and the extent to which the state might regulate the Indians' exercise of those rights.

The question of the geographic extent of the Ojibwa's treaty rights would remain pivotal during litigation in the 1980s. Throughout the litigation, the state attempted to persuade the court to limit those rights to public lands within the ceded territory. The LCO, by contrast, claimed that the geographic extent of the treaty rights should be determined by whether non-Indian use of a particular plot of land was compatible with Indian hunting, fishing, and gathering activities, not by a strict public or private land distinction. The "use" criteria applied to navigable waters and to both public and private lands. The tribe made a strategic choice not to claim access to all lands, both public and private, within the ceded territory. According to James Jannetta, the lead counsel for the LCO during the *LCO I* appeal, "The modern implication of a hunting right on all private lands was just too socially unpalatable to obtain court recognition and would have jeopardized what was already (at the time) viewed as a fragile claim."[8] The tribe limited the geographic scale of its definition of the treaty rights, arguing that tribal members could exercise their treaty rights only on those public and private lands where non-Indians were hunting or gathering.

On March 8, 1983, the Court of Appeals for the Seventh Circuit amended its original *LCO I* decision, ruling in favor of the state by restricting LCO rights to public lands. The court defined the rights as being "limited to those portions of the ceded lands that have not passed into private ownership." The court added that the Ojibwa who signed the treaties understood that "their rights could be limited if the land were needed for white settlement." A year later, Judge Doyle upheld the Seventh Circuit's decision, saying that the rights were "limited to those portions of the ceded lands that were not privately owned as of March 8, 1983."[9]

The state contested the court's definition of what constituted public lands. In its appeal of Doyle's judgment, the state advanced a geographic definition of the tribes' treaty rights based upon a theory of patent extinction. The state interpreted strictly the Seventh Circuit's statement that the treaty rights were "limited to those lands that have not passed into private ownership" to mean that only those "ceded lands which have *continuously* remained part of the public domain are subject to the treaty rights."[10] In northern Wisconsin, a large percentage of public lands (state and county forests) were once privately owned; these lands returned to the public domain beginning in the 1920s because of extensive tax delinquency.[11] If the court accepted the state's patent-extinction theory, approximately 3.8 million acres of state- and county-owned lands in the ceded territory would be declared unavailable for tribal members to exercise their treaty rights. The tribes, of course, disagreed with the state's view.

On April 25, 1985, in *LCO II*, the Seventh Circuit ruled against the state's patent-extinction interpretation and returned the case to Judge Doyle for a determination of the scope of the treaty rights.[12] The state was forced to pursue a different strategy. Its new assertion involved the meaning of *settlement*. The state argued that the Ojibwa understood they would have to "accommodate" to non-Indian use of resources and non-Indian "settlement" of the land in the ceded territory. *Settlement*, according to the state, should be defined not merely by property title or merely as private ownership of land but, more important, by non-Indian *use* of the public domain. At an absolute minimum, the state claimed that privately owned lands were settled lands and not subject to the treaty right. But the state's definition of *settlement* was broadened to include public lands also. The state argued: "All privately owned land is settled. [However] all settled land is not necessarily privately owned."[13] The state defined *settlement* so broadly as to include almost any type of public land found in northern Wisconsin, including school grounds, campgrounds, cemeteries, parks, roads and highways, county, state, and national forests. "Settlement," included, moreover, not only public lands but also the state's lakes, rivers, and other waters. This extension to include "waters" was directed against the tribal tradition of spearfishing and gill-netting on inland lakes.

The tribes, by contrast, offered a geographic definition of the rights based upon a distinction between public and private property and a determination of whether or not treaty-reserved usufructuary activities would be compatible with non-Indian use of a particular piece of property regardless of title. Citing *LCO II*, the tribes agreed that the treaty rights were limited to publicly owned land. All public lands, they said, except school grounds and hospitals, were available to the treaty right. The tribes also maintained that lakes, rivers, and streams were not "settled." According to the tribes, all navigable bodies of water within the state were subject to the public trust doctrine. Their beds were part of the public domain, and therefore they were subject to the exercise of the treaty-reserved rights. Moreover, forest crop lands and managed forest lands, private lands receiving property tax abatement in return for forest management and public access for hunting and fishing, were not "settled" and were, therefore, subject to the exercise of the treaty rights.[14]

In his February 18, 1987, *LCO III* opinion, Judge Doyle rejected the state's contention that settlement diminished or extinguished the treaty right. He said there was no evidence to suggest that the Ojibwa at treaty times understood that their rights could be diminished by the patenting of land into private ownership or by state and local government dedication of public lands to uses incompatible with the exercise of the tribes' usufructuary rights. He also concluded that the Ojibwa lost their reserved usufructuary rights on privately owned lands, unless this geographic diminution of their rights prevents "them from enjoying a modest living." Judge Doyle also ruled that private lands opened to public hunting and fishing, such as forest crop lands, managed forest lands, and woodland tax lands, were also open to the exercise of the treaty right.[15]

Boat-Landing Protests

While the litigation was in progress, the state and the tribes negotiated a number of interim agreements that allowed the tribes a limited exercise of off-reservation treaty rights. The tribes agreed to limit the exercise of their rights, and the state agreed not to arrest and prosecute Indians for such activity. Although tribal members harvest furbearers, small game, waterfowl, bear, deer, and many different species of plants, it is the off-reservation spring spearing of walleye and muskellunge that has drawn active, sometimes violent, protest.[16] Between 1985 and 1992 large crowds, sometimes 200–500 protesters, gathered at small public boat landings in order to protest the *Voigt* decision and pressure the Ojibwa to give up treaty rights. Most protests occurred in Vilas, Oneida, and Price counties and were directed at the spearers from the Lac du Flambeau Reservation, the most active spearfishers of the Wisconsin Ojibwa. Protesters screamed racial insults, threw rocks, and used wrist rockets. During their demonstrations they sang mock Indian chants, blew whistles, and made death threats, chanting "drown them, drown them," and told spearers, "We hope you brought your body bags." Spearfishers were also harassed on the water. Protest boats circled the spearers' boats, creating large and dangerous boat wakes that caused several swampings. Hoping to prevent the spearers from harvesting fish, protesters shined bright lights in spearer eyes, dragged anchors through the walleye spawning beds, and planted concrete walleye decoys.

As a result of the protests, large numbers of law enforcement personnel were stationed at the boat landings, canine units were utilized, and National Guard helicopters flew overhead. High-powered lights illuminated the dark landings, and police clad in riot gear patrolled snow fences used to separate protesters from spearers, their families, and treaty-rights supporters. The FBI investigated a number of incidents, including the placement of a pipe bomb at a landing in 1989. At the peak of the protests, in the spring of 1989, 109 protesters were arrested at a Trout Lake (Vilas County) boat landing after crossing a police line, sitting, and refusing to leave. One of the largest protests occurred at the conclusion of the 1989 spearing season at Butternut Lake in Ashland County. An estimated 215 law enforcement officers and 30 wardens' boats were required to keep the peace between an estimated 1,000 treaty-rights supporters and 700 protesters. The events that evening at Butternut Lake were unusual; for the first time, treaty-rights supporters outnumbered treaty-rights protesters. Public opinion about the issue seemed as strongly felt as it was polarized.

Treaties, Comanagement, and Environmental Protection

At the heart of the conflict between the Ojibwa, anti-treaty-rights organizations, and the state of Wisconsin is the question of who has sovereignty over the ceded territory of northern Wisconsin. More specifically, it is a question of who has the legitimate authority to control
the use and development of the natural resources of the ceded territory. For the Ojibwa, the exercise of their treaty rights may be viewed
as part of a larger struggle for political autonomy, economic self-
sufficiency, and a desire to preserve their cultural heritage. The Ojibwa
have struggled not only to gain territorial sovereignty or jurisdiction
over what happens on the reservations but also to expand the territorial reach of their sovereignty off-reservation to the ceded territory.

The *Voigt* litigation has resulted in tribal participation in the formu-
lation of state natural resource policy for the ceded territory. The Great
Lakes Indian Fish and Wildlife Commission (GLIFWC), an intertribal
natural resource management organization, works with the state and
federal government on numerous resource management projects. *Voigt*
litigation stipulations and orders require the DNR to recognize tribal
representatives as official members on any DNR committee involved
with resource management in the ceded territory. GLIFWC biologists serve as tribal representatives on 15 species advisory committees.
For example, GLIFWC biologists sit on the quail advisory committee,
the ruffed grouse and woodcock advisory committee, the fisher and
marten advisory committee, the wild rice management committee, and
the Inland Fisheries Technical Working Group. *Voigt* litigation stipulations and orders also require an approach that involves "all reasonable
efforts to reach consensus" in these committees.[17]

One goal of the tribes is to expand their authority over the natural
resources of the ceded territory. James Schlender, executive administrator of GLIFWC, has asked that the federal government "provide
tribes with the opportunity to participate fully with other governments on matters which have a substantial impact on tribal natural
resources." The tribes want government-to-government relationships
with local, state, and federal governments and seek recognition of
their sovereign status. They want to participate as "full partners" in
the management of the ceded territory and its resources.[18]

The tribes have been unsuccessful, however, in becoming full partners with the state in managing the natural resources of the ceded
territory. The state has rejected the tribes' comanagement proposal.
While the DNR and GLIFWC have successfully coordinated their activities regarding seasonal harvests and biological assessments, the

DNR is critical of the concept of comanagement. The federal district court recognized the DNR as having ultimate authority over the management of the natural resources of the ceded territory. Thus, sharing of resource management is seen by DNR officials as a "breach of its responsibilities towards its citizens." The DNR's position is that the tribes should be directly involved in the DNR's decision-making process. It endorses coordinated and cooperative management of natural resources by the state and the tribes, but not coequal management.[19]

The Ojibwa are especially concerned that some off-reservation activities permitted by the DNR, such as mining, may endanger the natural resources of the ceded territory. They seek a greater role in determining the permittance of such activities and in protecting the "treaty resources" from pollution and environmental degradation. Because the tribes have limited power to comanage, they have attempted to use their treaty rights as means to prevent environmentally harmful practices from taking place in the ceded territory. Treaty-protected harvesting rights, they argue, are meaningless if the resources have been damaged or destroyed because of habitat destruction. One implication of the *Voigt* decision, they claim, is that the resources are "treaty-protected" and that the tribes "have legal standing to stop or remedy" these environmental threats.[20] Although eventually unsuccessful, the LCO band attempted in 1989 to evoke their treaty rights to stop the construction of a large open-pit copper mine located south of their reservation near the town of Ladysmith. The LCO claimed that the state had improperly excluded them from full participation in the mine permitting process and had not considered the potential adverse effect mining might have on their treaty rights.[21]

The anti-spearing groups, PARR (Protect American Rights and Resources) and STA (Stop Treaty Abuse), favor a return to the status quo as it existed prior to the 1983 *Voigt* decision. They want the state of Wisconsin to exercise jurisdiction over both Indian and non-Indian resource users in the ceded territory. Their political agenda, however, goes beyond opposition to off-reservation treaty rights, and more directly challenges claims of tribal existence and sovereignty over *on-*reservation space. They seek a geopolitical solution to what they call "Red Apartheid" and the "crisis in Federal Indian Policy." Reservation Indians, they claim, pay no state or federal income or sales taxes, receive cradle-to-grave welfare benefits, free housing, and many other benefits paid for by federal tax dollars. The reservation system is described as a "prime example of unworkable socialism and a lesson in despair" and "a form of genocide where intelligent human beings are exiled for a lifetime and conditioned to be totally dependent on the handout of the federal government."[22] Wisconsin anti-treaty-rights groups favor the abrogation of Indian treaties, the termination of reservations. They believe that the federal government should promote a policy of Indian assimilation into non-Indian society.

Conclusion

The *Voigt* decision altered the social and political geography of northern Wisconsin. The Ojibwa now have access to the fish, game, and other harvestable species located on public lands in their former territories. They have a court-recognized share in these resources. Two types of harvesting now exist in the northern part of the state: a recreational or sport harvest, and a subsistence harvest of fish and game. But in addition to this changed allocation of resources, a more fundamental change has occurred. This is a change in how the state and tribes interact to manage the resources and natural habitat of the territory. While the tribes have not succeeded in achieving comanagement, they do have some voice in the decisions that affect their lives and their environment. The DNR and the Ojibwa now work cooperatively in the management of northern Wisconsin's fish and wildlife.

The *Voigt* decision has also affected the relations between the Ojibwa and their non-Indian neighbors. Prior to the *Voigt* decision, most non-Indians had very little contact with their Ojibwa neighbors and knew very little about their culture and history. Although PARR continues its fight against treaty rights, boat-landing protests are now a thing of the past. Today, greater interaction and cooperation characterize relations between the Ojibwa and their neighboring communities. Many who protested against treaty rights at boat landings, for example, now find themselves allied with the Ojibwa in their battle against Exxon's proposed zinc mine adjacent to the Mole Lake Reservation. Treaty rights, no longer considered a danger to the northwoods environment and economy, are seen today as a possible means to protect and preserve the region's natural and social environment.

Notes

1. Portions of this chapter appear in different form in S. Silvern, "Nature, Territory and Identity in the Wisconsin Treaty Rights Controversy," *Ecumene* 2 (1995): 268–286; and S. Silvern, "Scales of Justice: The Legal Geography of Indian Treaty Rights in Northern Wisconsin," *Political Geography* (forthcoming).

2. See E. Danziger, *The Chippewas of Lake Superior* (Norman, 1979); T. Pfaff, *Paths of the People: The Ojibwe in the Chippewa Valley* (Eau Claire, 1993); H. Hickerson, *The Chippewa and Their Neighbors: A Study in Ethnohistory* (New York, 1970).

3. R. Satz, "Chippewa Treaty Rights: The Reserved Rights of Wisconsin's Chippewa Indians in Historical Perspective," *Transactions, Wisconsin Academy of Sciences, Arts and Letters* 79 (1991): 13–82; P. Shifferd, "A Study in Economic Change, the Chippewa of Northern Wisconsin: 1854–1900," *Western Canadian Journal of Anthropology* 6 (1976): 16–40.

4. Satz, "Chippewa Treaty Rights," 79.

5. *Lac Courte Oreilles Band of Lake Superior Chippewa Indians et al. v. Lester P. Voigt et al.*, reported s. nom. *U.S. v. Ben Ruby*, 464 F. Supp. 1316–1376 (1978).

6. *Lac Courte Oreilles Band of Lake Superior Chippewa Indians et al. v. Lester P. Voigt et al.*, 700 F 2d 341–365 (1983).

7. 464 U.S. 805 (1983).

8. Personal communication, March 25, 1994.

9. *Lac Courte Oreilles Band of Lake Superior Chippewa Indians et al. v. Lester P. Voigt et al.*, 700 F 2d 365; unpublished order, U.S. District Court of the Western District of Wisconsin, March 1983.

10. State of Wisconsin, *Reply Brief of Defendant-Appellant* (United States Court of Appeals for the Seventh Circuit, July 5, 1984, p. 3; emphasis added).

11. I. Vogeler, *Wisconsin: A Geography* (Boulder, 1986), 112–115.

12. *Lac Courte Oreilles Band of Lake Superior Chippewa et al. v. State of Wisconsin et al.*, 760 F 2d 177 (1985), at 182.

13. Stenographic transcript of sixth day of court trial, U.S. District Court for the Western District of Wisconsin, December 16, 1985, vol. VI-B, p. 55.

14. Plaintiff's pretrial brief (U.S. District Court for the Western District of Wisconsin, September 11, 1985, p. 13).

15. *Lac Courte Oreilles Band of Chippewa Indians et al. v. State of Wisconsin et al.*, 653 F. Supp. 1420 (1987), at 1426.

16. For a survey of Ojibwa off-reservation harvesting, see: S. Erickson, *Season of the Chippewa* (Odanah, Wis., 1993). Tribal spearers, whose numbers have ranged from 194 to 426 (1986–1990), primarily target walleye. Between 1985 and 1993 they harvested an average of 18,562 walleye and 162 muskellunge per year. The state's 1.4 million licensed anglers took an estimated 623,000 walleye and 9,454 muskellunge per year for the same time period. United States Department of the Interior, *Casting Light upon the Waters: A Joint Fishery Assessment of the Wisconsin Ceded Territory* (Minneapolis, 1991), 60–61; *Masinaigan* (Odanah, Wis.) (Summer 1993): 2.

17. Stipulation for deer trial, Docket no. 1167, (1989), 18–19, *Lac Courte Oreilles Band of Chippewa Indians et al. v. State of Wisconsin et al.*; J. Schlender, "Testimony before the House Committee on Natural Resources, Subcommittee on Native American Affairs," February 18, 1993 (photocopy courtesy of Great Lakes Indian Fish and Wildlife Commission), 4.

18. Schlender, "Testimony . . . on American Affairs," 7–9.

19. C. D. Besadny, secretary, Wisconsin Department of Natural Resources, to Wisconsin state legislators, memorandum, November 6, 1989.

20. W. Bresette, "Treaties and the Environment," *Masinaigan* (July 1984): 14. Also see W. Bresette, "Indian Treaty Rights Point Way to Future," *Milwaukee Journal*, February 29, 1984; and A. Gedicks, *The New Resource Wars: Native and Environmental Struggles against Multinational Corporations* (Boston, 1993), 105.

21.   Despite the opposition of the LCO, the Sierra Club, and other environmental groups, on May 14, 1993, four years after its application for a mining permit, the Flambeau Mining Company started shipping ore, rich in gold and silver, out of state for processing. See N. Seppa, "Flambeau Mine Breaks Ground," *Wisconsin State Journal* (Madison), June 13, 1992; R. Smith, "Tons of Ore Being Dug Out of Flambeau Mine," *Milwaukee Sentinel,* May 14, 1993.

22.   "Native American Denounces Special Rights for Tribes," *Lakeland Times* (Minocqua, Wis.) July 13, 1984, 1–2.

25

The Wild Rice Harvest at Bad River
Natural Resources and Human Geography in Northern Wisconsin

Thomas E. Pearson

The story of wild rice at the Ojibwa's Bad River Indian Reservation in northern Wisconsin comes alive at the end of every summer (Figure 25.1). As the nights grow cool and the shadows of the midday sun grow long, Bad River Ojibwa men and women move their boats through the shallow waters of rivers, lakes, and sloughs, harvesting wild rice by hand from the dense natural stands which dominate the watery parts of this region (Figure 25.2). The material and symbolic importance of this annual ritual looms large in the minds of many Ojibwa at Bad River, and the persistence of this food-production strategy is readily apparent to any late-summer visitor to northern Wisconsin.

Archaeological evidence suggests that wild rice has been an important food for peoples in the upper Midwest of North America for over 2,500 years.[1] While the methods of the hand harvest are thought to be similarly ancient, the more modern methods and technologies which currently produce the majority of wild rice in North America are the result of much more recent developments. Only for the last three decades, since about 1970, have domesticated wild rice varieties been bred for a large-scale, mechanized, flooded-paddy agriculture. And the increase in harvest yields which has followed successful domestication has been dramatic.[2]

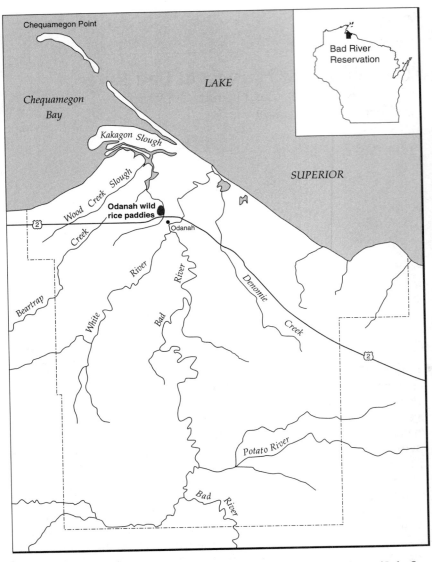

Figure 25.1. Map of the Bad River Indian Reservation, on the southern shore of Lake Superior

Figure 25.2. Mr. and Mrs. John Holmes participating in the late-summer harvest of wild rice in northern Wisconsin, ca. 1930 (Courtesy of the State Historical Society of Wisconsin, neg. number WHi [X3] 50182)

Despite the close cultural relationship between the Ojibwa and wild rice, the practice of mechanized paddy production has been almost exclusively a non-Indian affair. At the Bad River Reservation, however, a paddy project was begun during the late 1960s but abandoned after only two growing seasons. Why did this occur? What elements of wild rice paddy production have motivated its rejection by Native Americans, the people for whom this resource holds such historical, material, and symbolic importance? This chapter explores these questions in the broad context of human geography. Implicit within this context is a consideration of how a people's sense of cultural identity and relationship to their environment can influence choices about how they will make their living from the earth.

This chapter begins with an exploration of the ecology of wild rice and its harvest by hand and mechanized methods. A closer look at the experience of the Odanah paddies is then presented. An interpretive section follows, which further explores the dynamics of why particular decisions were made and the rationales for those decisions. This context reveals much more than a simple dichotomy between acceptance or rejection of technological change. A final section of conclusions closes the chapter and considers how issues of culture, identity, and perceptions of relationship to environment can influence some of the basic decisions of economic production.

Wild Rice Ecology

Wild rice is an annual emergent aquatic grass which grows widely in the eastern United States. It can be found along the Atlantic coast from Maine to central Florida and along the Gulf coast from central Florida to Texas. In Canada, wild rice grows naturally from southern Manitoba to the Atlantic coast and has been planted as far west as Saskatchewan. Wild rice is found in its greatest densities, however, in northern Minnesota and northern Wisconsin. Researchers estimate that some 8,000–10,000 hectares of wild rice exist in Minnesota and 2,600 hectares in Wisconsin. In Wisconsin, it is thought that some 2,000 of these hectares lie in the areas of territory ceded to the United States by bands of Ojibwa Indians in the treaties of 1837 and 1842.[3]

There is considerable confusion in the scientific classification of wild rice. Definitions of species and varieties have been based upon differences in morphology, including leaf and stalk size, both of which may vary within a single stand. However, the prominent species identified in the upper Midwest are *Zizania palustris* and *Z. aquatica*. *Z. palustris* has been the more important species for harvesting because of its larger seeds, and accordingly is the species of focus in this study.[4]

Soil and water factors are important in influencing wild rice prevalence and range. Wild rice is found in a wide variety of soils, but the densest stands are found in soft, alluvial, organic mucks. These soils provide necessary nutrients and may assist in seed lodging, root development, and nutrient uptake. Water depth and fluctuation are of particular importance to wild rice plants. Wild rice grows best in water depths between 15 centimeters and 1 meter, and when depths are stable or gradually receding during the growing season. Rapid rises in water level may either drown or uproot plants, particularly in the floating leaf stage, when root structures are unusually weak. The slowly moving waters of streams and sloughs are thought to be helpful for wild rice growth, perhaps because of the influx of water-borne nutrients and disturbance in the substrate, which can often favor colonizing annuals such as wild rice. Wild rice seeds mature in late August and early September and drop from the plants, lodging in lake or stream beds, where they remain during stages of dormancy until germination the following spring. Seed dispersal is usually quite limited, but water currents, ice, and animals (including humans) can increase dispersal range.[5]

Natural stands of wild rice rarely experience significant mortality from pests and pathogens. However, human activities have negatively impacted the prevalence of wild rice stands in a variety of ways. Large wakes from passing boats can uproot wild rice plants in their early stages of growth. Dam construction has flooded out large stands of wild rice plants in many instances in northern Wisconsin. Deliberate removal of wild rice in order to create boat landings or openings in waterways is a common practice. In addition, inappropriate harvesting, including harvest before seeds are ripe or the use of excessive force to dislodge seed, has also been an important human impact curtailing the prevalence of wild rice plants.[6]

Harvesting Wild Rice

Cultural History and the Hand Harvest

The harvest of wild rice is an ancient process which archaeologists believe has been practiced for at least 2,500 years. Methods of hand harvest as it is practiced today vary only slightly from those methods first described in written records from the seventeenth century. These harvest methods have persisted over such a substantial period of time and into the present, in part, because they generally prevent damage to rice stands, allow for reseeding in healthy stands, and minimize disturbance to local wildlife.[7]

The hand harvesting of wild rice is currently characterized by two persons in a canoe, one with ricing sticks (two wooden sticks 1 meter long, 3–4 centimeters in diameter) and one with a long pole to move the canoe through the rice bed (Figure 25.3). With one ricing stick, sometimes curved, the harvester bends the stalks over the canoe. Then with the other stick he or she gently strikes the heads of the plant to dislodge the ripened grain. This is repeated until the canoe is full. With a canoe full of wild rice, the harvesters return to shore to unload their yield for processing or for sale as unprocessed, or "green," rice.

This practice of hand harvesting wild rice in natural stands generally has a very low environmental impact. Studies have found that between 5 percent and 20 percent of the annual seed production was collected in areas which were harvested regularly during a three-week harvesting season. Researchers have argued that this leaves substantial seed for waterfowl, other wildlife, and reseeding. In addition, because

Figure 25.3. Mrs. Holmes "knocks" while Mr. Holmes poles their boat through a healthy stand of wild rice in northern Wisconsin, ca. 1930 (Courtesy of the State Historical Society of Wisconsin, neg. number WHi [X3] 50183)

small stands are rarely harvested, they are often completely protected from harvesting impacts. The hand harvest, with its low capital inputs, is also highly amenable to the heterogeneous environments of wild rice waters. The small boats and canoes are capable of reaching wild rice plants wherever they may grow, although the harvest is most efficient in healthy dense stands. In addition, this hand harvest, like many traditional agricultural contexts, allows the natural biodiversity of the wild rice plants to continue to coevolve with pests and pathogens in local environments. Evidence also suggests that hand harvesters have long participated in the conservation and management of wild rice biodiversity through the propagation and protection of wild rice waters.[8]

Technological Innovation and the Mechanized Harvest

In January 1951, a conference was held at the University of Minnesota to review the potentials and problems associated with human use of wild rice. This conference of 23 botanists and other specialists created an agenda for the development of the wild rice agricultural industry. Listed on this agenda were 12 items, including "investigation of methods for cultivation after the pattern of domestic [white] rice."[9] Mechanical harvesting of natural stands presented significant ecological problems, particularly in terms of stand regeneration, and, as such, this practice was made illegal in Minnesota, Manitoba, Ontario, and Wisconsin. Thus, the focus of mechanized harvesting turned from natural stands of wild rice to planted wild rice paddies. Early attempts at paddy production were begun during the 1950s but were relatively unsuccessful. Growers of paddy wild rice encountered problems with diseases, particularly fungi, and pronounced seed-head shattering or dehiscence of ripened seeds, a characteristic common among wild grasses and an important regeneration mechanism.[10]

During the early 1960s successful efforts in the domestication of wild rice began to occur. In 1963, two University of Minnesota agronomists, Paul Yagya and Erwin Brooks, working with a pioneer wild rice grower in northern Minnesota, isolated a strain of wild rice with increased indehiscence. This characteristic allowed ripened grains to remain on the seed head during the 10–14-day ripening period, and thus allowed growers using mechanical harvesting machines to harvest fields in one pass, rather than in multiple passes, which had previously been practiced. This development occurred while many large food companies were beginning to develop a market for wild rice among the larger North American society. When a routine failure in the harvest of hand-harvested wild rice occurred in 1965, one company, Uncle Ben's, Inc., bought an estimated 80 percent of the world supply. This event called attention to the limited and fluctuating supply of wild rice and spurred efforts to create more productive alternatives. Uncle Ben's began funding scientists and farmers to further the research and development of paddy production of wild rice. Additional funding came from the Minnesota state legislature in 1969 and 1971, and by 1972 wild rice had received the status of a minor domesticate.[11]

With the advent of crop breeding, agrochemical, and other technological developments, levels of production of wild rice from North American paddy operations increased dramatically during the 1970s and the 1980s. These projects were characterized by large flat fields below the floodplain, which were surrounded by constructed dikes and levees. Fields were flooded and mechanically planted each spring using hybrid seed and agrochemical fertilizers and pesticides, and drained and mechanically harvested each fall.[12]

This combination of botanic and economic developments facilitated dramatic growth in the overall production of wild rice in North America. In addition, average yields per acre increased significantly and consumer prices declined markedly. The average yield per acre of wild rice rose approximately 500 percent between 1964 and 1987. Overall North American yields increased from 0.7 million pounds of processed rice in 1968 to a peak of 15.8 million pounds in 1986. Prices rose between 1969 and 1978 from $2.60 to $5.15 per pound with increased demand for wild rice. However, as levels of paddy production began to exceed market demand, prices declined sharply from $5.15 in 1978 to $1.50 per pound in 1988.[13]

Prior to the development of paddy wild rice, hand-harvested wild rice was among the most important cash commodities of the Ojibwa people. However, the dramatic growth of the wild rice paddy industry, controlled primarily by non-Indian peoples, has greatly reduced the commercial importance of the hand harvest. In particular, the increases in annual yields of paddy wild rice have enabled large food corporations and national marketing firms to meet demand without dependence upon the often fluctuating and decentralized supply of hand-harvested wild rice. With great regularity, the low price and abundant supply of paddy-grown wild rice have enabled paddy agriculturalists to outcompete hand harvesters of wild rice in national and many local commercial markets. The loss of this commercial income has been significant for many Ojibwa people. With this loss, the primary use of hand-harvested wild rice has returned to subsistence and local trade, somewhat similar to the context prior to the arrival of Europeans in the upper Midwest.[14] Fieldwork from the present study suggests that in 1991 some 70 percent of hand-harvested wild rice was used for subsistence purposes.

Wild Rice at Bad River: The Odanah Paddy Project

The Bad River Indian Reservation was created in the treaty of La Pointe in 1854. Article II of this treaty designated reservation lands for many bands of Ojibwa people, including the La Pointe band. Three separate areas of land were set aside for the La Pointe band, including the Bad River and Red Cliff reservations and a small portion of Madeline Island particularly suitable for fishing. Bad River, which has a population of some 1,500 Ojibwa people, is the largest reservation in Wisconsin, with over 52,000 hectares of land, more than 25 kilometers of Lake Superior shoreline, and more than 145 kilometers of inland rivers and streams. The Bad River Reservation also encompasses over 3,000 hectares of wetlands, including the Kakagon Slough, where large natural stands of wild rice are prevalent.[15]

The Odanah Paddy Project

Despite the longstanding commitment to the hand harvest of wild rice at Bad River, a paddy operation was started in the late 1960s. This was during the period of early development of the paddy industry in the Great Lakes states, and the promise of high yields from mechanized harvesting, "improved" seed, and disease and pest control were appealing to many people in northern Wisconsin and Minnesota. Agricultural scientists from the University of Minnesota and the University of Wisconsin–Madison served as consultants for the paddy project at Bad River and at four other Ojibwa reservations (Mole Lake in Wisconsin, and Red Lake, Nett Lake, and White Earth in Minnesota). These scientists included many of the leaders in the early efforts to domesticate wild rice. Significant funding for the Bad River project was supplied by the Hunger Foundation of America, and funds were administered first by the Great Lakes Intertribal Council and then through the Bad River Tribal Council. In 1968, 55 hectares of diked paddy were constructed on the north side of U.S. Highway 2, on tribal land within the town limits of Odanah. The Odanah paddies were prepared and planted, and the first harvest was reaped in 1969. The second and final harvest came the following year. In 1971 a promising crop went unharvested, and thereafter the Odanah paddies were abandoned. These paddies still exist 20 years later as overgrown and visible reminders of the failed attempt of paddy production at Bad River.

The abandonment of paddy production at Bad River was clearly a contested decision. Significant time, money, and effort had been invested, and many people were interested in seeing the project become an economic and productive success. Yet other people were not supportive and felt that significant criticisms of the project were being dismissed. A confrontation arose at Bad River with technological innovation and economic development on one side and a varied and complex coalition of resistance and opposition on the other side.

Consultant Perceptions

Erwin Brooks, agricultural scientist at the University of Minnesota, served as the primary consultant to the Bad River Tribal Council concerning the Odanah paddy project from inception in 1964 to abandonment in 1971. Brooks produced an October 1964 report concerning current and potential wild rice production at Bad River, including recommendations for construction sites for future paddy production.

Brooks also advised the production operations of the Odanah paddy in 1969, 1970, and 1971. In my interview with Brooks in October 1991, he suggested that ecological problems were not significant in the failure of the project. Although no fertilizers or pesticides were used, the yields of 1969 and 1970 were good and relatively free from disease and pest problems. Brooks also indicated that economic problems were not the cause of the project failure. However, Brooks did suggest that the employment policy was somewhat unrealistic, with too many persons employed as full-time workers where only seasonal part-time persons were needed. Brooks felt that the cause of the failure was primarily political. A tribal election took place in 1971, and the tribal chair who had supported the project was replaced by one who did not. Brooks believed that people at Bad River were generally in favor of the paddy, but that political infighting and a lack of political coordination were the causes of the downfall.[16]

Dave Stuiber was another academic involved in the Odanah paddy. Stuiber was a professor in the Department of Food Science at the University of Wisconsin–Madison and was heavily involved with the development of wild rice–processing technology. Concerning the Odanah paddy, Stuiber also felt that lack of coordination and persistent infighting between the Great Lakes Intertribal Council and the Bad River Tribal Council represented the major problem. "It [the Odanah paddies] gave other tribes the opportunity to stick their noses into Bad River," which, Stuiber suggested, was not well-received by the Bad River Tribal Council. Stuiber cited additional problems including "unwise use of funds" and conflicts over management responsibilities. "They threw something away that had potential."[17]

Ervin Oelke was an agronomist at the University of Minnesota and a close colleague of Brooks during the life of the Odanah paddy. Although Oelke also felt that political problems were present, his understanding of the events at Odanah in 1971 reflects a perception different from those of his colleagues. "There was some opposition from within the tribe; they weren't sure whether they should do this [domestication] or not, [this] trying to change nature. There were problems with funding and marketing, but the real difficulty was getting that society to change from their ways to ours."[18]

Bad River Indian Reservation Perceptions

Division and conflict over the Odanah paddy project also existed within the Bad River communities themselves. While some people saw potential for economic growth and needed jobs for the reservation, others joined in opposition to the project, motivated by a large and varied array of sentiments.

Richard Ackley worked in the economic development department for the tribe during the time of the Odanah paddy project. In our interview, Ackley suggested that during the late 1960s the tribe had been looking for opportunities to generate jobs and income for the reservation, and the wild rice project was seen as a promising possibility. Ackley, like a number of Bad River Ojibwa, had been in favor of the project for many practical reasons but has since become more critical. As the project progressed he suggested, "People just didn't like the paddy. People didn't like it because they felt it undercut the traditional ricing. They [the Bad River Ojibwa] don't support traditional ricing just because it's the way we do things; they support it because they value that way of doing things, the communal harvest."[19]

Eugene Bigboy was the head of the tribal council during the late 1960s and early 1970s and held significant responsibility for the operations and management of the Odanah paddy project. When we talked about the project he suggested, "The paddy possibly could have been incorporated into the kinship networks after it had got up and going. Perhaps it wouldn't have had to remain a wholly commercial operation. The project was somewhat devoid of tradition, but the management attempted to make the project enjoyable and community-based by providing Friday afternoon cookouts for employees, their families, and friends. But this remained a problem."[20] Bigboy pointed to other problems with the project, including potential fertilizer and pesticide runoff, marketing and pricing problems, and concern over outcompeting hand-harvested wild rice in commercial markets. Although he felt the project had promise, Bigboy suggested that the economic and ecological problems began to look insurmountable.

Joe Rose was another Bad River tribal member living near Odanah during the time of the paddy project. In talking about the project, Rose suggested, "The Ojibwa here at Bad River looked at the project with a concern for social welfare. They tried to get paychecks to as many people and families as they could." Like many individuals at Bad River, Rose was aware of the concern over lack of income and jobs on the reservation. But Rose also articulated another perspective on the project: "Traditional [hand] harvesting is also about resistance to assimilation. It's an Indian thing; it identifies us as a people. It's a chance for people to participate in their cultural tradition. We have a different culture than white culture; we have a different history, different traditions, and different goals."[21]

The perspectives of many reservation residents suggest that the

Odanah paddy project and its variety of not-so-wild wild rice presented direct economic and symbolic competition for hand-harvested wild rice. In the views of a number of Bad River Ojibwa, the project was seen as a symbolic attack on a central cultural icon in contemporary Ojibwa society. "[Wild rice] is a sacred resource; their [paddy] rice tastes like mud. They grow it with chemicals and poisons and it doesn't even cook. I tried some one time. It's not the same," said Vince Bender, an elder at the Bad River Indian Reservation. Marvin Nellis similarly suggested, "People here didn't like that project; that wasn't wild rice." Sharon Nellis agreed: "They use all those pesticides; that's not ricing; ricing is what we do out in the sloughs when we fill up the boats; that's ricing!" JoeDan Rose, head of the tribe's natural resources department during 1991 suggested, "Northern Wisconsin is an economically 'depressed' area; however, the value of ricing and preserving such activities for one's children is more important than money."[22]

I later interviewed Peter David, a non-Indian botanist at the Great Lakes Indian Fish and Wildlife Commission (GLIFWC) in Odanah, whose primary focus is wild rice. After discussing some of the botanic and biogeographic aspects of wild rice, I asked David about his understanding of the cultural importance of this plant. He responded, "I've been working on wild rice ecology for five years, and I'm only beginning to appreciate its importance to these people [the Ojibwa]. I don't think there is any resource more important to these people. It's a sacred resource with an intrinsic value of its own."[23]

During the period of my fieldwork, I was told that at least 25 percent of the reservation residents participate in the end-of-summer hand harvest and that over half the reservation population receives some wild rice annually through kinship networks.[24]

Interpreting the Events at Bad River

The contested decision to abandon paddy production at the Bad River Indian Reservation was made in 1971. Since that time no other paddy projects have been attempted at Bad River, and the hand harvest of natural stands of wild rice has continued widely across the reservation. Paddy production has succeeded at other places on Indian and non-Indian lands in Minnesota and in the central valley of California, which now produces the majority of commercially available wild rice.

At Bad River, the conflicts over paddy production ended with the victory of those opposed to the Odanah project. But it is clear that this experience cannot be fairly viewed as a simple dichotomy of two sets of uniform perceptions. Persons arguing against the paddy project did so for their own reasons, which differed, sometimes significantly, from those of other people critical of the project. While many were concerned over the economic competition of a lower-priced alternative to hand-harvested wild rice, others were concerned over the potential fertilizer and pesticide runoff. Still others felt that wild rice and its hand harvest represent important symbols of local Indian culture, which were threatened by the presence of mechanized and controlled production on the reservation. And within this realm of cultural symbolism and identity, there was a variety of opinions across the reservation about which technologies represent threats and which do not. Other groups felt that Ojibwa identity and cultural survival were more threatened by passing up economic opportunities than by taking advantage of them. Despite the broad spectrum of opinions about the Odanah paddy project, a large and varied coalition of Bad River residents was sufficiently concerned with the project that its members came together to bring about its end.

Many writers who have investigated similar traditional agricultural contexts in less-developed areas around the globe have argued that local peoples are "selective, accepting only those items from other cultures that may have lasting value in their own."[25] For many Ojibwa at Bad River, the tradition of the hand harvest of wild rice has a lasting value which was not present in the paddy production context. Many people at Bad River consider the hand harvest of wild rice from natural stands to be a metaphor of realism and authenticity within their culture. In continuing to support the hand harvest while allowing less radical change and innovation in harvest technique and processing method, many Bad River Ojibwa people suggest that they are working to maintain some of their cultural traditions, and, perhaps more important, the principles behind their cultural traditions.[26] This balance between acceptance of certain technologies and opposition to others which they feel may compromise important principles and traditions has helped the Bad River Ojibwa maintain a strong yet flexible sense of cultural identity.

Conclusions

The development and selection of food-production strategies by cultures around the globe have long been central themes in human geography.[27] In searching to understand the origins and persistence of hunting, gathering, and agricultural strategies, geographers have frequently looked to cultural history and, to a lesser extent, to cultural dynamics for explanation.[28] The present study has adapted this perspective to explore some fundamental questions about the persistence and importance of the hand harvest of natural stands of wild rice in northern Wisconsin.

Although paddy production of wild rice has been successful in many areas of Minnesota and California, the Ojibwa at Bad River chose, after a brief experience with a paddy project begun in 1969, to reject this technology and practice only the hand harvest of wild rice on their lands. Many Bad River residents were opposed to the Odanah paddy project for a broad array of reasons. Some were concerned over economic competition; others were concerned over potential ecological degradation of waterways from paddy fertilizer and pesticide runoff; still others viewed paddy production on the reservation as a threat to the symbolic importance of wild rice as an icon of local Indian cultural identity. Despite their differences in motivation, a large and varied coalition of Bad River residents came together in 1971 to end production at the Odanah project.

Such decisions are, however, more than just isolated incidents. Through these decisions, people affect and change the ecological, economic, and ideological contexts in which they live. And it is decisions such as these, together with their uncounted predecessors, which sculpt the elemental shape of a people's relationship to their landscapes and to the earth itself.

Notes

This chapter was completed, in part, with support from a National Science Foundation graduate student fellowship.

1. R. I. Ford and D. S. Brose, "Prehistoric Wild Rice from the Donn Farm Site, Leelanau County, Michigan," *Wisconsin Archeologist* 56 (1975): 9–15; E. Johnson, "Archaeological Evidence for the Utilization of Wild Rice," *Science* 163 (1969): 276–277; E. Johnson, "Preliminary Notes on the Prehistoric Use of Wild Rice," *Minnesota Archaeologist* 30 (1969): 31–43.

2. P. M. Hayes, R. E. Stucker, and G. G. Wandrey, "The Domestication of American Wildrice (*Zizania palustris,* Poaceae)," *Economic Botany* 43 (1989): 203–214; T. Vennum, *Wild Rice and the Ojibway People* (Minneapolis, 1988); G. P. Nabhan, *Enduring Seeds: Native American Agriculture and Plant Conservation* (San Francisco, 1988); Y. G. Lithman, *The Capitalization of a Traditional Pursuit: The Case of Wild Rice in Manitoba,* Manitoba University Center for Settlement Studies, Series 5, Occasional Paper no. 6 (Manitoba, 1973); T. E. Pearson, "Wild Rice and the Ojibway People of Bad River: A Study of Production, Tradition, and Cultural Persistence," master's thesis, University of Wisconsin–Madison, 1992.

3. T. Andryk, *Wild Rice Wetland Inventory of Northwest Wisconsin,* Great

Lakes Indian Fish and Wildlife Commission Administrative Report 86-4 (Odanah, Wis., 1986); P. F. David, *Wild Rice Management Plan for the Northern Wisconsin Ceded Territories of 1837 and 1842*, Version 1.1, Great Lakes Indian Fish and Wildlife Commission (Odanah, Wis., 1989); G. Fannucchi et al., "The Effects of Traditional Wild Rice Harvesting on a Wild Rice Marsh," in G. Fannucchi, "Wild Rice in East Central Minnesota," 21–39, M.S. thesis, University of Wisconsin–Stevens Point, 1983; R. Libertus, *Wild Rice*, Minnesota Department of Natural Resources Commissioner's Report (St. Paul, 1981); Nabhan, *Enduring Seeds*; T. E. Pearson, "The Historical Biogeography of Wild Rice on Northern Wisconsin," typescript, Department of Geography, University of Wisconsin–Madison, 1991.

4. W. G. Dore, *Wild Rice*, Plant Research Institute, Ottawa, Ontario, Publication 1393 (Ottawa, 1969); P. F. Lee, *Northwestern Ontario Wild Rice Management Plan*, Ontario Ministry of Natural Resources (Kenora, 1976); S. Warwick and S. Aiken, "Electrophoretic Evidence for the Recognition of Two Species in Annual Wild Rice (*Zizania*, Poaceae)," *Systematic Botany* 11 (1986): 464–473.

5. David, *Wild Rice Management*; G. Fannucchi, W. Fannucchi, and S. Craven, *Wild Rice in Wisconsin: Its Ecology and Cultivation*, University of Wisconsin Extension Service, (Madison, 1986); Lee, *Northwestern Ontario*.

6. David, *Wild Rice Management*.

7. W. F. Aller, "Aboriginal Food Utilization of Vegetation by the Indians of the Great Lakes Region as Recorded in the Jesuit Relations," *Wisconsin Archeologist* 35 (1954): 59–73; David, *Wild Rice Management*; Fannucchi et al., "The Effects of Traditional Wild Rice Harvesting"; Ford and Brose, "Prehistoric Wild Rice"; Johnson, "Archaeological Evidence"; E. Johnson, "Preliminary Notes on the Prehistoric Use of Wild Rice," *Minnesota Archaeologist* 30, 2 (1969): 31–43; R. G. Thwaites, ed., *Jesuit Relations and Allied Documents: Travels and Explorations of the Jesuit Missionaries in New France 1610–1791* (Cleveland, 1897).

8. Aller, "Aboriginal Food"; David, *Wild Rice Management*; Fannucchi et al., "The Effects of Traditional Wild Rice Harvesting"; Vennum, *Wild Rice*; J. Weatherford, *Indian Givers: How the Indians of the Americas Transformed the World* (New York, 1988).

9. D. Lawrence, "Wild Rice," *Minnesota Naturalist* 2 (1951): 24–25.

10. Hayes et al., "The Domestication of American Wildrice"; Vennum, *Wild Rice*; J. E. Meeker, *Taming Wild Rice*, Wisconsin Department of Natural Resources Publication (Madison, 1988).

11. C. Prescott-Allen and R. Prescott-Allen, *The First Resource: Wild Species in the North American Economy* (New Haven, 1987); E. H. Winchell and R. P. Dahl, *Wild Rice: Production, Prices, and Marketing*, University of Minnesota, Agricultural Experiment Station Miscellaneous Publication 29 (St. Paul, 1984).

12. E. A. Oelke et al., *Wild Rice Production in Minnesota*, University of Minnesota Agricultural Extension Service, Extension Bulletin 464 (St. Paul, 1982); E. A. Oelke et al., *Commercial Production of Wild Rice*, University of Minnesota Agricultural Extension Service, Extention Folder 284 (St. Paul, 1973).

13. R. N. Nelson and R. P. Dahl, *The Wild Rice Industry: Economic Analy-*

sis of Rapid Growth Implications for Minnesota, Department of Agricultural and Applied Economics, Staff Paper P86-25 (St. Paul, 1986); Prescott-Allen and Prescott-Allen, The First Resource; Winchell and Dahl, Wild Rice.

14. T. A. Steeves, "Wild Rice–Indian Food and a Modern Delicacy," Economic Botany 6 (1952): 107–142; Vennum, Wild Rice.

15. E. J. Danziger, The Chippewa of Lake Superior (Norman, 1978); Great Lakes Indian Fish and Wildlife Commission, Annual Report (Odanah, 1990); Vennum, Wild Rice.

16. E. Brooks, personal communication, 1991.

17. D. Stuiber, personal communication, 1991.

18. E. Oelke, personal communication, 1991.

19. R. Ackley, personal communication, 1991.

20. E. Bigboy, personal communication, 1991.

21. J. Rose, personal communication, 1991.

22. V. Bender, M. Nellis, S. Nellis, J. Rose, personal communication, 1991.

23. P. David, personal communication, 1991.

24. Pearson, "Wild Rice and the Ojibway People of Bad River."

25. Nabhan, Enduring Seeds.

26. For a thorough discussion of traditions and principles behind traditions, see J. Vansina, Paths in the Rainforests: Toward a History of Political Tradition in Equatorial Africa (Madison, 1990).

27. C. O. Sauer, Land and Life: A Selection from the Writings of Carl Ortwin Sauer, ed. J. Leighly (Berkeley, 1963); W. L. Thomas et al., eds., Man's Role in Changing the Face of the Earth (Chicago, 1956); B. L. Turner II et al., eds., The Earth as Transformed by Human Action: Global and Regional Changes in the Biosphere over the Past 300 Years (Cambridge, 1990).

28. H. Brookfield, "Questions on the Human Frontiers of Geography," Economic Geography 40 (1964): 283–303; K. W. Butzer, "Cultural Ecology," in G. L. Gaile and C. J. Willmott, eds., Geography in America (Columbus, 1989); W. Denevan, "Adaptation, Variation, and Cultural Geography," Professional Geographer 35 (1983): 399–405; B. L. Turner II, "The Specialist-Synthesis Approach to the Revival of Geography: The Case of Cultural Ecology," Annals of the Association of American Geographers 79 (1989): 88–100.

26

Wisconsin Is Almost Anywhere
Generic Places and the Routinization of Everyday Life

Robert Sack

Much of this book has been devoted to those qualities that make Wisconsin different from other states or even unique as a specific place. These traits pertain to both its physical environment and its settlement and cultural patterns; these are the characteristics usually recounted to others when describing what it is like to be in Wisconsin. They have appeal for residents and nonresidents alike. Consider, though, that any place on the earth has a mix of attributes that make it both different from and at the same time similar or almost identical to other places. Madison, for example, is different from other American cities in many ways. It is situated on a scenic isthmus, which helps mold the city's unusual urban morphology, and Madison's political climate is unusually liberal. The Driftless Area is also strikingly different from the rest of its surroundings, with its water-weathered sandstone and limestone hills and valleys. But these two places are also like others. In many respects Madison is a typical university and administrative center, with a high proportion of white-collar workers and a low rate of unemployment; the Driftless Area, apart from its location, shares properties with other sedimentary landscapes that are exposed to the elements and that support a rural, agricultural economy. Even the state of Wisconsin as a whole is similar to other states: in its relationship to the federal government; in its blend of agriculture, industry, and trade; in its mix of rural and urban, rich and poor, Asian, African, and Caucasian.

The unusual, unique, or specific (as it is called in geography) forms one end of a continuum, with the usual, similar, or generic on the other end. Place is a mix of both. Some places, however, are intentionally constructed to be more generic than specific; the most important and numerous examples of generic places are those that organize our everyday lives: our rooms, apartments, houses, streets, buses, offices, banks, restaurants, and schools. In contrast with the specificity of Wisconsin Dells or New Glarus, these everyday places are at the other end of the continuum, being as close to generic types as can be found.

These generic places arise for two related reasons. First, every society must have standardized places in order for people to feel at home. If all places were extremely different from one another, it would take enormous effort to adapt to them, and little else would be accomplished. Having generic places for everyday activities helps people to concentrate on the activity, not on the place. For the most part, when such places work well, they blend into the background, providing the support needed to get on with the tasks at hand, to the extent that we may even become unaware of place while we perform our activities. This unobtrusive quality is essential then for the routinization of life.

A second reason that generic places occur, a reason that also helps explain the multiplication of such places throughout the landscape over the last century or so, is that the world is becoming more interconnected economically and culturally. Wisconsin is part of an integrated American economy and a global economic system. People from other states and countries arrive here for business, for pleasure, and to live. This means that they too must feel at home, so the generic places in Wisconsin will be like those in other states and even abroad. This push toward generic places will continue as the world becomes still more interconnected.

These generic places are important to examine because they are spreading over an ever larger portion of the Wisconsin landscape and occupy an increasingly large part of our daily lives. Indeed, one can argue that they mold more of our experience of being in Wisconsin than do much more specific places such as the Dells, New Glarus, and the Driftless Area. For a good deal of the time, Wisconsinites may not be aware that they are in the Badger State or in any particular county or community. This lack of awareness is a result of the generic types of place that nevertheless are essential parts of being in Wisconsin or anywhere else.

The ubiquity and power of such generic places can perhaps be seen most directly by considering a typical day in a Wisconsin family's life. Where in Wisconsin we envision the family does not matter much, for the generic places make the experience typical. The parents and children wake up to a crisp autumn Monday morning. The bedroom, with its bed, chest of drawers, and desk; the bathroom, with its tub, toilet, sink, and Kohler fixtures; the kitchen, with its countertops, cupboards, drawers, and half-open cereal boxes—all are familiar not only to this family in its daily routine, but also would be to any North American family and even to many well-to-do families around the globe. Only a minimum amount of consciousness is needed to move from bed, to bathroom, to kitchen: it is almost pure routine. What happens this morning in this home is happening in countless others throughout the state.

Since it is a weekday, everyone must be out of the home before 8:00 A.M. to go either to school or to work. The children wait at the corner school-bus stop, which is marked by the same yellow street lines found everywhere in North America. They climb into the regulation yellow school bus, which moves along streets of standard width, divided by uniform pavement markings, lined with regular sidewalks and traffic signs, then stops and discharges its passengers at a building that has the look of countless other post-1950s single-story elementary and secondary schools. Here students enter familiar corridors, deposit their belongings in regulation lockers, and seek their first-hour class in a classroom that must be distinguished from hundreds of others by a number. Lunch is in a cafeteria (or converted gymnasium) furnished with long, rectangular, tan, formica-covered tables and aluminum folding benches or chairs. (Even the cafeteria food is similar across the state.) The day is punctuated by school bells of a similar pitch nationwide, and while most classrooms have windows, intended to admit light, students are admonished not to look out them and thereby be distracted from their tasks. When the last bell rings, students return home, many on the same school buses and along the same routes, to play or relax, watch T.V., or begin doing homework. This vast movement of children from home to school and home again occurs throughout the state and the nation at virtually the same time during the day, the week, and the season. These places are soon taken for granted by the children, and a student who finds himself in a different town and school, not knowing another child or teacher, will quickly discover that the journey on the bus and the generic layout of the school are remarkably familiar and comforting.

The parents, too, move from one generic place to another. The car, itself serving as a place, is like countless others, regardless of its make;

the similarities of the many brands of automobiles far outweigh their differences. The trip to work follows the regular streets and highways with their standard signs and signals, so that the journey can be negotiated without paying much attention to where one is or should go next. Thoughts can be on almost anything other than the details of the journey.

At about the same time the children arrive at school, the parent pulls into the open-air parking place attached to the large multistory office building which houses her office. Parking in the standard space and then passing the large double-glass doors to the entrance hall with its commercial-grade carpeting, she waits in front of the building's central column containing its elevators. Upon arriving at the fourth floor, she enters a small eight-by-eight cubicle, furnished with a standard gray steel desk with a lighter gray formica top, a computer, phone, phone book, and framed photograph of the present governor. She works in a state agency, and because of the size of her office and its furnishings, she and any other visitor are aware of her status. For her, the day is as routine as it is for her children, and the places in which she works, eats, and meets people are as generic.

Work ends a bit later than usual. To feed everyone at a reasonable time, on the trip home she selects a meal from among a dozen or more nationwide fast food restaurants on the highway strip near her home. At 5:30 she parks the car, enters the house, and calls the children to the table, where they consume the food, knowing beforehand exactly how it will taste. If there is time after homework, the T.V. may be turned on to watch a favorite program. This, too, is being broadcast everywhere in Wisconsin and the United States, so that this family is virtually together with millions of others, seeing the same thing.

Even in these most generic of places, the specific does intrude to some extent. After the syndicated sit-com or drama, for example, the T.V. may be tuned to the local news station, where the weather, sports, and even some of the political news is about Wisconsin. But, even here, the details are often part of a larger national or global picture; the time allotted to the local is dwarfed by the time given to the generic.

Generic places help with routine, and routine is not a bad thing. Even the most creative individuals require routine. Immanuel Kant stayed in Königsberg for most of his life, and his daily activities were so predictable that people claimed they could set their watches by the timing of his evening walk. Although Kant was curious about virtually every facet of life and the world and was one of the first to lecture on geography, he was reluctant to leave Königsberg because travel requires effort and time to adjust to new surroundings, and these commitments he thought would be a less efficient means of learning about the world than to read about others' travels and to have travelers visit him. But if mass transportation and multinational motel chains had been available to Kant in the eighteenth century, even he might have traveled.

Generic types of place foster routine and mundane activities, but they also encourage unusual and even creative ones. An ordinary high school science class encourages a child's mind to stretch across the globe, even to the edges of the cosmos. And the modern information era with its electronic media allows us to be in contact with people and events anywhere. Much of the world can be literally brought within the confines of each of these places. Computers and modems place libraries and encyclopedias at our fingertips. This availability means that what goes on in generic places is not isolating, regardless of their location, and hardly need be dull.

Even so, we often want to add something special to generic places so that they call attention to themselves. The house of our particular Wisconsin family may be a mass-produced, prefabricated, three-bedroom suburban ranch; yet it is painted a bright red, and has a flowering crabapple next to the front entrance. A child's small regular-shaped room has its walls plastered with posters of rock groups. The school classroom is decorated with portraits of famous physicists and a collection of Far Side cartoons. And the wall of the mother's office displays the children's school artwork.

Even greater efforts are made to break out of the generic. Museums, churches, and even some governmental units commission novel architectural designs. And great architects, like Frank Lloyd Wright, create truly original and beautiful structures. Such unique constructs may also be functional and even enhance one's ability to concentrate on other things, like the Wright-designed Johnson Wax Wingspread retreat near Janesville; but often they call so much attention to themselves that seeking them out and enjoying them for their unique qualities turns out to be their principal function.

Such arresting and beautiful structures are sought mostly during our leisure: after hours, on weekends and vacations. The same is true of natural scenery. People often use their leisure to drive into the country, along the scenic lake shorelines, into the Driftless Area, to fish or backpack in the woods, all to encounter places with qualities different from (and more "natural" than) the generic ones of their daily lives. Because time is so short and so much of it takes place in the generic type of landscape, we want to make sure our trips count—that we see what is indeed different.

Specific places at all scales come to satisfy this demand for the different and the unique. Different buildings, communities, scenery, and the entire state can be presented as meeting these needs. Boosterism, which is what such presentations are often called, is a common feature of American life. Such claims about uniqueness may also be made about places that have become extensively transformed precisely to attract the attention of vacationers and tourists. These places push themselves forward by altering their appearance to conform to some idea of the specific or unique things people would like to see.

This alteration of place can build on what was already there, but many of their landscapes are fabricated. New Glarus and Stoughton originally contained significant numbers of Swiss and Norwegians, respectively. But now these groups are in the minority. Still, by staging festivals and adding architectural veneers, these towns are now presented in some senses as re-creations of this Swiss or Norwegian past, even though the majority of their residents are now non-Swiss and non-Norwegian. Most of the residents of these communities enjoy participating in these presentations, which not only help foster community bonds but also provide destinations for those seeking some place different from their own. Not least, such presentations afford these communities significant sums in tourist money.

As business ventures, these places attempt to sell themselves. Their attributes and scenery become commodities. However, as with all other commodities, these places run the risk of losing the very uniqueness they have carefully constructed. Places that trade on tourism can become too "touristy," too much like tourist traps. They soon contain fast food restaurants, motel chains, and mass-produced souvenirs (often manufactured in Mexico or China). In other words, places that are different, and attempt to capitalize on it, run the risk of becoming generic.

This tendency toward the generic occurs to an even greater extent when places are created from scratch as tourist attractions. The Wisconsin Dells, for example, have unusual natural beauty, but the nearby roadsides are covered with generic tourist spots such as Bible Lands and Frontier Towns. These theme parks are indeed different from the homes, schools, and offices of our daily lives, but they are by no means unique to the Dells or to Wisconsin. They can be found in every state.

The continuum from generic to specific thus presents several tensions. Every place contains elements that are specific and generic, but some have more of one than another. The uniformity and predictability that come from the generic are essential for us to feel at home and to accomplish our tasks. The emphasis on the generic aspects of any given place will continue to increase as our tasks and lives become more integrated into a global culture and economy, or as the artifacts in that place become mass produced, or when the place itself is promoted as a tourist attraction. Generic places are the ones in which we spend much of our time and which mold a good deal of our consciousness.

Even though we must have the uniformity and predictability provided by the generic, we often yearn for the specific or unique. We accentuate the few specific qualities of our generic environments, and we also want to experience places that are on the other end of the continuum, those containing more evident specific qualities. This desire for the specific appears to increase as our daily landscapes become more alike.

Modern life leaves little free time, so we want the specific to be rewarding and accessible. But this premium on the specific sets in motion exactly those forces that help push these places closer to the generic. The increasing number of people who now wish to visit these unique places would also like to feel at home in them. These visitors and tourists encourage the construction of such familiar features as motels and fast food establishments. Turning these places into commodities makes them look more like other tourist attractions, so they may try even harder to become specific.

Wisconsin is both types of place. Certainly its place-name—Wisconsin—naturally brings to mind a single and specific place with unusual qualities, and Wisconsin has these in abundance. But living and working in this state involve experiences that are not very different from ones in other states, and these similarities are increasing. To maintain the specificity of place and our identities as Wisconsinites, we often resort even to fictional differences: we become Badgers. This identity can bond us together during certain rituals like football games. But even this bond is porous enough to allow others who enter the state immediately to join the totem. Such an affiliation is not unique to Wisconsin; Minnesota has its Gophers, Michigan its Wolverines, and Iowa its Hawkeyes. Current interest in specific places and landscapes, then, whether it be Frank Lloyd Wright's architecture, Stoughton's Norwegian past, the Wisconsin Dells' allure, or Madison's football team, must be understood partly as a reaction to an overall increase in the role of the generic in the place we call Wisconsin.

Postscript

Wisconsin: Place, Time, Model

Yi-Fu Tuan

"Where are you from?" a New Yorker or Texan asks. "Wisconsin" is the answer we Wisconsinites normally give rather than the name of our hometown—Madison or Milwaukee—and, in this country at least, the answer will invoke a nod of recognition. But not necessarily in a foreign country, not necessarily when we reply to a foreigner, who may have heard of California, New York, or even Florida, but registers a blank at the mention of Wisconsin or any other state in America's heartland. How can this be? How can the heartland—the core and essence of a country—not command instant recognition? This is not as paradoxical as it sounds, for words such as *heartland* and *core* suggest home or homeplace; and home's virtues, which are many, do not include striking or sparkling images that catch attention. A tart adage has it that New York City is a good place to visit, but who would want to live there? As for the heartland (Wisconsin included), it is a good place to live in, but who would want to visit it? Now, we Wisconsinites, while we appreciate the compliment, do not appreciate the idea that we have so few outstanding features worthy of a visit. In any case, it is not true that we lack them. Wisconsin does have a distinctive personality. It attracts many visitors and tourists, and indeed tourism is a flourishing business, a major source of the state's wealth.

Still the question remains, How would we describe Wisconsin to outsiders or, for that matter, to ourselves? Self-understanding is an important part of being human, and there can be no such understanding unless, sometime, we have made an effort to understand the habitat and habit that constitute our being—the hills and valleys, the peoples and institutions, that have made us, sometimes directly and more often indirectly (and very subtly), into the sorts of persons we are. Obviously, each of us will have his or her own experience and appreciation of Wisconsin. The deeper and more extensive such experiences and appreciations are, the greater will be their overlap, for we shall have learned to see beyond our own neck of the woods, and feel, however tentatively, other people's lives—their aspirations, failures, and achievements.

To reach these levels of depth and overlap, most of us need help. The chapters in *Wisconsin Land and Life* have provided me such help: they have added to my knowledge, expanded my mental horizons, and, besides that, better equipped me to say what Wisconsin is and what it means. Unlike the other authors in this volume, I am not an expert on Wisconsin's geography. Perhaps like most readers, I am a resident who has a fondness for the state and some knowledge of it too, but knowledge that has been informally acquired. What follows, then, is one resident's account, the purpose of which is not to provide new insights and knowledge, but rather to stimulate other readers of the book to give their own answers to the questions, What is Wisconsin like? What does it mean to me?[1] I will divide my answer into three parts—place and location, time and narrative, and model.

Place and Location

The personality of Wisconsin owes much to its central location on a continent. Consider first two important aspects of the state's physical character—climate and topography. Is it too fanciful to say that Wisconsin's physical character is somewhat schizophrenic in the sense of containing two contradictory elements? Its climate, as we know, is extreme—bitterly cold in winter, surprisingly hot (and humid) in summer. To insiders, at least, Wisconsin is no place for wimps; the severe and long winters keep others out. Topography, by contrast, is undulating—lacking in extremes, except here and there, locally, along riverbanks. We don't have towering mountains and endless plains. Our landscapes, as Wisconsinites will admit, are pretty rather than breathtakingly beautiful or sublime. Early travelers and boosters had no difficulty assigning Edenic attributes to Wisconsin's landscape. Another ideal—the pastoral (admired since classical antiquity)—also well applies to southern Wisconsin's partly grass-covered, partly wooded, gently rolling hills. Climate is more problematic. To find a place for it in this Eden, travel writers have to play down winter's severity and play up the swing of the seasons—nature's theater in four acts. We have indeed four seasons, and they are for many Wisconsinites a source of satisfaction and even of pride. Southern California, lacking such seasons, is considered monotonous. Wisconsin's seasons do not, however, merge into one another, least of all winter into spring. After the siege of winter, one expects spring to come gently in wafts of balmy air, at long last liberating people from their bondage to heavy overcoats. Not so. Spring is the time of violent conflict between polar and tropical air masses. Thunderstorms, squalls, crushing hail, and the roar of a tornado are not uncommon, and can seem even more frequent than they are, thanks to conscientious weather reports that interrupt regular radio programs to warn listeners of nature's rage.

Wisconsin's central location in North America, unprotected by latitudinal mountain ranges, explains in large part the state's climatic extremes. Location is also a factor in the state's restrained topography. At one time or another during the last million years, much of Wisconsin lay in the path of continental ice sheets, which dampened vertical topographic differences by removing the obtruding heights and filling up the valleys. "Among the ten thousand undulations," the chancellor of the new state University of Wisconsin intoned, "there is scarcely one which lifts its crown above its fellows, which does not disclose to the prophetic eye of taste a possible Eden, a vision of loveliness, which time and the hand of cultivation will not fail to realize and to verify."[2] I find in these words, pronounced so early in the history of the state, a minisermon in stone. The chancellor couldn't have used climate, could he, to lace his sermon? Extremes of cold and heat, violently conflicting air masses, are hardly suitable models for a future Edenic society. But topography suits his purpose well: "ten thousand undulations," yet "scarcely one . . . lifts its crown above its fellows."

Location, then, accounts for some of the principal characteristics of Wisconsin's natural environment—climate and topography, but also vegetation and soils, and how these elements form interacting systems. Let me turn now to Wisconsin's human personality. We may again start with location by taking another look at the heartland theme. The heartland is all-American in a way that coastal states—especially large coastal cities—are not. This may be dismissed as the bias and boosterism of inland regions, but it is not entirely that; and even if it is largely that, the bias is deep-seated, has a basis in fact, and is by no means held only in America. Coastal cities, in general, are more likely to be engaged in international trade, exposed to foreign influences in their purest form, than cities in a continental interior are. Along the coasts, foreign cultures can be set up undiluted, a reason for this being that, despite the uprooting from their countries of origin and the months of separation from them, these cultures have been carried across ocean space sealed, as it were, in the self-contained worlds of vessels. Consider now the settlement of Wisconsin. Migrants entered the territory in the 1820s and 1830s in two main streams: one stream from Missouri and Illinois to the lead mining areas of the southwest, the other from such regions as New England, New York, and Ohio to the rich agricultural lands of the southeast. These people, though they no doubt retained many traits from the Old World, entered Wisconsin as Americans. A different situation arose in the second half of the nineteenth century, when immigration reached a flood tide, and many settlers came straight from foreign countries. Still, *straight* is not quite the right word: traveling across the land, even if it is on a train or a riverboat, is never quite as sealed off from circumambient cultural influences as traveling on an ocean-crossing ship is. This, of course, is especially true if the overland journey occurs in stages, with intermediate stops of varying duration. In short, by the time migrants reached their destinations in the continental interior, a little bit of their foreignness had been rubbed off.

The reputation of the heartland for all-Americanism has another basis—fertile soil and productive agriculture. Location again plays a role. The core of a large land mass may well be arid; Australia is an outstanding example. But the core of North America is not arid, a reason being its accessibility to moisture-laden tropical air. And the core of North America is exceptionally fertile, thanks in large part to deposits laid down by continental ice sheets and their meltwaters. Agriculture, since the founding of the republic, has not been just food production but also a mythic way of life, idealized and romanticized over and over again in the rhetoric of politicians and writers. In the myth, cornfields are the Real America; so when Khrushchev wanted to see that America, he was obliged to stop in Iowa. Cows and dairy barns also have mythic resonance. Wisconsin can claim to be many things—the land of white pine and sparkling streams, beer and heavy manufacture. But it prefers to be known as America's dairyland and so garner for itself an image of the pastoral idyll, wholesomeness, nurture, and perhaps even the immemorial prestige of Mom.

Wisconsin is a place. I have tried to show how its special character has something to do with its location at the core of a continent. But adjoining states have equal and perhaps even better claim to membership in the heartland. For Wisconsin to seem unique, appeals to other factors must be made. One such appeal is to the past, to historical happenings and initiatives. I will take up history in the next section. But, first, I would like to touch on one type of human initiative that had and still has the most direct impact on the realization of Wisconsin as place. It is politics—political imagination and ideal. Wisconsin is first and foremost a political entity. Before 1848, when Wisconsin became a state, or perhaps before 1836, when Wisconsin became a territory, with its own legally recognized boundaries, its own form of government— that is, its own machinery for the promotion of regional consciousness—Wisconsin had no existence in any individual's or group's mind. Of course, one could still talk of glaciation and air masses, white pine and sandpiper, settlers struggling to make a living, and so on, but these and other elements did not add up to Wisconsin, which, to exist, had first to be invented, then to be recognized by both the local people and outsiders.

The invention of Wisconsin as a modern political entity required, early on, the establishment of a sense of boundedness. Clearly drawn borders have the psychological effect of making people within them more aware of who they are, of how they and their part of the world differ from others. Negotiating a boundary with a neighbor in itself promotes place awareness. Where does Wisconsin end and Michigan begin? Wisconsin postponed statehood, in part because it wished to incorporate the Upper Peninsula of Michigan within its borders—a wish that went unfulfilled. Once a state has acquired a firm area and shape, information of all sorts, from weather and vegetation to population and forms of government, can be gathered and propagated as though everything ends at the border—a procedure, backed by bureaucratic organization, that clearly promotes group identity. Listen, for example, to Wisconsin Public Radio's daily report on the weather. We are told about an impending storm in the northern part of the state, sunshine in Dane County, possible rain along the Mississippi River, strong wind near Lake Michigan. And that's about it. The fate of people in Minnesota and Iowa, whether they will be able to enjoy their weekend, is not our concern.

Establishing borders is a first step. More important to Wisconsin's mature political image is that which is eventually created within them —a political ethos sufficiently powerful and innovative to permeate other areas of life, one that gives the state distinction. Politically, Wisconsin had little to be proud of in the first 50 years of its founding: it had its share of mob violence, lynching, spectacular bribery and fraud, the works. Even a century later, it could produce a Senator McCarthy and McCarthyism, which cast a shadow over the entire American political landscape. Happily for Wisconsin, these warts and shadows tend to be forgotten. Prominent in our mind, and in the nation's mind, is Wisconsin's purity and progressivism, its significant contributions to America's political discourse, outstandingly, the role of government in promoting fairness and justice, and its contributions to the articulation of environmental ideals and the means to achieve them. In effect, Wisconsin has come to stand for, and indeed is, a distinctly mid-American realization of the Good Life, though how it has done so is far from clear.

Time and Narrative

A region takes on a humanized character as a result of the peoples who have made their homes there. Each people has a separate story to tell; each subregion has its own narrative. Do such stories add up to Wisconsin? Do they give credence to the popular notion that there has been progress—that, however difficult the earlier stages of settlement, Wisconsin is now a Good Place in the experience and view of most of its citizens? I have no answer to these difficult questions. Their degree of difficulty is suggested by the following three narrative sketches: (1) Wisconsin from the perspective of the Native American experience; (2) the story of northern Wisconsin; (3) the story of southern Wisconsin.

Wisconsin was the name of a river to Native Americans. Its transmutation into the name of a state occurred at their expense. As part of that process of transmutation, their right to their own land was extinguished; they were forced to move beyond the newly drawn state borders or confined to reservations within them. Suppose we forget Wisconsin as a political entity and consider simply the region. What happened here in pre-Columbian times? People came in successive waves and settled down; some groups then left, rather suddenly. What happened? Was the cause of departure war or the threat of war? Disease? Hunting or crop failure in combination with ill omens? It may be that large gaps in the record make the Indian story seem exceptionally episodic. However, the gaps also tempt one to construct retrospectively, and perhaps deceptively, stages of cultural change. Shifts of scene certainly occurred. They may be characterized as follows: First, there was the arrival of big-game hunters, who had the skill to track down and kill mammoth, mastodon, caribou, and musk ox, and whose cultural repertoire included the drawing of animal and human figures on cave walls. In the thousands of years that followed, waves of peoples moved into Wisconsin, some of whom were probably ancestors of the modern Menominee and Winnebago. Later arrivals were hunters and gatherers, agriculturalists who also worked metal (copper), and mound builders. Indian mounds, even today, are a distinctive feature of the south Wisconsin landscape. Earlier mounds were rounded graves of modest size. Later ones, built between A.D. 700 and 1300 by a people called the Mound Builders, were not only larger but also often molded into the shape of animals hundreds of feet in extent. The sophistication of prehistoric Indian culture was manifest in other ways as well: for instance, by the Oneota gardeners, who grew new vegetables like pumpkins, squash, corn, and beans, and by the gardeners and pyramid builders of the stockaded village of Aztalan, an outpost of the Cahokia mound culture of southern Illinois. The Aztalan site was occupied for some 200 years and then inexplicably abandoned.

Pre-Columbian reality thus appears to be made up of many components which do not add up to a unified story or picture that can then be contrasted with post-Columbian reality. Folks with different ways of livelihood entered at different times and from different directions. As contemporaries, they no doubt learned from one another and engaged in trade to mutual benefit. But they also intruded on each other's territory and fought. Indeed, Indians in enmity continued to fight long after contact with Europeans. Confrontation with the Other in this case failed to produce a unified We. Nevertheless, that contact was a radical turning point for all Indians. Native populations, unable to resist diseases brought over by Europeans, declined precipitously.[3] Native cultures, for the first time, had to submit to major change. Indians knew how to cope with cultural change, which was a part of their own historical experience. They could adjust to European ways, provided these were not too alien to their tradition. Thus the Menominee and the Winnebago were able to modify or abandon their diversified economy in favor of hunting, that is, fur trapping, in order to plug into the Euro-American economy. Unhappily for the Indian peoples, Euro-American culture would not stay put. It shifted from fur trapping and lumbering to agriculture and manufacturing, to high-technology industries and information services, moving all the while further and further from the common ground of negotiating directly with "land and nature." What to do remains the Native American dilemma: the challenge is daunting, for it lies in locating a middle ground between security and pride in one's cultural heritage and security and pride in contributing to a Western scientific-technological culture that changes constantly.

Let us turn to the Euro-American story as it unfolds in Wisconsin. I would like to tell it as the biographies of two subregions—north and south—using "progress" as a possible theme. My thesis is that progress doesn't work well as a theme for the north. It works better for the south and perhaps for the state as a whole.

Since Wisconsin became a state, the north offered two primary attractions to Euro-Americans: (1) lumber and iron ore, and, as these approached exhaustion, (2) the dream of agricultural prosperity based on land. Boom and bust, the grandest visions followed by rudimentary achievements, or no achievements at all, were rather typical stories. Throughout the 1890s, Wisconsin was a leading lumber producer in the nation. Yet by 1900, the most sought-after wood, the white pine, already showed evident signs of decline. Rather than give up the region in discouragement, boosters of the north (lumber and rail interests, the state legislature) quickly persuaded themselves that the removal of one source of wealth was preparation for the exploitation of another— the soil. They were able to conjure a rosy picture of farms and towns scattered among woodlots in the former forest. However, neither soil nor climate would cooperate. Farms never made much headway; by the 1920s many had to be abandoned. As for that other major resource, iron ore, the story is similar, though more localized and short-lived. Iron mining became serious business only when railroads reached the ore sites in the 1880s. So many claims were made in the Gogebic Range that, in the year 1887, Milwaukee had close to 200 stock offerings. Late that year, scores of companies had to declare bankruptcy, and in 1889, only 15 remained. With the exploitation (from the 1890s on) of the vast iron deposits of the Mesabi Range in Minnesota, mining in the Gogebic Range declined in importance, though some operations were kept open till the 1960s. Predictably, towns in the north fluctuated in size as economic opportunities came and left. Great hopes were boosted by a combination of the advantages of Great Lakes and rail transportation and the quite unrealistic appraisals of hinterland agricultural resources. Images of metropolises bordering Lake Superior at first seemed plausible. Ashland, which was connected to the railroad system in 1877, expanded rapidly with the Gogebic iron ore boom. Superior's rise in urban stature came later and was far more astonishing: a small town in 1890, it became the state's second largest city 10 years later. But it could not maintain its position. In the first century of statehood, no town or city in the north could boast a record of continuous growth, as could Milwaukee, Madison, and a number of diversified industrial towns in the southern and eastern parts of the state.

Among America's most popular myths are (1) a wholesome way of life in a beautiful rural setting, and (2) progress. Both appear to have gone beyond myth to reality in southern Wisconsin. Pioneering agriculture there was difficult, as were such beginnings all over the world, though in southern Wisconsin they were mitigated by fertile soil, benign (relatively speaking) climate, and good access to rail transportation after the 1850s. What would one see in 1870, at the end of the pioneer period? In the southwestern part of the state (the most beautiful topographically) one would still see much wilderness, but also scattered farms; small irregular fields producing such crops as wheat, corn, and potatoes; improved log cabins, more and more wood-frame houses, and even a few made of brick or stone. In the course of the next 50 years (1870–1920), farms expanded in number and size; their chief commercial product shifted from wheat to milk. Dairy farming introduced new features into the landscape, the most distinctive among which were the large dairy barn and the silo. The farmstead itself became a handsome two-storied building with shade trees, shrubs, flowers, and even a front lawn, all of which strongly suggest that life had gone far beyond the earlier struggles to subsist. The next 50 years saw the introduction of the internal combustion engine and electricity, two forces that radically altered agricultural life and landscape, slowly at first, then rapidly after 1940. Roads were built to accommodate mechanized transport, including tractors and field implements moving from one part of the farm to another, tank trucks carrying milk, and yellow buses carrying school children; farms were consolidated for more efficient organization; already-large dairy barns became bigger, though they were overshadowed by huge silos and towering steel structures designed to provide air-sealed preservation of green chopped corn. In recent decades, as mechanization continues apace, farms have become efficient family corporations, farmers have turned themselves into managers of a complex business, and they and their family live in a style little different from that of well-to-do suburbanites.

Here is a success story—one of hard work, adaptability, ingenuity, technical prowess. True, some Wisconsinites may feel that technology has gone too far in transforming agricultural life, that that life has lost much of its charm, once evident, if not in the split-rail fences and log cabins of the pioneer period, then in the handsome wood-frame houses with sheltering trees, dairy barn, and silo of the late nineteenth and early twentieth centuries. Yet, few will deny that even today southern Wisconsin presents a landscape of agricultural wealth and beauty. Even the story of nature is not one of continuous withdrawal, for parts of southern Wisconsin, especially the steeper slopes, are better wooded and more ecologically diverse today than they were in the pre-Columbian past.

Model

Every human settlement can be seen as a microcosm, with its own struggles, aspirations, and achievements reflecting those of humanity at large. However, not every human settlement—not every humanized region—is equally suitable as a possible model, as an idealized yet plausible projection of humanity's future. Does Wisconsin qualify? I think it does, not so much because it is already a decent place to live (although it is), but more because it is a place in which steady and fairly rapid progress has occurred. The idea of progress is often dismissed as naive by academics and intellectuals; however, to dismiss it utterly not only distorts history but also dismisses hope.

Wisconsin has had a past of despoiling nature. The lumber story is not pretty. Yet once the mistake was recognized, as early as 1897, the state began a serious effort to conserve and restore its forests. Today, the extent of the forest cover in northern Wisconsin is about what it was two centuries ago. As for southern Wisconsin, I already noted that some parts are better wooded now than in pre-Columbian times, thanks to farms and other human works that checked the spread of prairie fire. Wisconsin remains less urban than the United States as a whole. Although more and more people now live in metropolitan fields and corridors, many still live in medium-sized towns, and this means that they are blessed with ready access to rural landscapes and nature. Perhaps such direct experience has served to remind Wisconsinites that their privileged region is vulnerable and that dedicated efforts are necessary to protect it against mindless urban encroachment and desecration. The environmentalist movement is strong here; it was a Wisconsin senator who in 1970 introduced Earth Day to the American public. Wisconsin, to its credit, learned a lesson from its early brutal treatment of nature—a lesson, let it be said, that has to be learned over and over again. Wisconsin is also fortunate in having had charismatic teachers, outstandingly John Muir (1838–1914) and Aldo Leopold (1886–1948). Another major reason for preservationist sentiment is the recognition that nature is an aesthetic resource that brings in tourists and money. As early as the 1860s, the Wisconsin Dells attracted tourists. Pamphlets and, above all, the artful photographs of H. H. Bennett promoted the idea that the Dells offered a unique combination of sublimity and picturesqueness. Maybe so—at the Dells. However, even loyal Wisconsinites will admit that the state as a whole has far more of the picturesque than of the sublime. The great northern forest is indeed awesome, but it is not a sublime view that can be taken in at any point: what visitors really enjoy are the good hunting and fishing, the good cross-country skiing and snowmobiling—the pleasures of outdoor life in a seemingly natural environment. As for the rest of the state, there are certainly individual features of striking appearance, for example, in the rugged Driftless Area and along the limestone coast of Door County; but in general the appeal of central and southern Wisconsin lies in rolling hills and valleys, woodlots and lakes, farms and charming small towns. Wisconsinites show mature sensibility to the extent that they no longer insist on being entertained or overwhelmed by nature. Subdued beauty—one that is not wholly natural, one that even clearly shows the human imprint—is widely recognized as having its own rewards.

Now, isn't this an inspiring model for the world at large? Isn't it inspiring to know, for example: that even gross environmental mistakes can be rectified if caught in time and truly regretted? that the forest cover can expand impressively even as the human population shoots up by five and a half times, as it did in Wisconsin within the 100-year period 1865–1965? that people can learn to care about the environment above quick monetary returns? that they can love nature in their own backyard—nothing exotic, just what they see when they leave the edge of town? Wisconsin today is obviously not representative of the current world, which is mostly poor and environmentally despoiled. But isn't it plausible to say that the world should strive for the sort of environmental quality that now obtains in Wisconsin rather than, as eco-extremists would wish, for a return to some pristine wilderness inhabited by hunters and gatherers?

Let us turn to the human-cultural scene. The world is ethnically diverse. So is Wisconsin. The world can boast centers of cosmopolitan excitement and prosperity, which, historically, are either great capital cities or flourishing seaports. Wisconsin can boast one small cosmopolitan city—Madison, the state capital—thanks in large part to the university and its substantial contingent of foreign students, visiting scholar-scientists, and faculty. As for "seaport cosmopolitanism," in Wisconsin it was best exemplified a century ago in the frontier lakeports and the milling and mining towns of the north. I have deliberately omitted Milwaukee, for it has always been more a conglomeration of ethnic neighborhoods than one great cosmopolitan center in which people and cultures mix indiscriminately. For the world as a whole, ethnic-cultural diversity implies human vitality and wealth, but that diversity has also been and remains today a seedbed of the most bitter and bloody conflicts. Now, how will Wisconsin play as a parable or model for the world? Its ethnic diversity is the result of successive waves of immigration. People came, sometimes over epic distances, in search of a better life. They were thus predisposed to find what they envisaged and yearned for. That's one factor in favor of success and social harmony. Another factor is this: ethnic groups could choose (obviously within limits defined by nature and practical social considerations) where to locate in Wisconsin; they were not, in any case, compelled to live next door to an arrogant social elite or to a people they found, for historical and other reasons, uncongenial, as they sometimes were in the Old World. The third factor is distinctive to the settlement history of the United States and so is not, presumably, applicable to the larger world. When Europeans of very diverse cultural and social backgrounds entered Wisconsin in increasing numbers from the 1830s onward, they found already, first, a Territory, then, a State. I capitalize these nouns to emphasize their political standing. Although sparsely settled and very "raw" as a human environment, Wisconsin already had in place a political machinery—a way of administering public affairs that, for all its ineffectiveness and corruption in the early period, was intended to encourage such American ideals as freedom, individual rights, and the pursuit of happiness. Last but not least is economic-technological development, which has enormous power to transform life and values, smooth out the differences, corrode ethnic-communal bonds, liberate—but also isolate—individuals.

Wisconsin was and, to some extent, still is a colorful patchwork of ethnic groups and cultures. The meaning of these entities has changed, and this to me is a cause for hope. The direction of change in Wisconsin is one that I would like to see happen in the world as a whole. And it is in this sense that, again, I see Wisconsin as a model. What is the direction of change? It is from a humorless, passionately self-engaged, and righteous attitude toward one's kin group and culture to an attitude that is warmly affectionate, loyal yet gently ironic and playful. For most nineteenth-century immigrants, the earlier attitude is born of the need to survive: in a strange and often-threatening frontier environment, people must be able to count on instant, unquestioning help that only kinsfolk can dependably provide; moreover, in the midst of newness and strangeness there naturally arises a profound need for the familiar, and this involves language, custom, and, above all, religion. I say religion above all because it is the one area of life where, paradoxically, the familiar (words and gestures that have been repeated over and over) becomes momentarily sublime. Not all settlers in Wisconsin, however, fitted the model of a "besieged" community. Those Americans who moved in from the 1820s onward probably did not feel besieged, no matter how hard life actually was for many of them. In Wisconsin, they soon established themselves as members of a political and social upper crust: their way was, after all, the American way. The Germans, especially the educated and well-to-do migrants that came after 1848, formed strong communities in southeastern Wisconsin—communities so strong and so self-confident that they hardly felt besieged as the underdog, except, resentfully, vis-à-vis the entrenched old-stock Americans. Germans were passionately loyal to their language, their schools and institutions, their way of life: in the decade before America entered World War I, they could be openly arrogant toward later and poorer arrivals from eastern and southern Europe, whom they regarded as inferior.

The nineteenth-century forebears of Wisconsin's current white population lived in communities that were often ethnically and culturally distinct. There was little social contact among them. Not only did the Dutch live apart from the Swedes, but the Catholic Dutch lived apart from the Protestant Dutch; and, for that matter, the Bavarian Germans lived apart from the Prussian Germans, and so on for the dozen or so other European groups. However, cultural uniqueness and separateness progressively lost ground. Even the Germans, for all their numeric strength and cultural pride, could not resist the path to prosperity, that is, to modernization and Americanization. Much intermarriage and merging of peoples occurred in the twentieth century, especially after 1940. Wisconsin is indeed a melting pot of Euro-Americans and, increasingly, of peoples from all parts of the world. Wisconsin is recognizably an entity—a state, region, people—not just congeries of communities and cultures. In recent times, the metaphor "melting pot" has fallen out of favor, for it projects an image of domination by a single set of values—an image of self-satisfied blandness. In favor is ethnic-cultural diversity. In favor is difference: Swiss Americans are not Swedish Americans. Just about every ethnic group now tries to resurrect its old customs and language; it does so playfully in the form of festivals and storehouse museums, but also with genuine pride and affection. The community it attempts to re-create is not introverted and defensive. On the contrary, festivals and museums openly invite outsiders to come and admire. The belief that outsiders will want to come and admire presumes the existence of transcultural commonality and, in the broader view, universal commonality. That is a reason for optimism and hope.

Unfortunately, not every ethnic group in Wisconsin can engage in affectional-playful ethnicity. That luxury is possible only for people who are already economically "well off," who can live just about anywhere they want. Recent arrivals in Wisconsin, such as the Hispanics and the Southeast Asians, suffer the disadvantages of all recent arrivals: they continue to live in introverted, besieged communities. The hope is that, as their economic well-being improves, they too will be able to live in playfully reconstituted ones. As a group, black Americans are the outstanding exception to the story of upward mobility. They migrated and continue to migrate into Wisconsin, seeking a better life, like everybody else. But unlike Wisconsinites of Latin American and Asian origin, they are not new arrivals from foreign places; indeed, they are among the oldest Americans (Indians excepted) in the nation. Yet success eludes them; the melting pot seems to reject them. Why? Not even Wisconsin has found a solution to what Gunnar Myrdal calls an American Dilemma.

Notes

1.   For the sake of clarity, and also because this is a personal essay rather than a scholarly paper, I will generally refrain from source notations. For my sources, turn to the chapters themselves and their ample notes. I would, however, like to draw the general reader's attention to a small book that I have found especially useful: R. N. Current, *Wisconsin: A Bicentennial History* (New York: 1977).

2.   J. H. Lathrop, "Wisconsin and the Growth of the Northwest," *De Bow's Review* 14 (March 1853): 230.

3.   The vexing question of just what was the pre-Columbian population of North America has been raised many times, most recently by D. R. Snow, "Microchronology and Demographic Evidence Relating to the Size of Pre-Columbian North American Indian Populations," *Science* 268, 16 (June 1995): 1601–1604.

Notes on Contributors

Arnold R. Alanen has a B.A. in architectural studies and a Ph.D. in geography from the University of Minnesota. He has been a professor in the Department of Landscape Architecture at the University of Wisconsin–Madison since 1974 and has written extensively on the settlement history, immigrant groups, and cultural landscapes of the Lake Superior region.

Timothy Bawden is a Ph.D. candidate in the Department of Geography at the University of Wisconsin–Madison. His research interests focus on the historical geography and environmental history of the upper Great Lakes. His dissertation examines how different groups of people envisioned and shaped the landscape of northern Wisconsin during the early twentieth century.

Waltraud A. R. Brinkmann has been a member of the University of Wisconsin–Madison geography faculty since the middle of the 1970s. Her teaching and research focus is on climatology, climatic change, and natural hazards. Brinkmann's wide-ranging science has included investigations of the historical changes in the levels of the Great Lakes and the temporal variability in the length of the growing season in Wisconsin.

Michael P. Conzen is professor of geography at the University of Chicago. He earned his Ph.D. from the Department of Geography at the University of Wisconsin–Madison in 1972 and has published widely on aspects of American historical geography. His books and edited anthologies include *Frontier Farming in an Urban Shadow; Boston: A Geographical Portrait; Chicago Mapmakers;* and *The Making of the American Landscape.*

William Gustav Gartner is a Ph.D. candidate in the Department of Geography at the University of Wisconsin–Madison. Long interested in the anthropology and archaeology of Wisconsin, Gartner brings the expertise of the physical geographer to his dissertation work on the pre-European ridged fields—elevated agricultural plots—at Hulbert Creek near Wisconsin Dells.

Duane Griffin is a Ph.D. candidate in the Department of Geography at the University of Wisconsin–Madison. His primary research interest is biogeography and plant ecology, with dissertation work on biological diversity in a naturally fragmented landscape. Griffin's expertise is broad, including vegetation change, natural hazards, environmental conservation, quantitative methods, and Africa.

Steven Hoelscher received his Ph.D. from the Department of Geography at the University of Wisconsin–Madison in 1995 and is currently assistant professor of geography at Louisiana State University. Hoelscher's dissertation explored the process by which an "ethnic place" was created at New Glarus, Wisconsin. A native of Racine, he is also codirector of the Cultural Map of Wisconsin and is working on a new book, *H. H. Bennett's Wisconsin Dells: The Creation of an American Landscape*.

Francis D. Hole is a retired professor in both the Departments of Soil Science and Geography on the University of Wisconsin–Madison campus. Known widely for his technical expertise in soils, his breadth of interests in Wisconsin geography, and his humane approach to education, Hole continues an active intellectual life. He is the author of *Soils of Wisconsin* and is the dean of soil science in the state.

John C. Hudson, currently on the faculty of Northwestern University, was a member of the University of Wisconsin–Madison Department of Geography in the late 1960s and early 1970s. His teaching and research interests span the discipline from quantitative methods to regional geography. Hudson's books include *Making the Corn Belt; Plains Country Towns;* and *Crossing the Heartland.*

Judith T. Kenny earned her Ph.D. at Syracuse University and is now associate professor of geography at the University of Wisconsin–Milwaukee, where she teaches urban geography. Her main interests are in cultural geography with particular attention given to dominant societies' efforts to control urban environments.

Anne Kelly Knowles received her Ph.D. from the Department of Geography at the University of Wisconsin–Madison in 1993. Knowles currently teaches historical and economic geography through the medium of Welsh at the University of Wales, Aberystwyth. Her most recent research focuses on the ways in which culture has influenced the development of industrial capitalism in Britain and the United States. She is the author of *Calvinists Incorporated: Welsh Immigrants on Ohio's Industrial Frontier.*

Ann Marie Legreid received her Ph.D. in geography at the University of Wisconsin–Madison in 1985. Ann is currently professor of geography at Central Missouri State University, where she pursues research on nineteenth-century Swedish and Norwegian immigration to her native Wisconsin and the American Midwest.

David M. Mickelson completed an undergraduate degree in geography at Clark University before moving on to a Ph.D. in geology at Ohio State University. For 25 years, Mickelson has been a member of the Department of Geology at the University of Wisconsin–Madison and the campus authority on the glacial history of Wisconsin. An active teacher and researcher, he is a frequent participant in the education of physical geographers in the Department of Geography.

Eric Olmanson is a Ph.D. candidate in the Department of Geography at the University of Wisconsin–Madison. He is interested in the variety of ways that people view both the world and one another across space and over time. His dissertation is a historical geography of human interactions with the environment in the Chequamegon Bay region of northern Wisconsin.

Clarence W. Olmstead is emeritus professor of geography at the University of Wisconsin–Madison. From the mid-1940s until his retirement in 1981, Olmstead was the department's expert in the geography of agriculture, with an emphasis on comparative and changing systems and on the regional-historical geography of the United States, stressing land use and rural landscapes.

Robert C. Ostergren has been a member of the University of Wisconsin–Madison geography faculty since 1978. His research focuses on processes and patterns of immigrant settlement in Wisconsin and the upper Midwest. He is the author of *A Community Transplanted: The Trans-Atlantic Experience of a Swedish Immigrant Settlement in the Upper*

Middle West, 1835–1915 and codirector of the Cultural Map of Wisconsin.

Thomas E. Pearson is a Ph.D. candidate in the Department of Geography at the University of Wisconsin–Madison. His research interests focus on environmental geography and mapping. Pearson's current work involves the use of satellite imagery and geographic information systems to study patterns of resource use and environmental change in northern Wisconsin.

Randall Rohe received his Ph.D. at the University of Colorado and is currently professor of geography at the University of Wisconsin–Waukesha. He has codirected the excavation of three logging camps that operated in Wisconsin between the late 1860s and the early 1920s and has written extensively on lumbering in Wisconsin and the Great Lakes region.

Robert Sack is Clarence J. Glacken Professor of Geography and professor of integrated liberal studies at the University of Wisconsin–Madison. His interests are geographic theory and philosophy. His books include: *Homo Geographicus; Place, Modernity and the Consumer's World; Human Territoriality;* and *Conceptions of Space in Social Thought.*

Yda Schreuder received her Ph.D. from the Department of Geography at the University of Wisconsin–Madison in 1982 and is currently associate professor of geography at the University of Delaware. She has published extensively on the subject of Dutch immigration to the United States, including a book entitled *Dutch Catholic Immigrant Settlement in Wisconsin, 1850–1905.*

Steven E. Silvern received his Ph.D. from the Department of Geography at the University of Wisconsin–Madison in 1995 and currently teaches geography at the University of Wisconsin–Oshkosh. Focusing on a series of legal and developmental issues that surround the treaty rights of Native Americans in northern Wisconsin, Silvern has approached his research from a variety of angles, ranging from participant-observer to legal archivist. The results of his research have been published widely.

Yi-Fu Tuan has been, since 1985, the J. K. Wright and Vilas Professor of Geography at the University of Wisconsin–Madison. His scholarly interest, as reflected in his most recent book *Cosmos and Hearth,* lies in the area of cultural and human geography. Other recent books by Tuan include *Passing Strange and Wonderful* and *Morality and Imagination*.

Thomas R. Vale has been a member of the Department of Geography at the University of Wisconsin–Madison since 1973. He teaches and critically considers issues of physical geography, natural resources, and the American West. Vale's books include: *Plants and People; Time and the Tuolumne Landscape;* and *John Muir's First Summer in the Sierra*.

Ingolf Vogeler earned his Ph.D. in geography at the University of Minnesota. Vogeler is currently professor of geography at the University of Wisconsin–Eau Claire, where he teaches and writes about rural areas and cultural landscapes in North America. He is the author of *The Myth of the Family Farm: Agribusiness Dominance of U.S. Agriculture* and *Wisconsin: A Geography*.

Robin P. White completed her Ph.D. in the Department of Geography at the University of Wisconsin–Madison in 1987, submitting a dissertation on the South American wintering environments of Wisconsin birds. For nearly a decade, White has been active in the Office of Technology Assessment, a research team serving the U.S. Congress in Washington, D.C.

Susy Svatek Ziegler is a Ph.D. candidate in the Department of Geography at the University of Wisconsin–Madison. Her academic interests embrace physical geography broadly and the concept of geography as a discipline. Her talents as a teacher have earned her several awards. Ziegler's research focuses on forest ecology and vegetation dynamics in southern Wisconsin and in the forests of northern New York State and New England.

Jeffrey Zimmerman is a Ph.D. candidate in the Department of Geography at the University of Wisconsin–Madison. His research explores the process by which historical and cultural landscapes—or "symbolic economies"—are used in contemporary urban revitalization and gentrification schemes. Much of Zimmerman's work has focused on Milwaukee and Chicago.

Index

Vegetation (*continued*)
  between northern and southern, 101–7;
  fire as factor, 14, 15, 16; history, pollen
  in study of, 97, 101–2; Holocene history
  in Wisconsin summarized, 97–98; individualistic theory of, 108; in northern
  Wisconsin, 100; presettlement pattern
  of, 99, 356–57; in southern Wisconsin,
  14–16, 100; tension zone in, 99; wind as
  factor in, 100
Vernon County: Norwegians in, 301,
  305–6; mentioned, 366
Vertical development of soil, 66–69
Victory (town), 305
Vilas County: Finns in, 445; forest reserves in, 461, 465; and Ojibwa, 499;
  and rural zoning, 464; mentioned, 451,
  453, 458, 459, 462
Villages: of Native Americans, 334–35,
  343–44
Viroqua: Old Americans in, 305, 306
Voigt decision, 496, 500–502

Walworth County: dairying in, 411; mentioned, 38
Washburn: bird's-eye view of, 471, 472,
  478–79; harbor of, 470; population of,
  479
Waste repositories, 46
Watertown, 145, 203
Waud, Alfred R., 439
Waukesha: Danish settlement around,
  154, as manufacturing center, 149;
  platting of, 203; mentioned, 24
Waukesha County: Welsh in, 153; 283,
  285–95; mentioned, 19
Waupaca: Danish settlement around, 154;
  platting of, 203
Wausau: Germans in, 158; as sawmilling
  town, 148, 207, 224, 225; mentioned, 23,
  142, 149, 198, 450
Weather: and drought, 62; modification
  of, 61–62; seasonal characteristics of,
  51–60; variations in, 60–62. See also
  Climate
Webster, 213
Wellge, Henry: and bird's-eye viewmaking, 471–74, 476, 482–83
Welsh: and fundamentalist Protestant
  religion, 282–95; immigration and

settlement of, 150, 153, 158, 283; and
  language, 290–91
Welsh Synod of Wisconsin, 287
West Allis: foreign born in, 158; as manu-
  facturing center, 149
Westboro: as sawmilling town, 223
Westby, 366
Westerlies: in atmosphere, 19, 50
Western Upland: settlement of, 141
Wetlands: soil in, 73
Weyerhauser, Frederick, 207
Wheat: cultivation of, 140, 141, 143; rail
  shipment of, 208; transition from, 411;
  mentioned, 362
Whitewater, 208
Whitney, Daniel, 201
Whittelsey, Charles W., 244
Wild rice: commercial production of, 511–
  14; and cultural identity of Ojibwa, 507,
  517–18; ecology of, 508–9; harvest by
  Ojibwa, 509–10; mechanized harvest
  of, 510–12; Native American influence
  on, 338; Ojibwa reaction to commercial,
  514–16; and University of Wisconsin,
  513, 514
Williams, Catherine, 285
Williams, Daniel Jenkins, 289, 295
Wind: trees toppled by, 72, 88; as vegeta-
  tion factor, 100
Winnebago: creation myth of, 331–32. See
  also Ho-Chunk
Winneconne: as sawmilling town, 207
Wisconsin and Lake Superior Mining and
  Smelting Company, 245
Wisconsin Central Railway: construction
  of, 209; and iron shipments, 245; and
  promotion of cut-over lands, 456; and
  sawmilling towns, 223, 453; mentioned,
  485
Wisconsin Dells: early boosterism in,
  428–29; as Kilbourn City, 428; tourism
  in, 432–33; mentioned, 522, 526, 527
Wisconsin Dells Company, 444
Wisconsin Land Economic Survey, 4
Wisconsin Mirror, 429
Wisconsin Rapids: as sawmilling town,
  207, 222, 224, 225, 230; mentioned, 149,
  198
Wisconsin River: early survey of, 426–27;
  forest reserves at headwaters of, 461;

DISCARD

DATE DUE

DEC 6 2002			
NOV 2 3 2004			

Demco, Inc. 38-293